Sediment
and
Contaminant
Transport
in
Surface
Waters

Sediment and Contaminant Transport in Surface Waters

WILBERT LICK

Publishing

CRC Press
Taylor & Francis Group
Boca Raton   London   New York

CRC Press is an imprint of the
Taylor & Francis Group, an **informa** business

CRC Press
Taylor & Francis Group
6000 Broken Sound Parkway NW, Suite 300
Boca Raton, FL 33487-2742

© 2009 by Taylor & Francis Group, LLC
CRC Press is an imprint of Taylor & Francis Group, an Informa business

No claim to original U.S. Government works
Printed in the United States of America on acid-free paper
10 9 8 7 6 5 4 3 2 1

International Standard Book Number-13: 978-1-4200-5987-8 (Hardcover)

Library of Congress Cataloging-in-Publication Data

Lick, Wilbert J.
Sediment and contaminant transport in surface waters / author, Wilbert Lick.
p. cm.
"A CRC title."
Includes bibliographical references.
ISBN 978-1-4200-5987-8 (alk. paper)
1. Water--Pollution. 2. Sediment transport. 3. Contaminated sediments. 4.
Streamflow. 5. Limnology. 6. Environmental geochemistry. I. Title.

TD425.L52 2009
628.1'68--dc22

2008023728

Visit the Taylor & Francis Web site at
http://www.taylorandfrancis.com

and the CRC Press Web site at
http://www.crcpress.com

Contents

Dedication

To
Jim and Sarah

Preface

This book began as brief sets of notes prepared for a graduate class of students at the University of California at Santa Barbara (UCSB). The course emphasized the transport of sediments and contaminants in surface waters. The students were mainly from engineering, but there also were students from the departments of environmental sciences and biology. The course was later given twice as a short course (with the same emphasis) in Santa Barbara to professionals in the field, primarily to personnel from the U.S. Environmental Protection Agency and the U.S. Army Corps of Engineers but also to personnel from other federal and state agencies, consulting companies, and educational institutions.

Sediment and contaminant transport is an enormously rich and complex field and involves physical, chemical, and biological processes as well as the mathematical modeling of these processes. Many books and articles have been written on the general topic, and much work is currently being done in this area. Rather than review this extremely large set of investigations, the emphasis here is on topics that have been recently investigated and not covered thoroughly elsewhere — for example, the erosion, deposition, flocculation, and transport of fine-grained, cohesive sediments; the effects of finite rates of sorption on the transport and fate of hydrophobic contaminants; and the effects of big events such as floods and storms. Despite this emphasis, the overall goal is to present a general description and understanding of the transport of sediments and contaminants in surface waters as well as procedures to quantitatively predict this transport.

Much of the work described in this book is based on the research done by graduate students and post-doctoral fellows in the author's research group at UCSB and previously at Case Western Reserve University. For their work, inspiration, and input, I am enormously grateful. Because they are quite numerous, it is difficult to list them with their specific contributions here; hopefully, I have thoroughly referenced their contributions in the text itself. I am also grateful to June Finney, who did much of the typing and assisted in many other ways. Several researchers (Lawrence Burkhard, USEPA; Earl Hayter, U.S. Army Corps of Engineers; Doug Endicott, Great Lakes Environmental Center; and Craig Jones, Sea Engineering) have each reviewed two or more chapters of the text. Their comments and suggestions were of great help.

About the Author

Wilbert Lick is currently a research professor in the Department of Mechanical and Environmental Engineering at the University of California at Santa Barbara (UCSB). His main expertise is in the environmental sciences, fluid mechanics, mathematical modeling, and numerical methods. His present interests are in understanding and predicting the transport and fate of sediments and contaminants in surface and ground waters and the effects of these processes on water quality. This work involves laboratory experiments and numerical modeling with some fieldwork for testing devices and data verification. He has researched these problems in the Great Lakes, the Santa Barbara Channel, New York Harbor, Long Beach Harbor, the Venice Lagoon in Italy, and Korea.

Lick is the author of more than 100 peer-reviewed articles and is a consultant to federal and state agencies as well as private companies. Previous to UCSB, he taught at Harvard University and Case Western Reserve University, with visiting appointments at the California Institute of Technology and Imperial College, University of London. His Ph.D. is from Rensselaer Polytechnic Institute.

1 Introduction

The general purpose of this text is to assist in the quantitative understanding of and ability to predict the transport of sediments and hydrophobic contaminants in surface waters. For this reason, fundamental processes are emphasized and described. However, any attempt to understand and predict sediment and contaminant transport inevitably leads to mathematical models, occasionally simple conceptual or analytical models, but often more complex numerical models. Two of the main limitations in the usefulness of these models are (1) an inadequate knowledge of the processes included in the models and (2) an inadequate ability to approximate and/or parameterize these processes. For these reasons, approximately half of the text is a presentation and discussion of basic processes that are significant and that need to be quantitatively understood to develop quantitative and accurate models of sediment and contaminant transport; the other half of the text is the development, description, and application of these models. Throughout, the emphasis is on realistic descriptions of sediments (e.g., cohesive, fine-grained sediments as well as non-cohesive, coarse-grained sediments); contaminants (finite sorption rates); and environmental conditions (including big events).

The most obvious application of the work described here is to the problem of contaminated bottom sediments. These sediments and their negative impacts on water quality are a major problem in surface waters throughout the United States as well as in many other parts of the world. Even after elimination of the primary contaminant sources, these bottom sediments will be a major source of contaminants for many years to come. To determine environmentally effective and cost-effective remedial actions, the transport and fate of these sediments and associated contaminants must be understood and quantified. More generally, the transport and fate of sediments and contaminants are basic processes that must be understood for assessing water quality and health issues (toxic transport and fate, bioaccumulation); water body management (navigation, dredging, recreation); and potential remediation methods (environmental dredging, capping, natural recovery).

Examples of surface waters that are heavily impacted by contaminated bottom sediments are the Hudson River, the Lower Fox River, the Passaic River/Newark Bay Estuary, and the Palos Verdes Shelf. These sites are all similar in that they are locations of historical industrial discharges; large amounts of contaminated sediments are still present; and, because of the toxicity and persistence of the chemicals and the large amounts of contaminated sediments, there is considerable uncertainty on how best to remediate each site. For each site, extensive descriptions of the site; the contaminated sediment problem; and the scientific, engineering, and political issues that arise in the solution

1

of the problem are given on the Web by the U.S. Environmental Protection Agency (EPA) as well as other organizations. Brief descriptions of these sites, the problems due to contaminated sediments at these sites, and progress toward remediation are given in the following section. Most of this information is from EPA listings on the Web.

In the description of the transport of sediment and contaminant transport and in water quality models in general, numerous parameters appear; many of these parameters may not be well understood or quantified. When they cannot be adequately determined directly from laboratory or field data, they often are determined by parameterization (also called model calibration) — that is, by varying the value of each parameter until the solution, however defined, fits some observed quantity. Although quite useful, there are limitations to this procedure, especially when multiple parameters are involved. For this reason, a preliminary discussion of modeling, the determination of parameters needed in the modeling, and the associated problem of non-unique solutions are given in Section 1.2.

Big events, such as large storms and floods, have been shown to have a major effect on the transport and fate of sediments and contaminants and, hence, on water quality. This is a recurring theme throughout the text. An introduction to this topic is given in Section 1.3, and an overview of the entire book is given in Section 1.4.

1.1 EXAMPLES OF CONTAMINATED SEDIMENT SITES

1.1.1 HUDSON RIVER

The Hudson River is located in New York State and flows from its source in the Adirondack Mountains south approximately 510 km to Manhattan Island and the Atlantic Ocean. Much of the river is contaminated by polychlorinated biphenyls (PCBs), and the contaminated part of the river has been designated as a Superfund site. This site extends approximately 320 km from Hudson Falls to Manhattan Island and is the largest and most expensive Superfund site involving contaminated sediments. For descriptive purposes, the site has been further divided into the Upper Hudson (from Hudson Falls to the Federal Dam at Troy, approximately 64 km, Figure 1.1) and the Lower Hudson River (from the Federal Dam to Manhattan Island). The Upper Hudson contains the highest concentrations of PCBs.

The PCB contamination is primarily due to the release of PCBs from two General Electric Company (GE) capacitor plants in the Upper Hudson at Fort Edward and Hudson Falls. At these two plants, GE used PCBs in the manufacture of electrical capacitors from approximately 1947 to 1977. During this time, as much as 600,000 kg of PCBs were discharged into the Hudson. The use of PCBs was discontinued in 1977; since then, additional PCBs have leaked into the Hudson River from the Hudson Falls plant through cracks in the bedrock.

The primary health risk associated with PCBs in the Hudson River is the accumulation of PCBs in the human body through eating contaminated fish.

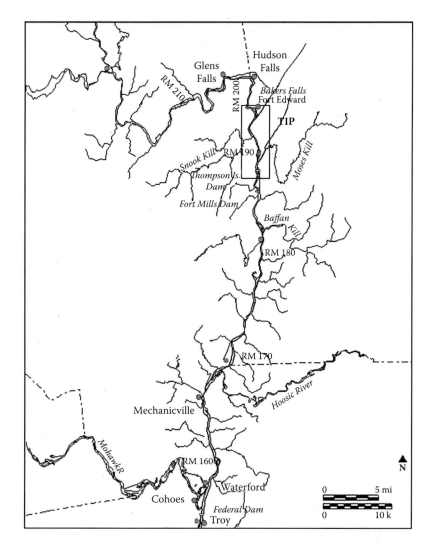

FIGURE 1.1 Upper Hudson River.

PCBs are considered probable human carcinogens and are linked to other adverse health effects such as low birth weight; thyroid disease; and learning, memory, and immune system disorders. PCBs have similar effects on fish and other wildlife.

Because of their hydrophobicity, PCBs sorb to sediments, are transported with them, and settle with them in areas of low flow (e.g., behind dams). Many of the sorbed PCBs settled initially behind the Fort Edward Dam just downstream of Hudson Falls (Figure 1.1). Because the dam was deteriorating and was in poor condition, the Niagara Mohawk Power Corporation removed the dam in 1973. During subsequent spring floods and other high flow periods, the PCB-contaminated sediments behind the dam were eroded, transported downstream, and again deposited in low-flowing and quiescent areas of the river (including areas behind downstream dams).

Areas with potentially high PCB concentrations were surveyed by the New York State Department of Environmental Conservation from 1976 to 1978 and again in 1984. Areas that generally had PCB concentrations greater than 50 mg/kg were defined as "hot spots." Approximately half of these hot spots were located in the Thompson Island Pool (TIP) behind Thompson Island Dam (Figure 1.1).

Since 1976, because of high levels of PCBs in fish, New York State has closed recreational and commercial fisheries and has issued advisories restricting the consumption of fish caught in the Hudson River. In addition, extensive investigations of PCB concentrations, their transport and fate, their uptake by organisms, and subsequent transfer through the food chain have been made by GE (e.g., QEA, 1999) and EPA by means of field measurements, laboratory analyses, and numerical modeling. In 2005, the Department of Justice and EPA reached an agreement with GE that required GE to begin remediation of contaminated sediments in the Thompson Island Pool. Proposed remediation is by a combination of dredging, capping, and natural recovery. Dredging is scheduled to begin in spring 2008. Approximately 2.5×10^6 m^3 of sediment and 7×10^4 kg of PCBs will be removed. The expected cost of the dredging is more than $700 million.

1.1.2 LOWER FOX RIVER

The Lower Fox River is in Wisconsin, is 63 km long, and runs from Lake Winnebago in the south to Green Bay in the north. It is similar to the Hudson River in that it is a Superfund site due to PCB contamination of the bottom sediments. However, compared to the Hudson, it is smaller in length, its volume flow rate is lower, and its sediments contain a smaller amount of PCBs; however, PCB concentrations are similar.

Over the distance from Lake Winnebago to Green Bay, the river drops by 52 m. To develop water power and improve navigation, the U.S. Army Corps of Engineers constructed a series of 17 locks and dams on the Fox; this construction was completed in 1884 and allowed navigation from the Great Lakes through the Fox to the Mississippi River. In the downstream direction from Lake Winnebago to Green Bay, the last dam in the series is at DePere (Figure 1.2). Dredging of the river below this dam for navigation purposes has caused the river in this area to be wide and deep. As a consequence, the flow velocities there are relatively low and sediment deposition is high. It is estimated that 75% of the present sediment deposition in the Lower Fox occurs between DePere Dam and Green Bay.

The combination of cheap water transportation and power attracted a large number of industries, including manufacturers of wood products and paper mills. This area is considered to have the largest concentration of paper mills in the world. In the pulp and paper manufacturing process (especially in the manufacture of carbonless copy paper), PCBs were used; the waste PCBs were discharged directly into the river until 1971. At that time, the use of PCBs was discontinued. It is estimated that more than 115,000 kg of PCBs were discharged into the Fox

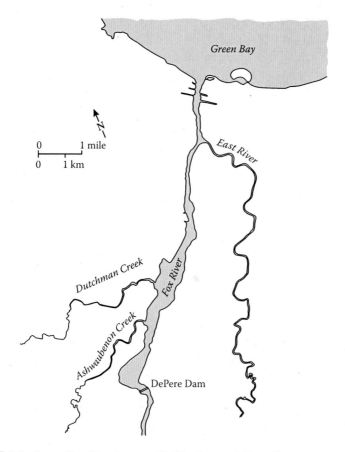

FIGURE 1.2   Lower Fox River between DePere Dam and Green Bay.

from 1957 to 1971 and that 31,000 kg of PCBs still remain in the sediments of the river. Of this latter amount, 87% (or 27,000 kg) are located in the sediments between DePere Dam and Green Bay. These sediments serve as a long-term source of PCBs to the overlying water in the Lower Fox and hence to Green Bay.

Below DePere Dam, most PCBs (80%) are buried below approximately 30 cm of relatively clean sediment. The average PCB concentration in these surficial sediments is 2.7 mg/kg; a few locations have concentrations up to 30 mg/kg. At depth, concentrations range from non-detect up to 400 mg/kg. The guidelines of the Wisconsin Department of Natural Resources suggest maximum PCB concentrations in the sediment of 0.25 mg/kg.

The major concerns are the flux of PCBs from the bottom sediments to the overlying water and the subsequent transport of PCBs down river to Green Bay. Possible remedial actions include dredging, capping, natural attenuation, or combinations of these actions for different parts of the river.

As an initial step in the cleanup of the Lower Fox, dredging of contaminated sediments began in an upstream region of the river, Little Lake Butte des

Mortes, in fall 2004. The entire cleanup of this area may last up to 6 years. In the region between DePere Dam and Green Bay, the plan is to remove approximately $6 \times 10^6 \, m^3$ of sediment and 2.6×10^4 kg of PCBs. Designs and plans for the cleanup of the rest of the river are being developed. The entire cleanup is expected to last 15 years.

1.1.3 PASSAIC RIVER/NEWARK BAY

The Passaic River/Newark Bay area (Figure 1.3) has been heavily industrialized since the 1800s and is recognized as the largest manufacturing and industrial center in the eastern United States (Endicott and DeGraeve, 2006). Because of this, the sediments in the river/bay are highly contaminated by dioxin/furans, PCBs,

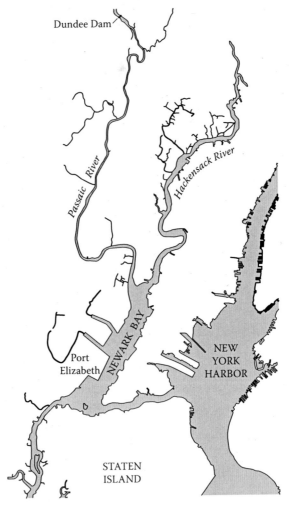

FIGURE 1.3   Lower Passaic River/Newark Bay area.

polycyclic aromatic hydrocarbons (PAHs), pesticides, and metals; dioxin is the major concern. A major source of this dioxin as well as a variety of pesticides is the Diamond Alkali site in Newark, a manufacturing facility for pesticides from mid-1940 to 1970. Although point sources of contaminants have been reduced, the bottom sediments continue to serve as a source of contaminants to the waters not only in the Passaic River and Newark Bay but also in the Hackensack River and New York Harbor and Bay. The Passaic River and Newark Bay areas are Superfund sites.

The most highly contaminated area of the Passaic River extends approximately 17 miles from the Dundee Dam downstream to Newark Bay. A study of this area (the Lower Passaic River Restoration Project) is being done by EPA in cooperation with the U.S. Army Corps of Engineers, the New Jersey Department of Transportation, and other government agencies. The purpose is to develop a comprehensive watershed-based plan for the remediation and restoration of the Lower Passaic River system.

The river, along with Newark Bay, is an estuary and is influenced by semi-diurnal tides up to Dundee Dam. Because of salinity variations from southern Newark Bay (sea water) to Dundee Dam (fresh water), salinity stratification in the vertical direction occurs and causes a reversal of currents between the top and bottom layers of the water column in the lower Passaic. This flow is superimposed on the tidal flow and the freshwater discharge. The combination transports contaminants from their main source in the Passaic throughout the river below Dundee Dam down into Newark Bay and upstream into the Hackensack River. To complicate matters further, the lower section of the Hackensack River includes a large area of tidal wetlands, the Meadowlands. These wetlands serve as a storage area for water and contaminants during tidal cycles and storms.

Large variations in contaminant concentrations occur throughout the sediments in the Passaic — as a function of distance along the river, across the river, and in the vertical direction. In the vertical direction, the maximum concentrations are typically found at a depth of about 2 m, with measurable concentrations down to 4 m or more, whereas the surface concentrations are mostly lower.

Newark Bay is a major commercial harbor. The surrounding area is heavily industrialized due to its proximity to Newark and New York City as well as many other metropolitan areas. The bay is also a tidal system with salt water moving in from the south through the Kill Van Kull and the Arthur Kill. Fresh water enters the bay from the north via the Passaic and Hackensack Rivers.

Plans for cleanup of the contaminated sediments in the Passaic River and Newark Bay are in their initial stages. Restoration plans for the lower 7 miles of the river are to be completed in 2008, and a feasibility study for the entire Lower Passaic River is to be completed in 2012.

1.1.4 Palos Verdes Shelf

From 1947 to 1983, Montrose Chemical Corporation manufactured DDT at its plant near Torrance, California. The plant discharged wastewater containing the pesticide into Los Angeles sewers that emptied into the Pacific Ocean near White's Point on the Palos Verdes Shelf (Figure 1.4). More than 1.5×10^6 kg of DDT were discharged between the late 1950s and the early 1970s. Several other industries also discharged PCBs into the Los Angeles sewer system that ended up on the Palos Verdes Shelf by way of outfall pipes. In 1994, the U.S. Geological Survey determined that an area of approximately 44 km^2 contained elevated levels of DDT and PCBs. EPA later expanded the study area to that shown in Figure 1.4.

The waters of the Palos Verdes Shelf have been used extensively by both sport and commercial fishers as well as for swimming, windsurfing, surfing, scuba diving, snorkeling, and shellfishing. Since 1995, fish consumption advisories and health warnings have been posted in southern California because of elevated DDT and PCB levels. Bottom-feeding fish are particularly at risk for high contamination levels.

In 1996, EPA initiated a non-time-critical removal action to evaluate the need and feasibility for actions to address human health and ecological risks. In 2000, EPA and the U.S. Army Corps of Engineers (COE) initiated a pilot capping project in which they placed clean sediment over a small area (1%) of the contaminated ocean floor. This pilot project provided an opportunity to evaluate cap placement methods and construction-related impacts. In 2002, EPA and COE

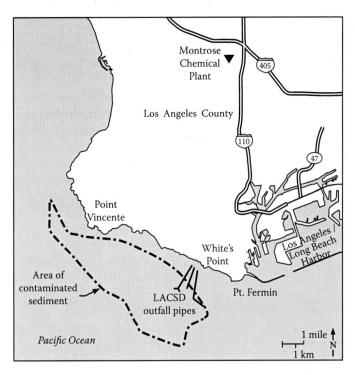

FIGURE 1.4   Palos Verdes Shelf.

concluded that cap construction would be technically feasible. However, they are currently conducting several studies (oceanographic, geotechnical, bioturbation, and resuspension) to better understand sediment fate and transport and the stability of the cap.

In December 2000, the U.S. Department of Justice (representing EPA and several natural resource trustee agencies) and the California Attorney General announced a $73 million settlement with Montrose Chemical as well as other companies that either owned or operated DDT manufacturing plants in Los Angeles County. This increased the total amount available for cleanup of the Palos Verdes Shelf to $140 million.

1.2 MODELING, PARAMETERIZATION, AND NON-UNIQUE SOLUTIONS

1.2.1 MODELING

In water quality investigations, mathematical models are used to summarize and efficiently describe laboratory and field data, to quantify sediment and contaminant transport and fate processes, and also to quantify and predict the future water quality in aquatic systems. In general, these models should be constructed to answer specific questions. In predicting the effects of contaminated sediments on water quality, the questions are usually similar to the following.

1. What will the water and sediment quality be if nothing is done, that is, if natural recovery is assumed? Time periods of interest are 5 years up to as long as 100 years.
2. What will the water and sediment quality be if something is done? For example, possible remediations could consist of some combination of dredging and/or capping in some areas and natural recovery in the remaining areas. Time periods of interest are also 5 to 100 years.
3. Implicit in the answers to these questions is the consideration of the high variability of nature, that is, the effects of large storms and floods. For example, are there contaminated sediments that are now buried that are likely to become exposed, resuspended, and transported due to big events such as large storms and floods?

There are an enormous number of processes (physical, chemical, and biological) that can affect water quality to some degree. What processes to include in a model and the amount of detail necessary in their description should be determined by the problem and the questions asked. Processes necessarily must be included if they significantly contribute to the answers to these questions. They should not be included if they do not. To determine whether processes are sufficiently important to include in a model, simple estimates can usually be made *a priori* without the use of a full, complex model.

In general, mathematical models should not be constructed "with an excruciating abundance of detail in some aspects, while other important facets of the problem are misty or a vital parameter is uncertain to within, at best, an order of magnitude. It makes no sense to convey a beguiling sense of 'reality' with irrelevant detail, when other equally important factors can only be guessed at" (May, 2004). More concisely, Einstein stated "models should be as simple as possible, but not more so."

1.2.2 PARAMETERIZATION AND NON-UNIQUE SOLUTIONS

In water quality models, many parameters are not well understood or quantified. In the absence of other information, they are often determined by parameterization, that is, by varying the values of each parameter until the solution, however defined, fits some observed quantity. For example, in problems of sediment transport, erosion rates and deposition rates are independently varied until the calculated values for the suspended sediment concentration agree with the observed values. In more general water quality models, other parameters such as sediment-water fluxes of PCBs are determined in a similar way.

Although this parameterization procedure is often used in many practical applications, there are serious limitations to this procedure. To illustrate this, consider a simple case of sediment transport where there is a local steady-state equilibrium between the erosion and deposition of sediments. Denote the erosion rate by E and the deposition rate by pw_sC, where p is the probability of deposition, w_s is the settling speed of the particles, and C is the suspended solids concentration. Local steady-state equilibrium then implies that

$$E = pw_sC \qquad (1.1)$$

Rearranging, one obtains

$$C = \frac{E}{pw_s} \qquad (1.2)$$

From this, it is easy to see that a numerical model can "predict" the observed value for C with an almost arbitrary value of E, as long as pw_s is changed accordingly, that is, such that $E/pw_s = C$. For example, a particular value of C can be obtained by high values of erosion and deposition or by low values of erosion and deposition, as long as they balance to give the observed value of C. That is, for a predictive model, unique values for both E and pw_s cannot be determined from calibration of the model by use of the suspended solids concentration alone. It should be emphasized that the quantity of interest in a contaminated sediment problem is usually the depth of erosion (governed by E) and the subsequent exposure of buried contaminants and not the suspended sediment concentration. Accurately predicting suspended sediment concentration, although necessary, is not sufficient for an accurate prediction of sediment erosion and deposition or contaminant flux, exposure, and transport.

The above problem is similar to that of fitting a polynomial through a set of points. For n points, a polynomial of order $m = n - 1$ ($y = a_0 + a_1 x + \ldots + a_{n-1} x^{n-1}$) with n coefficients can be determined that will go exactly through all n points. For $m < n - 1$, it is generally not possible to determine a polynomial that goes exactly through all n points. However, for a polynomial of order $m > n - 1$, an infinite number of solutions can be found that go exactly through all n data points; that is, the solution is inadequately constrained. For most water quality problems, the parameters and hence the solution are inadequately constrained and an infinite number of solutions is therefore possible.

As a practical illustration of this problem, consider the sediment and contaminant transport modeling in the Lower Fox River (Tracy and Keane, 2000). Two groups independently developed transport models. Each group calibrated their model based on suspended sediment concentration measurements. Each group believed that the parameters used in their model were reasonable. However, the results predicted by the two models were quite different, in the transport of both sediments and contaminants. As an example, the amount of sediment eroded at one location at a shear stress of 1.5 N/m² (a large, but not the maximum, shear stress in the Fox) was predicted by one group to be 11.3 g/cm² (on the order of 10 cm), whereas the other group predicted 0.1 g/cm² (or 0.1 cm), a difference of two orders of magnitude!

This difference, of course, has a direct impact on the choice of remedial action. Small or no erosion at high shear stresses indicates that contaminants are probably being buried over the long term and natural recovery is therefore the best choice of action. Large amounts of erosion indicate that buried contaminants may be uncovered, be resuspended, and hence will contaminate surface waters and bottom sediments elsewhere; dredging and/or capping is therefore necessary. The differences in the model estimates by the two groups make it difficult to decide on the appropriate remedial action.

The problem of non-unique solutions is compounded when additional processes and parameters are considered. For example, if erosion and deposition rates are determined incorrectly, the sediment-water flux of contaminants due to erosion/deposition is, of course, also determined incorrectly. The sediment-water flux of contaminants due to non-erosion/deposition processes (sometimes called "diffusion" processes, i.e., molecular diffusion, bioturbation, groundwater flow, etc.) is usually determined as the difference between the measured total flux and the flux due to erosion/deposition. If the latter is incorrect, then the diffusive flux is also incorrect. Because the sediment-water flux due to erosion/deposition and the sediment-water flux due to "diffusion" processes act differently over time, long-term predictions of sediment-water fluxes of contaminants and their effects on sediment and water quality will be incorrect.

It is sometimes argued that, in order to make a solution unique, additional constraints can be imposed on the solution by the collection of data on additional variables. In some cases and for some variables, this is certainly true. For example, in sediment transport problems, other measured parameters in addition to suspended sediment concentration might be erosion rates or changes in sediment bed thickness. If this is done accurately throughout the body of water, then non-uniqueness of solutions can be eliminated or at least minimized, consistent with Equation 1.2 and subsequent arguments. However, when some additional variables (such as PCB concentrations) are introduced, additional parameters (such as PCB partition coefficients, sorption rates, and sediment-water fluxes) are also introduced. As a result, if these additional parameters are not adequately determined, the solution is then no more constrained than before, and non-unique solutions are again possible and, indeed, seem to be the rule. This is discussed further in Sections 8.6 and 8.7.

To eliminate, or at least minimize, this non-uniqueness of solutions, the significant processes and parameters must be determined independently and quantitatively. For this as well as other reasons, the emphasis throughout this book is on a thorough and quantitative understanding of the processes that are significant in sediment and contaminant transport.

1.3 THE IMPORTANCE OF BIG EVENTS

In recent years, it has been realized that big events, such as large storms and floods, have a major effect on the transport and fate of sediments and contaminants and hence on water quality. Much of our early understanding of the significance of big events derived from geologists who were concerned with somewhat different, but related, events — that is, events that could be interpreted from the stratigraphic record. In this case, there has been a long-standing controversy between the uniformatarionists and the catastrophists (Ager, 1981). The uniformatarionists believe in the "gentle rain from heaven" theory; that is, sedimentary conditions and rates are uniform with time, and the stratigraphic record can be interpreted from a knowledge of the present-day conditions and rates. Catastrophists, on the other hand, believe that the sedimentary record is primarily determined by large episodic events separated by long periods of time when very little occurs.

Ager (1981), in his book entitled *The Nature of the Stratigraphical Record*, is a strong and persuasive proponent of the importance of big events in the interpretation of the stratigraphic record. In the book, he presents information on numerous catastrophic events and their effects on the stratigraphic record and emphasizes the spasmodic nature of sedimentation. He concludes that "Nothing is worldwide, but everything is episodic."

Of course, as far as pollution in surface waters is concerned, the spatial and temporal scales of concern are generally smaller than those described by Ager. Nevertheless, a careful examination of present-day sediment dynamics at the smaller spatial and temporal scales of interest in pollution problems also leads to recognition of the importance of the large and rare event in the transport and fate of sediments (Lick, 1992). More importantly, it follows that the big event is also of major significance in the transport and fate of contaminants and in the resultant exposure of organisms to these contaminants. It is now generally agreed that large episodic events such as storms on lakes, large runoffs and floods in rivers, and hurricanes in coastal areas, despite their infrequent occurrence, are responsible for much, if not most, of the sediment and contaminant transport in surface waters. A corollary to this is that, over a long enough period of time, exceptional events are not the exception but the rule.

As an example of the high variability that might be expected in a river, consider Figure 1.5, which shows the flow rate in the Saginaw River as a function of time for the 50-year period from 1940 to 1990 (Cardenas et al., 1995). The average flow rate is about 50 m^3/s. However, this average flow is punctuated by high flows of relatively short duration. Twice during the period, flow rates of almost 2000 m^3/s have occurred. For this river, the ratio of the largest flow to the average flow is about 40; the ratio of the once-in-5-year flow to the average flow is about 24. For comparison, the latter ratio is about 60 for the Buffalo River and approximately 4 for the Fox River, which is strongly controlled by a series of dams. Large variations in currents, wave action, and sediment transport also occur in lakes and near-shore areas of the oceans, as is indicated, for example, in Figure 1.6, which shows the turbidity and bottom shear stress at the Oregon water intake in the western basin of Lake Erie in 1977 (Paul et al., 1982).

To qualitatively understand the effects of large variations in winds, currents, and wave action on sediment and contaminant transport, consider as a specific

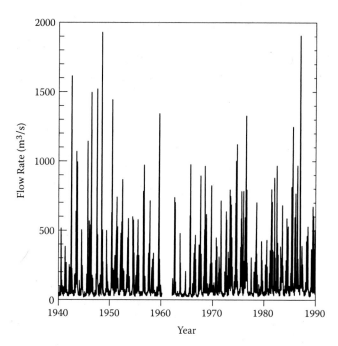

FIGURE 1.5   Saginaw River: flow rate as a function of time from 1940 to 1990. (*Source:* From Cardenas et al., 1995. With permission.)

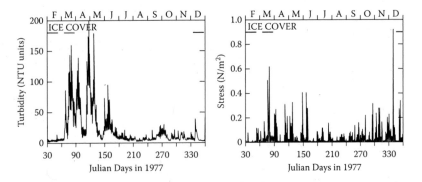

FIGURE 1.6   Oregon water intake in Lake Erie in 1977: turbidity (measured) and bottom shear stress due to wave action (calculated) as a function of time. (*Source:* From Paul et al., 1982. With permission.)

example the forcing of currents and/or wave action in a lake by a wind with speed U. The magnitude of the steady-state wind-driven currents and/or wave action is approximately proportional to the wind speed. It is also well known that the bottom shear stress τ due to currents and/or wave action is approximately proportional to the square of the magnitude of the currents and/or wave action. From this it follows that τ is proportional to the square of the wind speed; that is,

$$\tau \sim U^2 \tag{1.3}$$

Experimental evidence indicates that the amount of sediment resuspended, ε, is a nonlinear function of the shear stress, perhaps proportional to as much as the cube of the shear stress; that is,

$$\varepsilon \sim \tau^3 \tag{1.4}$$

From the above equations, it follows that the amount of sediment resuspended for a particular wind speed is approximately proportional to the sixth power of the wind speed; that is,

$$\varepsilon \sim U^6 \tag{1.5}$$

For example, when the wind speed is doubled, the amount of sediment resuspended is increased by a factor of 64; when the wind speed is increased by 4, the amount of sediment resuspended is increased by a factor of 4096. In practice, this increase will generally be limited by the consolidation and armoring of the sediments; nevertheless, it can be readily seen that the resuspension of bottom sediments is a very nonlinear and rapidly increasing function of the wind speed. The amount of sediment transported, of course, depends on the amount of sediment resuspended, as well as the currents, and is therefore also a very nonlinear function of the wind speed.

To substantiate these very qualitative arguments, calculations of sediment transport in Lake Erie were made for a variety of wind speeds and directions as well as the 1940 Armistice Day storm, probably the largest storm on Lake Erie in the last century (Lick et al., 1992). For purposes of verification, calculations of sediment deposition were then compared with ^{210}Pb and ^{137}Cs data at an Eastern Basin core location. The data indicate that deposition at this site was very non-uniform with time, with infrequent large depositions caused by major storms that were separated by long periods of time in which very little deposition occurred. The results of the calculated deposition for the Armistice Day storm as well as those for other storms were consistent with this idea. This is discussed in more detail in Section 6.5.

From the above and similar calculations, it was shown in general that winds with average and below-average speeds caused negligible deposition at this site compared to the 1940 storm or any equivalent large storm. As the wind speed increases, the probability of an event with this speed decreases, but the net transport of sediment caused by the total of all events with a similar wind speed increases. On the basis of these calculations, it was hypothesized that the largest storms, despite their infrequent occurrence, cause the most transport of any class of wind events and are responsible for more of the transport in Lake Erie than the total of all of the lesser storms and wind events.

Additional examples to substantiate this hypothesis are as follows. First, consider Figure 1.7, which shows the cumulative percentage of suspended sediment yield as a function of the percentage of total time for the period from 1972 to 1979 for the River Creedy, Devon (Walling and Webb, 1981). It can be seen that 90% of the sediment transport occurs in less than 5% of the time. As another example, Figure 1.8 shows the huge increase in sediment discharge and suspended sediment concentration in the Susquehanna River in 1972 due to Tropical Storm

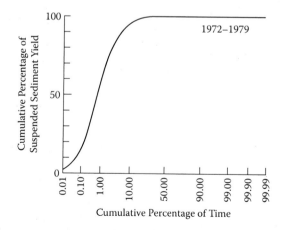

FIGURE 1.7 River Creedy, Devon, for the period from 1972 to 1979: cumulative curve of suspended sediment yield as a percentage of total time. (*Source:* From Dyer, 1986. With permission.)

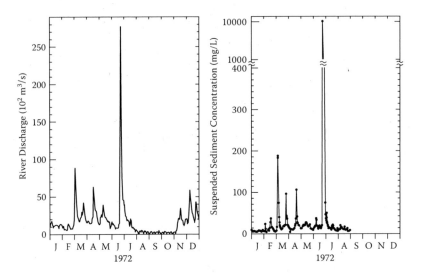

FIGURE 1.8   Susquehanna River during 1972: river discharge and suspended sediment concentration as a function of time. (*Source:* From Dyer, 1986. With permission.)

Agnes (Schubel, 1974). In 1 week, the Susquehanna discharged more sediment than in the previous 50 years. It has been estimated (Zabawa and Schubel, 1974) that more than 50% of the sediment deposited in the upper Chesapeake Bay in the last 100 years was due to Tropical Storm Agnes and another flood event in 1936.

A generalized and very approximate hypothesis that follows from these studies is that, during any specified period of time (whether it is 1, 20, or 100 years), the largest events (either storm or flood) expected during that period are responsible for more of the sediment and contaminant transport than all the other events during that period. It follows that, in attempting to predict the trend of water quality in surface waters over long periods of time, it is the large floods and/or storms that must be considered in the modeling and prediction of the transport and fate of sediments and contaminants. This, of course, is contrary to many investigations that emphasize the uniform deposition of sediments and the chemical flux from the bottom sediments due to steady-state diffusion. In some situations, these may be valid approximations. Nevertheless, the above studies illustrate and emphasize the dynamic nature of sediment and contaminant transport and also the importance of large events in this dynamics. There is no steady state. Because of this and the nonlinearity of the processes, an average state is difficult to define and may not be meaningful.

As stated by Ager (1981) when writing about the stratigraphic record, "The history of any one part of the Earth, like the life of a soldier, consists of long periods of boredom and short periods of terror." In the following, it is emphasized that sediment and contaminant transport are also like the life of a soldier with long periods of moderate conditions where little sediment and contaminant transport occurs and short periods of severe storm and flood conditions where much or most of the transport occurs.

1.4 OVERVIEW OF BOOK

As stated previously, the purpose of this book is to assist in the quantitative understanding of and ability to predict the transport of sediments and hydrophobic contaminants in surface waters. Chapter 2 is concerned with general properties of sediments, properties that will be needed and referred to throughout the rest of the book. Properties of individual particles (sizes and size distributions, settling speeds, and mineralogy) are discussed first. However, most particles do not exist as individual particles but are aggregated together into flocs. An introduction to flocculation is then given because flocculation affects sediment erosion, settling, deposition, consolidation, and, of course, the transport of sediments and contaminants. As sediments deposit on the bottom and are buried by sediments depositing above them, the sediment aggregates are compressed by the weight of the overlying sediments; the water and gas in the pores of the aggregates are forced upward toward the surface, and the bulk density of the bottom sediment (including particles, water, and gas) generally increases with time and with depth in the sediments. These processes and measurements of sediment density are discussed in the final section of Chapter 2.

The most significant process in the modeling of sediment transport is sediment erosion. The reason for this is that erosion rates control sediment and contaminant transport and are also highly variable in space and time — often by orders of magnitude. The erosion rate strongly depends on many sediment parameters (e.g., particle size, bulk density, mineralogy, volume of gas, organic content, bioturbation, consolidation time) and hence is very difficult to predict. No theory of sediment erosion rates exists that is uniformly valid over a wide range of parameters. Because of this, erosion rates must be measured. Chapter 3 describes devices for measuring erosion rates as well as results of field and laboratory measurements of erosion rates. These results are then used to develop approximate and useful equations for the initiation of motion and for erosion rates for a wide range of sediments. The effects of the slope of the sediment bed on erosion also are discussed.

In addition to erosion, knowledge of the settling and deposition of particles/flocs and their subsequent consolidation in the bottom sediments is necessary. Most sediments that are contaminated tend to be finer-grained. Because of this, the flocculation of these sediments occurs and greatly affects settling, deposition, and consolidation. Chapter 4 first discusses experimental and theoretical work on the flocculation of suspended sediments; this is followed by a discussion of the settling speeds of flocs, the modeling of flocculation, the deposition of particles and flocs, and the consolidation of sediments.

In the modeling of sediment and contaminant transport in surface waters, an understanding of and ability to predict the hydrodynamics (both currents and wave action) are essential. For many reasons, there have been extensive investigations of the hydrodynamics of rivers, lakes, estuaries, and oceans; much of this work has been incorporated into numerical models, many of which are useful for our present purposes. All hydrodynamics are essentially described by the three-dimensional, time-dependent Navier-Stokes equations, supplemented by other equations such as an energy equation, an equation of state, and equations for various parameters appearing in these equations. However, the full set of equations is much too complex and cumbersome for practical use, so the equations are generally approximated by, for example, reducing the number of space dimensions, eliminating time as a variable and only considering steady-state solutions, or using turbulence models to approximate small-scale hydrodynamic processes. Chapter 5 briefly discusses some of these models, their applications, and their limitations.

Numerous models of sediment transport also exist. They differ in the number of space dimensions, the inclusion of time variation, and how they approximate various sediment transport processes. If the processes discussed in Chapters 2, 3, and 4 were well understood, well quantified, and easy to measure, then the modeling of sediment transport would be relatively easy. As will be apparent, this is not true. All of the processes are not understood and have not been quantified as well as might be desired; in addition, sediment properties are generally highly variable throughout an entire system and are therefore difficult to adequately measure. Because of this, semi-empirical approximations, or even parameterizations of some transport processes, are necessary in some cases. However, the emphasis here is on accurately describing and approximating the fundamental processes and hence minimizing parameterization. Chapter 6 presents applications of models to sediment transport in rivers, lakes, bays, and estuaries. The advantages and limitations of different models are also discussed.

Hydrophobic organic chemicals (HOCs) are of concern because they are often toxic and do not degrade readily. Because of their hydrophobic nature, they sorb to particles and are transported with them. When these sorbed contaminants are deposited as part of the bottom sediments, they may later desorb and serve as a large source of contamination to the overlying water, to benthic organisms, and to organisms in the overlying water. The adsorption and desorption of these chemicals to sedimentary particles and aggregates is time dependent, with time scales (days to years) that are often comparable to or greater than the transport times of interest (seconds to days). Because of this, the assumption of chemical equilibrium between the chemical sorbed to the sediment particles and that dissolved in the water, a common assumption, may not be a valid approximation. Chapter 7 discusses the time-dependent sorption and partitioning of HOCs to sedimentary particles and aggregates. The emphasis is on HOCs such as PCBs and hexachlorobenzene (HCB). However, numerous chemicals, such as flame retardants (e.g., PBDE), surfactants, and synthetic musks, as well as a variety of metals (e.g., mercury) are hydrophobic and act in a similar manner to the HOCs emphasized here.

The flux of chemicals between the sediments and the overlying water is primarily due to sediment erosion/deposition, molecular diffusion, bioturbation, and groundwater flow. Gas transport also may be important in some cases. All these processes are significantly influenced by sorption and especially the finite rate at which HOC sorption occurs. These fluxes are often not well understood or quantified so that they are often parameterized with little reference to the actual flux processes. Chapter 8 first discusses models of the transport of HOCs as affected by sediment erosion/deposition. This is followed by a description of a simple diffusion approximation for the sediment-water flux of a dissolved chemical. Experimental results and models of the sediment-water flux of HOCs due to molecular diffusion and bioturbation are presented in Sections 8.3 and 8.4. A critical comparison of these results with a more-or-less standard mass transfer approximation is then made in Section 8.5. Sediment and contaminant transport models are generally complex, and it is often difficult to determine from the model results how accurate the model is, how significant each of the processes in the model is, and whether all the significant processes have been included. For these purposes and as an example, Section 8.6 presents a relatively simple study of sediment and contaminant transport for a particular case. Section 8.7 revisits the topic of water quality modeling, parameterization, and non-unique solutions.

2 General Properties of Sediments

Certain properties of sediments are basic and common to many of the topics in the remainder of this book and are presented here. In Sections 2.1, 2.2, and 2.3, essential characteristics of individual sedimentary particles (particle size, settling speed, and mineralogy) are described. Although individual particles are the basic building blocks for suspended as well as bottom sediments, sedimentary particles generally do not exist in the form of isolated particles. In the overlying water, finer-grained particles are generally present as aggregates of particles, that is, as flocs; in contrast, coarser-grained particles usually do not aggregate and are present as individual particles. Properties of a floc, such as size and density, are often significantly different from those of the individual particles making up the floc. An introduction to flocculation and the properties of flocs is given in Section 2.4. As the suspended particles or flocs settle out of the overlying water, they are deposited on the sediment bed, where they may be buried by other depositing particles. As this occurs, flocs are compressed and formed into larger and denser aggregates. Due to the weight of the overlying sediments, interstitial water and gas are forced out from between the solid particles/aggregates in the sediment bed and are transported toward the sediment-water surface; differential settling and transport of different size particles may also occur. In addition, gas may be generated within the sediments due to the decay of organic matter. The bed consolidates with time due to these processes and also due to chemical reactions in the bed. In Section 2.5, procedures for measuring the bulk densities of these bottom sediments are discussed, and illustrations of the variations of bulk densities with depth and time are given.

2.1 PARTICLE SIZES

2.1.1 CLASSIFICATION OF SIZES

The most obvious property of a sediment is the sizes of its particles. Because most particles are irregular in shape, a unique size or diameter is difficult to define. However, a conceptually useful and unique diameter can be defined as the diameter of a sphere with the same volume as the particle. Nevertheless, in practice, effective particle diameters are usually defined operationally, that is, by the technique used to measure the particle sizes. Because of this, small differences in particle sizes as measured by different techniques may occur. For a typical sediment of interest here, particle diameters generally range over several orders of magnitude, often from less than a micrometer to as much as a centimeter. Because a particular measuring technique is only valid for a limited range of particle sizes, more than one measurement technique may be necessary to measure the complete size distribution of a sediment.

Particle sizes often are classified using the Wentworth scale (Table 2.1). This scale standardizes and quantifies the definitions of terms commonly used to describe sediments (e.g., clay-size, silt, sand). The basic unit of the scale is 1 mm; different size classifications follow by multiplying or dividing by two. For example, very coarse sand is defined as particles with diameters between 1 and 2 mm, and very fine sand is defined as particles with diameters between 1/8 and 1/16 mm. Broader classifications that will be used extensively are: clay-size particles with $d < 1/256$ mm; silts, $1/256 < d < 1/16$ mm; sands, $1/16 < d < 2$ mm; and larger particles (granules, pebbles, cobbles, and boulders), $d > 2$ mm. In units of micrometers (μm), these sizes are clay-size ($d < 3.91\ \mu$m), silts ($3.91 < d < 62.5\ \mu$m), sands ($62.5 < d < 2000\ \mu$m), and larger particles ($d > 2000\ \mu$m). For convenience, a scale with other metric units is also given. Because the Wentworth scale proceeds in

TABLE 2.1
Classification of Particle Sizes

Wentworth scale	Particle diameter		
	mm	metric	phi units
Boulder			
	256		−8
Cobble	128	10 cm	−7
	64		−6
Pebble — Large	32		−5
Pebble — Medium	16	1 cm	−4
Pebble — Small	8		−3
Pebble — Very small	4		−2
Granule	2	2 mm	−1
Sand — Very coarse	1	1 mm	0
Sand — Coarse	1/2		1
Sand — Medium	1/4		2
Sand — Fine	1/8		3
Sand — Very fine	1/16	63 μm	4
Silt — Coarse	1/32		5
Silt — Medium	1/64		6
Silt — Fine	1/128	10 μm	7
Silt — Very fine	1/256	4 μm	8
Clay — Coarse	1/512		9
Clay — Medium	1/1024	1 μm	10
Clay — Fine	1/2048		11
Clay — Very fine	1/4096		12

powers of two, the phi scale is sometimes used (usually by geologists), where $\phi = -\log_2$ (particle diameter in mm), or $d = 2^{-\phi}$. The negative logarithm is used so that particles smaller than 1 mm (the most frequently encountered particle diameters) have positive values of ϕ. The phi scale also is shown in Table 2.1.

2.1.2 MEASUREMENTS OF PARTICLE SIZE

Various methods are used to measure particle sizes. The most common are sieving, sedimentation, and the use of light diffraction instruments. Sieving is done by means of a series of stainless steel sieves with aperture sizes at 1/2 phi intervals. The sediments generally are dried, disaggregated, and passed through a series of these sieves; the residue on each sieve is then weighed. Because fine sieves tend to become clogged rather rapidly, sieving generally is useful only for coarse sediments with $d > 63$ μm.

For fine- and medium-size particles, various types of sedimentation procedures have been popular in the past and are still used to some extent. These procedures depend on the fact that different-size particles settle at different speeds and use Stokes law (Section 2.2) to deduce particle size from a measurement of settling speed. There are numerous difficulties with these procedures, including time-dependent flocculation of fine-grained particles as they settle, hindered settling due to the upward movement of water when sediment concentrations are high, and the slow dissipation of the turbulence due to the initial mixing of the sediment-water mixture. Because of this, these sedimentation procedures are not used extensively at present.

The most accurate and convenient procedure for measuring particle sizes over a wide range is by means of light diffraction instruments. In these instruments, light is passed through low concentrations of the sediment-water mixture and is diffracted by the particles. The particle size distribution is then determined from the diffraction pattern. The procedure is non-invasive and, with different lenses, can measure particle sizes from about 0.1 to 2000 μm. For example, the Malvern Particle Sizer (Mastersizer X) can measure particle sizes from 0.1 to 80 μm with one lens, from 0.5 to 180 μm with a second lens, from 1.2 to 600 μm with a third lens, and from 4.0 to 2000 μm with a fourth lens. Only sieving for large particles and the use of a light diffraction instrument for all other particles are recommended for accurate particle sizing.

2.1.3 SIZE DISTRIBUTIONS

In a natural sediment sample, particle sizes are generally not uniform or even close to uniform; a wide distribution of sizes generally is present. As an example, the particle size distribution for a sediment from the Detroit River in Michigan is shown in Figure 2.1(a), where the fraction of particles by mass in a particular size class is plotted as a function of the logarithm of the diameter of that size class. A wide asymmetric distribution is evident, with a significant fraction of particles less than 4 μm (clays) and also greater than 62.5 μm (sands). However, even when it is known, the detailed distribution of sizes usually generates more information than is needed or can be used. Because of this, the distribution often is described in terms of various statistical parameters (e.g., the mean, standard deviation, skewness, and possibly higher moments). However, only the simplest parameters (the mean, median, and mode) are used here.

The mean (also generally known as the average) is defined as the center of gravity of the area under the distribution curve on a linear scale; that is,

$$d(\text{mean}) = \sum_i x_i d_i \qquad (2.1)$$

where d_i and x_i are the diameter and fraction of particles by mass in size class i, respectively. The median particle diameter is defined as that diameter for which 50% of the sediment by mass is smaller and 50% is greater, whereas the mode is defined as the diameter of the size fraction with the largest fraction of particles within it. For the size distribution shown in Figure 2.1(a), the mean, median, and mode are 18, 6.0, and 2.7 μm, respectively.

In some cases, the size distribution (on a log-diameter scale) may be similar to a Gaussian distribution; this is termed "log-normal." However, there is no known theoretical reason why this should be true. In fact, upon close examination, most sediment size distributions are not log-normal. As a specific example, a log-normal distribution (with a mean of 18 μm and a standard deviation the same as that of the Detroit River sediments) is shown in Figure 2.1(a) and can be compared there with the Detroit River size distribution. The Detroit River distribution is clearly not log-normal.

As additional examples, size distributions for three sediment samples from the Detroit River (different from the previous sample), the Fox River in Wisconsin, and the Santa Barbara Slough in California are shown in Figure 2.1(b) (Jepsen et al., 1997). The means are 12, 20, and 35 μm, respectively.

The general character of the size distributions shown in Figures 2.1(a) and (b) is more or less typical of most sediments with a wide range of particle sizes and with one dominant size fraction. A more unusual distribution is shown in Figure 2.1(c), where measurements of two subsamples of a single and larger sample from Lake Michigan (McNeil and Lick, 2002a) are plotted. Both measurements were made by a Malvern particle sizer and show close agreement with each other. Two peaks in each distribution are obvious, indicating that the sediment may be a mixture of sediments from two different sources. For both subsamples, the means (66.5 μm) and medians (22.0 and 20.7 μm) are in good agreement with each other. However, the modes (149.2 and 2.1 μm) are not. The reason for this is that the modes depend in a very sensitive manner on the relative heights of the two peaks, which are slightly different for the two subsamples. In this particular case, neither the mean, median, nor mode is a useful parameter for characterizing the particle size distribution. Quite clearly, the distributions shown in Figures 2.1(b) and (c) are also not log-normal. Herein, log-normality will not be assumed for any size distribution.

Because transport and chemical sorption properties often and significantly depend on particle size, some representation of the distribution of particle sizes is often necessary. For modeling purposes, this is usually done by separating the entire size distribution into three or more intervals, with the corresponding mass

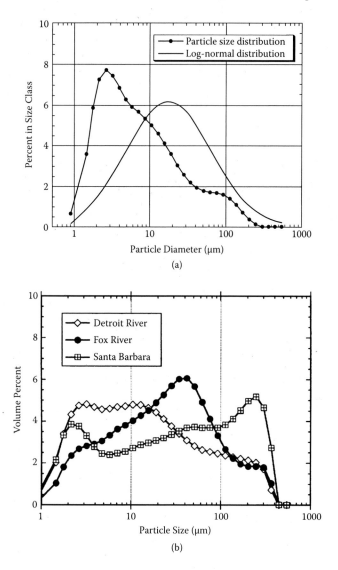

FIGURE 2.1 Particle size distributions. Percent by volume (mass) as a function of diameter: (a) sediments from the Detroit River. Also shown is a log-normal distribution with a mean and standard deviation the same as the Detroit River sediments; (b) sediments from the Detroit River, Fox River, and Santa Barbara Slough (From Jepsen et al., 1997. With permission.)

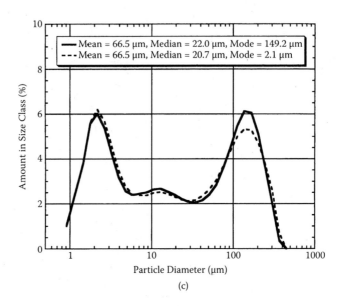

(c)

FIGURE 2.1 (CONTINUED)   Particle size distributions. Percent by volume (mass) as a function of diameter: (c) sediments from Lake Michigan near the Milwaukee bluffs. (*Source:* From Jepsen et al., 1997. With permission.)

fraction of particles given in each interval — for example, 20% fine (0–10 μm), 50% medium (10–64 μm), and 30% coarse (>64 μm). The number of intervals (or size classes) and the range of particle sizes within each interval depend on the sediment and the application. Examples are given in Chapter 6.

2.1.4 Variations in Size of Natural Sediments throughout a System

Sediment properties, including particle size, often vary greatly throughout a sediment bed, in both the horizontal and vertical directions. Sediment types can change rapidly in the horizontal direction from coarse sands (where currents and wave action are strong) to fine-grained muds (where currents and wave action are small). They also can change rapidly in the vertical direction. For example, this can happen due to strong storms that may deposit layers of coarse sands on top of finer sediments. During quiescent conditions, this coarse layer may then be covered by fine sediments.

As an example of horizontal variations in particle size, consider the surficial sediments of Lake Erie. The bathymetry of the lake is shown in Figure 2.2(a). Lake Erie is a relatively shallow lake and can be conveniently separated into three basins: (1) the Western Basin, with an average depth of 6 m; (2) the Central Basin, with an average depth of 20 m; and (3) an Eastern Basin, with an average depth of 26 m. Mean particle sizes (on a phi scale) of the disaggregated surficial sediments are shown in Figure 2.2(b) (Thomas et al., 1976) and range from greater than medium sand (d > 1/2 mm, ϕ < 1) to less than coarse clay-size particles

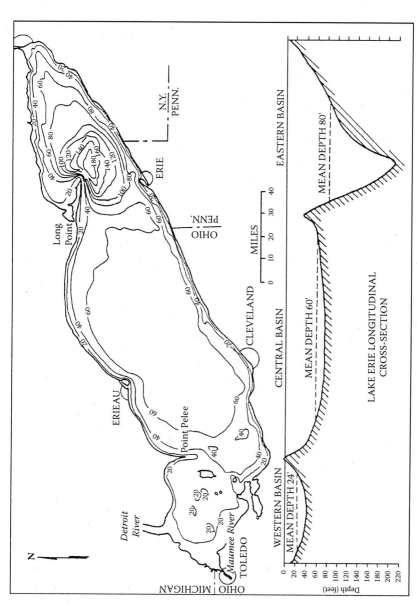

FIGURE 2.2(a)   Lake Erie: bathymetry, depth in feet (*Source:* From Thomas et al., 1976. With permission.)

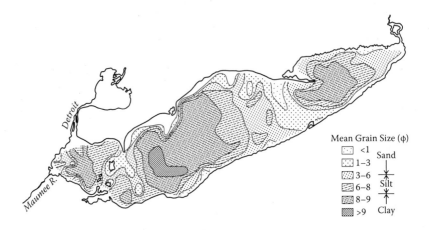

Mean Grain Size (φ)

<1 } Sand
1–3
3–6 } Silt
6–8
8–9 } Clay
>9

FIGURE 2.2(b) Lake Erie: mean grain size of surficial sediments. (*Source:* From Thomas et al., 1976. With permission.)

(d < 2 μm, φ > 9). Particles tend to be coarser in the shallower parts and finer in the deeper parts of each basin. This variation in particle size depends on the primary source of the sediments (major primary sources for Lake Erie are the Maumee River, the Detroit River, and shore erosion, especially along the northern shore of the Central Basin); differential erosion of the bottom sediments due primarily to wave action (greater in shallow waters than in deep); transport of the suspended sediments by currents and wave action; and differential settling of the particles. This is discussed more quantitatively in Chapter 6.

As an example of vertical variations in particle size, Figure 2.3 shows the mean diameter, d, as a function of depth for a sediment core from the Kalamazoo River in Michigan (McNeil and Lick, 2004). An irregular variation of d from 85 μm at the surface to 200 μm at a depth of 10 cm, to 70 μm at 15 cm, and to 180 μm at 19 cm is shown. Although vertical variations in d can be relatively small, the rapid and irregular variation shown here is not atypical.

Also shown in Figure 2.3 is the variation of sediment bulk density with depth. A close correlation between bulk density and particle size is evident. However, other quantities in addition to particle size also influence the bulk density so that the correlation is not exact. The increase in particle size and density between 8 and 14 cm is probably due to a large flood transporting coarse sediments from upstream; these then deposit near the end of the flood as the flow velocity decreases. After this flood, the flow velocity is relatively low, and only finer sediments can be transported from upstream and deposited in this area.

In general, the mean size and size distribution of suspended sediments will be different from those of the bottom sediments at the same location. This is due not only to transport of the suspended sediments from other locations where sediments have different sizes but also to the dependence of suspended particle size on the magnitude of the shear stress causing the resuspension. This is illustrated in Figure 2.4, which gives the disaggregated particle size distribution for

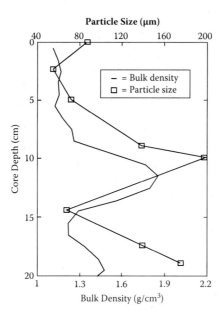

FIGURE 2.3 Mean grain size and bulk density as a function of depth for a sediment from the Kalamazoo River. (*Source*: From McNeil and Lick, 2004. With permission.)

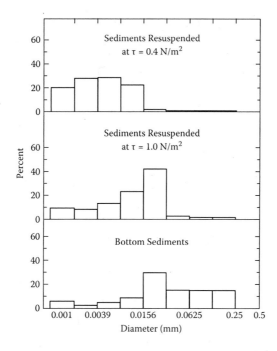

FIGURE 2.4 Size distributions of a bottom sediment and the same sediment resuspended at shear stresses of 1.0 and 0.4 N/m². (*Source*: From Lee et al., 1981. With permission.)

well-mixed bottom sediments and for these same sediments when they have been resuspended at shear stresses of 1.0 and 0.4 N/m² (Lee et al., 1981). It can be seen that the suspended sediments have a smaller average particle size than the bottom sediments and that this average size depends on the shear stress — that is, the greater the shear stress, the greater the average suspended particle size.

2.2 SETTLING SPEEDS

When a particle is released into quiescent water, there will be an initial transient in the vertical speed of the particle until a steady state is reached. This steady state is due to a balance between the gravitational force, F_g, and the drag force, F_d, on the particle. The distance until the steady state is reached depends on the properties of the particle and of the fluid but, for a typical sedimentary particle in water, will generally be on the order of a few particle diameters.

The steady-state speed is known as the settling speed, w_s. For a solid particle, w_s will depend on the size, shape, and density of the particle as well as the density and viscosity of the fluid. For aggregates of particles, w_s will depend not only on these same parameters for the aggregate and the fluid but also on the permeability of the aggregate.

Because of the complex dependence of w_s on the above parameters, w_s generally cannot be determined theoretically but must be determined experimentally. However, in certain limiting cases, theoretical determinations of w_s can be made. A most important case is the simplest case, that is, a solid spherical particle with diameter d, volume $V = \pi d^3/6$, and density ρ_s settling at constant speed in a fluid of density ρ_w and kinematic viscosity μ. In this case, the gravitational force is just due to the immersed weight of the particle, or

$$F_g = V(\rho_s - \rho_w)g$$

$$= \frac{\pi d^3}{6}(\rho_s - \rho_w)g \qquad (2.2)$$

where g is the acceleration due to gravity and is equal to 980 cm/s². For slow, viscous flow past a sphere, the drag force has been determined by Stokes and is (Schlichting, 1955):

$$F_d = 3\pi\mu d w_s \qquad (2.3)$$

By equating the gravitational and drag forces as given by the above two equations, one obtains

$$w_s = \frac{gd^2}{18\mu}(\rho_s - \rho_w) \qquad (2.4)$$

which is known as Stokes law. The formula is valid for settling speeds such that the Reynolds number (Re $= \rho_s w_s d/\mu$) is less than about 0.5. For solid sedimentary

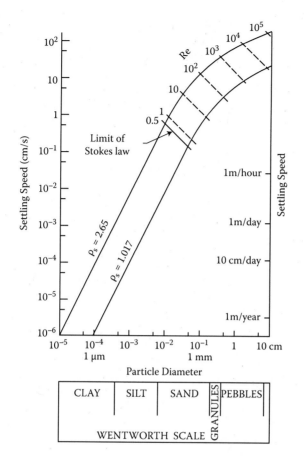

FIGURE 2.5 Stokes law for settling speed as a function of diameter.

particles with a density of 2.65 g/cm³, this corresponds to a diameter of about 100 μm. Stokes law for settling speed as a function of diameter is shown in Figure 2.5. Straight-line relations are shown for densities of 2.65 g/cm³ (an average density for mineral particles) and 1.017 g/cm³ (an immersed density of 0.017 g/cm³ or 1/100 that of the immersed density of a mineral particle and similar to the effective densities of many phytoplankton and flocculated sediments).

For a particle with a density of 2.65 g/cm³ in water at 20°C ($\mu = 1.002 \times 10^{-2}$ g/cm/s, $\rho_w = 0.9982$ g/cm³), the above equation can be written as

$$w_s = a\, d^2 \qquad (2.5)$$

where a = 0.898×10^4/cm-s and the units of d and w_s are cm and cm/s, respectively. For d in μm and w_s in μm/s, $w_s = 0.898\, d^2$, or

$$w_s \cong d^2 \qquad (2.6)$$

This is a convenient and easy-to-remember equation for $w_s(d)$, especially for fine-grained sediments whose diameters are often expressed in micrometers.

In the derivation of Stokes drag (Equation 2.3), it is assumed that the flow is steady and laminar and that there is no wake behind the body. This is correct for $Re < 0.5$; for $Re > 0.5$, the flow separates behind the sphere and vortices are shed. As Re increases beyond 0.5, these vortices are first shed periodically and symmetrically, then periodically and asymmetrically, and then randomly (Schlichting, 1955). As a result, Stokes law is no longer valid for $Re > 0.5$. In this case, the drag force is a function of the Reynolds number, a dependence that has not been determined theoretically but has been determined experimentally. It is generally reported as:

$$F_d = C_d A \frac{\rho_w}{2} w_s^2 \qquad (2.7)$$

where $A = \pi d^2/4$ and is the cross-sectional area of the particle, and C_d is a drag coefficient. This coefficient is a function of the Reynolds number, must be determined experimentally, and varies as follows (Schlichting, 1955). For $Re < 0.5$, $C_d = 24/Re$. As Re increases, the deviation of C_d from this relation continually increases; for Re greater than about 10^3, C_d approaches a limiting value of 0.4. Once C_d is known, w_s can be determined by equating the gravitational and drag forces as before. The effect of Re on w_s can be seen in Figure 2.5. For large Re, when $C_d \cong 0.4$, it follows from Equations 2.2 and 2.7 that w_s is then proportional to $d^{1/2}$.

Of course, most sedimentary particles are irregular in shape and are not spheres. Because of this, they often spin and oscillate as they settle. The flow around them is no longer smooth or steady, and their hydrodynamic drag is greater than for smooth, steady flow. As a result, settling speeds for most particles are somewhat lower than those given by Stokes law. On the basis of experimental data for natural particles, Cheng (1997) has developed an equation that more closely approximates $w_s(d)$; it is given by:

$$w_s = \frac{v}{d} \left[\left(25 + 1.2\, d_*^2 \right)^{1/2} - 5 \right]^{3/2} \qquad (2.8)$$

where

$$d_* = d \left[\left(\rho_s - 1 \right) \frac{g}{v^2} \right]^{1/3} \qquad (2.9)$$

and $v = \mu/\rho_w$. This relation is compared with Stokes law in Figure 2.6.

In the above, it has been implicitly assumed that $\rho_s > \rho_w$, that particles move vertically in a downward direction, and that w_s is positive. Particles with densities less than that of water are buoyant and will move vertically upward, and

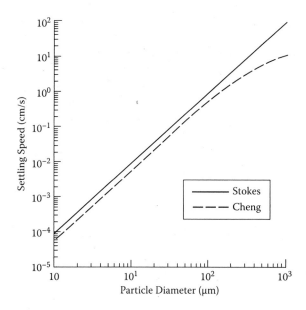

FIGURE 2.6 Settling speeds as a function of diameter as calculated from Cheng's equation and Stokes law.

their settling speed is therefore negative. Unless specified otherwise, it will be assumed hereafter that $\rho_s > \rho_w$ and therefore that settling speeds are positive.

Despite all the modifying factors and approximations, Stokes law is a very good approximation to $w_s(d)$ for nonaggregated particles when $d < 100$ μm (more accurately, when Re < 0.5). Because of this, Stokes law will be used frequently throughout this text as a first approximation to $w_s(d)$. For a quick estimate of $w_s(d)$, Equation 2.6 can be used. When accurate results are desired and/or necessary, improved determinations (including experimental measurements) may be necessary to take into account the nonspherical nature of particles and nonlinear effects when Re > 0.5. However, the greatest effect on settling speeds of fine sedimentary particles is that most of these particles do not exist as individual particles but are in the form of flocs. These flocs have very different diameters and densities and hence different settling speeds from those of the individual particles. An introduction to this problem is presented in Section 2.4 and discussed more thoroughly in Chapter 4.

2.3 MINERALOGY

Sediment particles are primarily inorganic particles and are generally coated with metal oxides as well as viable and nonviable organic substances. Most inorganic particles are silicate minerals that can be further subdivided into silica minerals (quartz and opaline silica); the clay minerals (usually kaolinite, illite, montmorillonite, and chlorite, but others as well); and other silicates (feldspars, zeolites). The mineralogy and size of an inorganic particle tend to be related. Large particles with a diameter greater than 62.5 μm are defined as sands and gravels and consist primarily of quartz and feldspar grains. Small particles with a diameter less than 4 μm are defined as clay-size particles and generally consist of clay minerals.

The basic structure of the silicate minerals consists of silicon and oxygen atoms arranged in some regular fashion. For example, quartz consists of closely packed tetrahedra formed by four oxygen atoms at the corners of the tetrahedron and a silicon atom at the center. The resulting chemical formula for quartz is SiO_2.

The clay minerals consist of sheets of silicon–oxygen tetrahedra (structured such that the silicon–oxygen ratio is four to ten) separated by layers of alumina or aluminum hydroxide. Different clay minerals differ by the type and number of layers forming the structure. For example, kaolinite has a two-layer structure of a silica tetrahedra sheet and an aluminum hydroxide layer, whereas montmoril-lonite is a three-layer structure with an aluminum hydroxide layer between two silica tetrahedron layers. This three-layer structure, or "sandwich" as it is some-times called, is connected to other sandwiches by a layer of water. Other minerals can replace aluminum in the lattice and produce a large variation in the properties of clay minerals.

Small particles have high surface-area-to-mass ratios compared to large particles and are therefore significantly influenced by surface chemical–electrostatic effects. Because of these chemical–electrostatic effects, small particles (especially the clay minerals) tend to aggregate together; this occurs in the overlying water, where the particles form flocs, and in the bottom sediments, where they initially form a rela-tively open and low-density structure that then deforms and compacts into increas-ingly larger and denser aggregates and clumps after burial and with time.

As will be seen later, this aggregation has significant effects on the settling and deposition of the suspended sediments and on the consolidation and erosion of the bottom sediments. The cohesive properties of a sediment depend to a great extent not only on the amount of clay present but also on the type of clay mineral. In this context, montmorillonite and illite are most significant, the other clay minerals are less so, and quartz and the other silicate minerals are least significant.

As an example of sediment composition, the mineralogy of a sediment sample from the Kalamazoo River in Michigan is shown in Table 2.2. The mineralogy is given for the bulk sediment, for the 2- to 15-μm size fraction, and for the less-than-2-μm size fraction. The bulk sediments consist primarily of clay minerals (30%) and quartz (45%), with the remainder (25%) being other minerals. The size fraction less than 2 μm consists primarily of clay minerals (93%) and contains relatively little quartz (3%). Kaolinite is the dominant clay mineral for all size classes.

The subject of clay mineralogy is quite diverse and is a fascinating area of research for many scientists. However, it should be realized that natural sedimen-tary particles generally consist of inorganic particles coated with aluminum, iron, and manganese oxides; nonviable organic substances such as proteins, carbohy-drates, lipids, and humic substances; and viable organic substances such as bacte-ria and phytoplankton. The coatings may cover all or at least a significant fraction of the outer surface as well as the interior surfaces of a particle. These coatings

TABLE 2.2
Mineralogy of a Composite Sediment Sample from the Kalamazoo River

Mineral	Bulk (%)	Sample 2–15 µm (%)	<2 µm (%)
Smectite	1	1	3
Illite	7	20	15
Kaolinite	20	35	70
Chlorite	2	3	5
Total Clay	**30**	**59**	**93**
Quartz	45	22	3
K-Feldspar	4	2	0.5
Plagioclase	3	3	0.5
Calcite	4	0	0
Dolomite	6	1	0
Weddelite	5	10	0
Talc	1	2	2
Other	2	1	1
Total	**100**	**100**	**100**

Note: Mineralogy is shown for the bulk sediment, for the 2- to 15-µm size fraction, and for the less-than-2-µm size fraction.

are significant because they may greatly modify the surface forces between particles and hence modify the amount of flocculation (floc size and effective density) as well as the rate of flocculation. In addition, many contaminants (especially nonpolar, nonionic organic contaminants) are known to partition to these organic coatings. The nature and properties of these coatings are not well understood.

2.4 FLOCCULATION OF SUSPENDED SEDIMENTS

Much of the suspended particulate matter in surface waters exists in the form of flocs. These flocs are aggregates of smaller solid particles that may be inorganic or organic. Floc sizes range from less than a micrometer up to several centimeters; their concentrations range from less than 1 mg/L in open waters during quiescent conditions to more than several thousand mg/L in rivers, estuaries, and near-shore areas of oceans and lakes during floods and storms; and their effective densities range from that of the solid constituents of the flocs (approximately 2.65 g/cm³) down to densities very close to that of water (1.0 g/cm³) for very open flocs. These effective densities affect the settling speed and hence transport of a floc. For large, low-density flocs, the settling speeds may be several orders of magnitude less than that of a solid particle of the same size as the floc and with the density of a particle in the floc. However, for this same floc, the settling speed is much greater than the settling speed of a single particle making up the floc.

Because flocs are porous, there will be flow through the floc; this also will affect the settling speed. Settling speeds of flocs generally cannot be predicted theoretically and hence must be measured. This can be done, for example, by means of settling tubes and rotating disks (see Section 4.3).

The size and density of a floc depend on the particle size and mineralogy, the suspended sediment concentration, the fluid shear, and the salinity of the water. In addition, flocs are dynamic quantities, and their properties change with time due to aggregation and disaggregation. Under steady flow conditions and constant concentration, the floc size distribution will reach a steady state as a dynamic equilibrium between aggregation and disaggregation. As examples of different flocs produced under different conditions, photographs of some typical unflocculated and flocculated inorganic particles from the Detroit River and Lake Erie (Lick et al., 1993) are shown in Figure 2.7. Figure 2.7(a) is a photograph of disaggregated sediments. The solids concentration is about 100 mg/L, and the median diameter of the particles is about 4 μm. Figure 2.7(b) shows numerous smaller flocs and a larger floc approximately 100 μm in diameter. These flocs were produced at a sediment concentration of 100 mg/L and a fluid shear of 100/s. For the largest floc, note the open structure and relatively low average density. Figure 2.7(c) shows several flocs, the largest of which is approximately 400 μm in diameter. These flocs were produced at the same shear but at a lower concentration (10 mg/L) than the previous ones and generally are larger and less dense. Figure 2.7(d) is a photograph of a floc approximately 600 μm in diameter. This floc is much less dense and much more fragile than the flocs in the previous two photographs. This type of floc tends to be transient, is generally formed early during a shear experiment, and then disappears and is replaced by denser flocs.

Figure 2.7(e) shows a much larger floc produced by differential settling; that is, no externally applied fluid shear is present. The diameter of this floc is about 2 mm. Because the median diameter of the particles making up this floc is about 4 μm, the number of particles in this floc must be on the order of 10^6 or more. Figure 2.7(f) is a photograph of a relatively large floc. It was produced by differential settling at a sediment concentration of 2 mg/L, and its horizontal diameter is about 2 cm. The floc is very open and fragile. With increasing settling time, this type of floc will become more like a teardrop in shape and will become denser as it continually sweeps up smaller particles and incorporates them into its structure. The flocs produced by differential settling are generally different in appearance from the flocs produced by fluid shear, with the former generally being larger and denser as well as being influenced and shaped by the flow around the floc as it settles through the water.

The times for flocs to aggregate or disaggregate can be on the order of seconds up to many days and hence can be comparable to, or sometimes much longer than, characteristic transport times for flocs in the overlying water. Because of this, to quantitatively understand the transport and fate of fine-grained sediments and their associated contaminants, the rates of aggregation and disaggregation as well as the steady-state sizes and densities of flocs must be understood and quantified. Flocculation and its dynamic nature are discussed in more detail in Chapter 4.

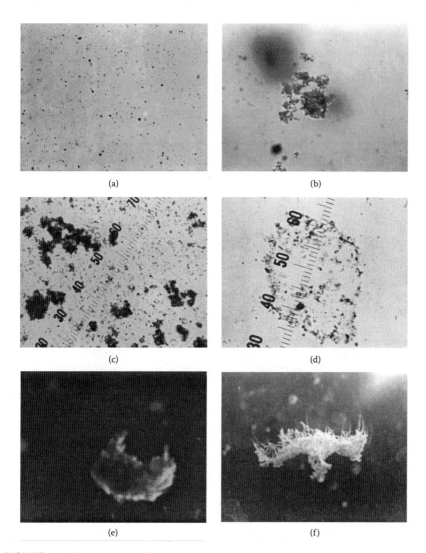

FIGURE 2.7   Photographs of flocs: (a) disaggregated sediments at a sediment concentration of 100 mg/L; (b) floc approximately 100 μm in diameter, produced at a sediment concentration of 100 mg/L and a shear of 100/s; (c) flocs produced at a sediment concentration of 10 mg/L and a shear of 100/s; largest floc is about 400 μm in diameter; (d) low-density, transient floc produced during a shear test; it is approximately 600 μm in diameter; (e) floc produced by differential settling, approximately 2 mm in diameter; and (f) floc produced by differential settling at a sediment concentration of 2 mg/L. It is approximately 2 cm wide. (*Source*: From Lick et al., 1993, *J. Geophys. Res.* With permission.)

2.5 BULK DENSITIES OF BOTTOM SEDIMENTS

Bottom sediments are a mixture of solid particles, water, and gas. The assumption often is made that no gas is present, but the gas volume fraction, especially for organic-rich sediments near the sediment-water interface, can be relatively large (typically 2 to 6% but occasionally as much as 40%) and must therefore be considered in the determination of the bulk density of a sediment. For a bottom sediment consisting of solid particles, water, and gas, the bulk density is defined as

$$\rho = x_s \rho_s + x_w \rho_w + x_g \rho_g$$

$$\cong x_s \rho_s + x_w \rho_w \qquad\qquad (2.10)$$

where ρ_s, ρ_w, and ρ_g are the densities and x_s, x_w, and x_g are the volume fractions of the solid particles, water, and gas, respectively. From their definitions, it follows that $x_s + x_w + x_g = 1$. The second equality in Equation 2.10 is a valid approximation because ρ_g is much less than ρ_s and ρ_w.

The water content W of a sediment is defined as the mass of the water contained in the sediment divided by the total mass of the sediment, or

$$W = \frac{m_w}{m_w + m_s + m_g} \qquad\qquad (2.11)$$

where m_w, m_s, and m_g are the masses of the water, solid, and gas contained in the sediment. With the above definitions for the densities and volume fractions, this equation can be written as

$$W = \frac{x_w \rho_w}{x_w \rho_w + x_s \rho_s + x_g \rho_g}$$

$$\cong \frac{x_w \rho_w}{x_w \rho_w + x_s \rho_s} \qquad\qquad (2.12)$$

If there is no gas present in the sediments, $x_s + x_w = 1$. The above equation then simplifies to

$$W = \frac{x_w \rho_w}{x_w \rho_w + (1 - x_w)\rho_s}$$

$$= \frac{x_w}{2.6 - 1.6 x_w} \qquad\qquad (2.13)$$

where it has been assumed that $\rho_s = 2.6$ g/cm^3 and $\rho_w = 1$ g/cm^3. By solving the above equation for x_w, one obtains

$$x_w = \frac{2.6\,W}{1 + 1.6\,W} \qquad\qquad (2.14)$$

TABLE 2.3
Densities (g/cm³) of Various Sediment Components

Humus	1.3–1.5	Anorthite	2.7–2.8
Clay	2.2–2.6	Dolomite	2.8–2.9
Kaolinite	2.2–2.6	Muscovite	2.7–3.0
Orthoclase	2.5–2.6	Biotite	2.8–3.1
Microcline	2.5–2.6	Apatite	3.2–3.3
Quartz	2.5–2.8	Limonite	3.5–4.0
Albite	2.6–2.7	Magnetite	4.9–5.2
Flint	2.6–2.7	Pyrite	4.9–5.2
Calcite	2.6–2.8	Hematite	4.9–5.3

Source: From Kohnke, 1968.

From the definition of x_w, it follows that $x_w \rho_w = W\rho$, or $x_w = W\rho/\rho_w$, and therefore

$$\rho = \frac{2.6}{1 + 1.6\,W} \qquad (2.15)$$

This latter formula often is used to relate sediment bulk density to water content, but it should be emphasized that it is only valid when no gas is present.

In the previous three equations, it has been assumed that $\rho_s = 2.6$ g/cm³. However, the densities of different sediment components vary from this value, as shown in Table 2.3. As a result, ρ_s for different sediments may differ somewhat from 2.6 g/cm³. Nevertheless, because sediments are often dominated by sand (quartz) with an average density of 2.65 g/cm³, whereas silts and clays tend to have densities close to 2.6 g/cm³ (Richards et al., 1974), a density of 2.6 g/cm³ is a reasonable approximation for the average density of the solid particles in sediments.

Organic material is generally much lighter than the other solid particles in sediments. Because of this, it may be desirable in some cases to correct Equation 2.15 for the change in density due to organic content. If it is assumed that organic material has a density of about 1.0 g/cm³, then Equation 2.15 becomes (Hakanson and Jansson, 2002)

$$\rho = \frac{2.6}{1 + 1.6(W + IG)} \qquad (2.16)$$

where IG is the loss on ignition expressed as a percent of the total mass. This correction is usually relatively small.

2.5.1 MEASUREMENTS OF BULK DENSITY

In the usual procedure for measuring the bulk density of bottom sediments (hereafter called the wet-dry procedure), sediment cores are frozen, sliced into 1- to 4-cm sections, and weighed (m_{wet}). They then are dried in an oven at approximately 75°C for 2 days and weighed (m_{dry}). Because $m_w = m_{wet} - m_{dry}$ and $m_w + m_s = m_{wet}$, the water content can be determined from Equation 2.11 and, in terms of experimental quantities, is

$$W = \frac{m_{wet} - m_{dry}}{m_{wet}} \qquad (2.17)$$

If no gas is present in the original sediment sample, then Equation 2.15, with W from the above equation, can be used to calculate the density. However, if gas is present, Equation 2.15 is not valid and the density cannot be determined in this manner.

Because of the way the sediments are treated in the wet-dry procedure, any gas present in the sediments is released during the procedure. In this case, Equations 2.15 through 2.17 can only be used to determine the density of the sediments in the absence of gas, defined as ρ^*. Other limitations of the wet-dry procedure are that the resolution is relatively coarse (1 to 4 cm) and the core is destroyed as samples are taken.

More recent devices use the attenuation of electromagnetic radiation to measure bulk densities in a nondestructive manner; they have higher resolution, are more convenient, and are more accurate. One such device, called the Density Profiler (Gotthard, 1997), is described here. It is similar to the x-ray device described by Been (1981) but uses a gamma radiation emitter, ^{137}Cs, as a source and measures the attenuation of the radiation as it is transmitted horizontally through the sediments. Once the transmitted radiation is measured, the density of the sediments in a core can be determined from

$$N = N_0 e^{-\mu\rho x} \qquad (2.18)$$

where N is the number of counts that pass through the core sample, N_0 is the number of counts with no sample present, μ is the mass absorption coefficient, ρ is the density, and x is the sample width (typically about 10 cm). By solving for ρ, one obtains

$$\rho = -\frac{1}{\mu x} \ell n \frac{N}{N_o} \qquad (2.19)$$

For ^{137}Cs and for the apparatus used by Gotthard (also see Roberts et al., 1998), μ was determined to be approximately 0.0755 cm^2/g by calibration with water and sediments of known density. However, μ depends somewhat on the mineralogy of the sediments, and its value therefore must be adjusted by a calibration for different types of sediment. Once this is done, the accuracy of the measurement for ρ is estimated at ±0.01 g/cm^3.

The Density Profiler measures the amount of horizontally transmitted radiation as the emitter traverses the core in a vertical direction. For most measurements, this traverse speed has been set at 3.3×10^{-3} cm/s (2 mm/min). A radiation rate meter gives an output in counts per minute every 2 s. Because of statistical fluctuations of the radiation, this output is averaged over time (or distance traversed). In most cases, the data are averaged over 1 cm to reduce the fluctuations in the data and also to illustrate trends more simply. For better resolution (at the expense of traverse time), the data can be averaged over smaller distances, as small as 0.1 cm, the lowest resolution obtainable in the Density Profiler.

The Density Profiler measures the actual density of the sediments, including solids, water, and any gas present. In contrast, the standard procedure for measuring sediment density measures the density of the sediments in the absence of gas, ρ^*, and ignores the presence of gas. By measuring both densities, it is possible to determine the gas volume in the sediments. From their definitions, $\rho^* = m_{sw}/V_{sw}$ and $\rho = m_{sw}/(V_{sw} + V_g)$, where m_{sw} is the mass of the solids and water, V_{sw} is the volume of the solids and water, and V_g is the volume of the gas within the sediments. The total volume, V, is given by $V_{sw} + V_g$. From these definitions, the fractional gas volume, $x_g = V_g/V$, can be shown to be

$$x_g = \frac{\rho^* - \rho}{\rho^*} \tag{2.20}$$

2.5.2 Variations in Bulk Density

For noncohesive sediments, the way in which sedimentary particles are arranged affects the density, stability, and hence erosion rate of the sediment. This packing of particles in water with no gas present has been investigated theoretically by numerous authors. In particular, the packing of uniform size spheres has been investigated by Graton and Fraser (1935). They define and evaluate four different, but regular, ways of packing with solid fractions, x_s, ranging from 0.52 to 0.74; that is, the porosity, x_w, varies from 0.48 to 0.26. For $\rho_s = 2.6$ g/cm^3, this corresponds to bulk densities from 1.83 to 2.18 g/cm^3. In addition, Allen (1970) has investigated the packing of prolate spheroids, and Scott (1960) has analyzed the random packing of uniform-sized spheres. Mixtures of nonuniform-sized spheres will have greater densities than those of uniform-sized spheres because the smaller spheres can occupy spaces between the larger spheres. Because of its importance in the properties of materials, there has been considerable interest in the packing of particles with uniform but more irregular shapes than spheres (usually prolate and oblate ellipsoids); for example, see Donev et al. (2004). Because of vibrations and the subsequent rearrangement of particles, the bulk densities of bottom sediments consisting of noncohesive particles may increase by a small amount with time.

For beds made by deposition of cohesive, fine-grained particles, the situation is somewhat different. In this case, the particles will generally be deposited in the form of flocs. Because these flocs are buried by other settling particles, the weight of the overlying particles will cause the flocs to deform and compress, and the bed will compact with time. A freshly deposited bed will have a very low density and weak structure and will be easily erodible. As burial increases, further deformation and compression of the structure occur, and the resistance to erosion increases. These changes occur less rapidly as time increases. Due to compaction, the interstitial water and gas will be forced out of the sedimentary structure and toward the surface. The generation of gas due to the decay of organic matter will modify this process. Fine-grained sediments tend to have significantly lower bulk densities (1.1 to 1.6 g/cm³) than coarse-grained sediments.

The vertical transport of water, solids, and gas is space and time dependent. Because of this, the bulk density of a bottom sediment can be a complex function of space and time. To illustrate this, two examples are given here. The first example is a reconstructed (well-mixed) sediment from the Detroit River (Jepsen et al., 1997). These sediments had a median disaggregated particle size of 12 μm, an organic carbon content of 3.3%, and a mineralogical composition (in decreasing order) of quartz, muscovite, dolomite, and calcite. The sediments were well mixed and seven duplicate 40-cm cores were prepared. Using the wet-dry procedure, the bulk densities of these cores were measured as a function of depth at times after deposition of 1, 2, 5, 12, 21, 32, and 60 days. No gas was observed during this time.

These results are shown in Figure 2.8. The densities are generally between 1.4 and 1.5 g/cm³. They increase with time and usually increase with depth; this is generally expected because the pore waters are forced upward and out of the bottom sediments due to the weight of the overlying sediments. The increase with time is most rapid initially, but then the rate of increase decreases as time increases. At the end of the experiment at 60 days, it is not obvious that a steady

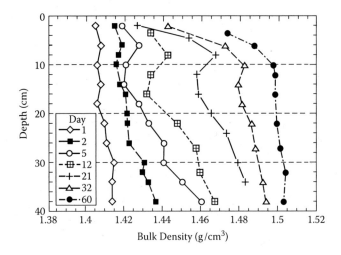

FIGURE 2.8 Bulk density as a function of depth for a Detroit River sediment core at consolidation times of 1, 2, 5, 12, 21, 32, and 60 days. (*Source*: From Jepsen et al., 1997. With permission.)

state has been reached, or is even near. It also can be seen that there are deviations from a monotonic increase of density with depth. These deviations tend to
appear between 5 and 21 days and are due to increases in water content (seen as
a decrease in bulk density with depth) at a particular depth as the waters above
this layer are transported upward slower than the waters below this layer. This
trapping effect disappears by 60 days. After 60 days, gas pockets began to form.
Because only the wet-dry procedure was used and this procedure cannot account
for the presence of gas, no measurements of bulk densities were made for times
after deposition greater than 60 days.

The second example is a relatively undisturbed field core from the Grand River
in Michigan (Jepsen et al., 2000). For these sediments, most bulk properties were
reasonably uniform with depth, the mean particle size was 50 μm, and the organic
content was between 5 and 6%. Gas bubbles were present in the cores, and gas per-
colated from the sediments during coring. Immediately after coring, bulk densities
were measured as a function of depth by both the Density Profiler and the wet-dry
procedure. Results for ρ and ρ^* as a function of depth are shown in Figure 2.9(a).
It can be seen that ρ fluctuates but has an average value of about 1.24 g/cm^3; the
density, ρ^*, (as measured with 1 to 2 cm slices) is always greater than ρ and also
shows less variability, primarily because of the coarser averaging. From these two
densities and Equation 2.20, the gas volume fraction can be determined, is shown
as a function of depth in Figure 2.9(b), and is generally about 0.03 to 0.04.

The density, ρ, shown in Figure 2.9(a) was determined from a 1-cm averaging
of the data from the Density Profiler. The same data, when averaged over only

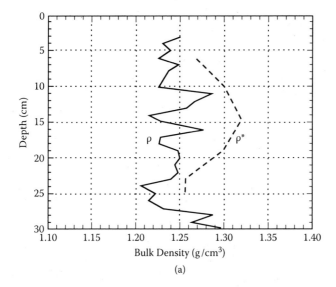

(a)

FIGURE 2.9 Densities and gas fraction as a function of depth for a sediment core from
the Grand River: (a) bulk densities as measured by gamma attenuation (ρ) and by the wet-
dry procedure (ρ^*). (*Source*: From Jepsen et al., 2000. With permission.)

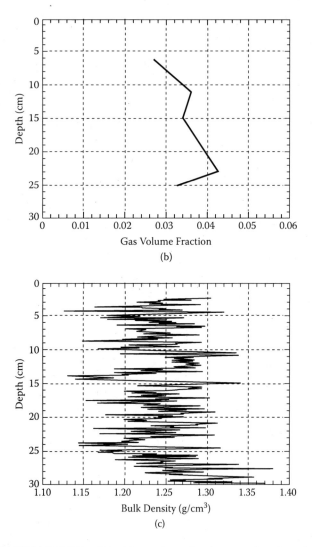

FIGURE 2.9 (CONTINUED)   Densities and gas fraction as a function of depth for a sediment core from the Grand River: (b) gas fraction and (c) bulk density (ρ) with 0.1 cm resolution. (*Source*: From Jepsen et al., 2000. With permission.)

0.1 cm, are shown in Figure 2.9(c) and can be seen to be much more highly variable with more rapid and larger excursions. It was estimated that more than half of the variability was because of the presence of gas pockets, whereas the remainder was from inhomogeneities in the solid-water matrix.

   As can be seen, gas generation and the subsequent transport of this gas can have significant effects on the time-dependent consolidation of sediments; this also will affect erosion rates. Sediment consolidation, including effects of gas, is discussed more thoroughly in Section 4.6.

3 Sediment Erosion

The erosion rate of a sediment is defined as the total flux, q (g/cm^2/s), of sediment from the sediment bed into the overlying water in the absence of deposition. This flux is generally due to shear stresses caused by currents and wave action. Because of their activity, benthic organisms and fish also can contribute to this flux, but their effect is usually small. Propwash and waves from large ships as well as smaller recreational boats can cause localized erosion.

Once eroded, sediments can go into, and be transported as, suspended load or bedload. The resuspension rate of a sediment is defined as the flux of sediment into suspended load, again in the absence of deposition. Particles in suspension in a horizontally uniform flow move horizontally with the average fluid velocity, whereas their vertical motion is governed by gravitational and turbulent forces; collisions between particles are usually negligible in modifying the transport. As a result of these forces, the concentration of the suspended particles typically varies in the vertical direction, with the concentration being largest near the sediment-water interface and decreasing approximately exponentially in an upward direction from there. The length scale of this exponential decay depends on the settling speed, w_s, and the eddy diffusivity due to turbulence, D_v. From a steady-state balance of the fluxes due to gravitational settling and turbulent diffusion and as a first approximation, this can be shown to be given by D_v/w_s.

As the settling speed increases and turbulence decreases, this length scale decreases. When it is on the order of a few particle diameters, collisions between suspended particles and between suspended particles and particles in the sediment bed become significant. In this limit, the particle transport is known as bedload. Bedload generally occurs in a thin layer near the sediment bed with a thickness of only a few particle diameters. In bedload, particle concentrations are relatively high and the average speed of the particles is generally less than the speed of the overlying water. Suspended load is usually dominant for fine-grained particles, whereas bedload is more significant for coarser particles.

For the quantification of sediment transport, the total sediment flux as well as the individual fluxes into suspended and bedload are generally necessary. These fluxes depend not only on hydrodynamic conditions (the applied shear stress due to currents and waves) but also on the bulk properties of the sediment bed. These bulk properties vary in both the horizontal and vertical directions in the sediment, and their variations can cause changes in the fluxes by several orders of magnitude.

At present, no uniformly valid, quantitative theory of erosion or resuspension rates is available, and experiments are therefore needed to determine these rates. In the following section, devices for measuring sediment resuspension/erosion (the annular flume, the Shaker, and Sedflume) are described and compared; advantages and limitations of these as well as other devices are discussed. In Section 3.2, some results of erosion measurements using Sedflume with relatively undisturbed sediments from the field are presented. These results illustrate the rapid and large variations of erosion rates often found in the sediment bed. To better understand and be able to predict the effects of sediment bulk properties on erosion rates, laboratory experiments with sediments with well-defined properties have been done. Results of these experiments are described in Section 3.3. In modeling erosion rates and the initiation of sediment motion, a useful parameter is a critical shear stress for erosion. Semi-empirical equations to approximate this parameter have been developed based on experimental data; these are presented in Section 3.4. For sediment transport models, approximate equations to quantify erosion rates as a function of shear stress and the bulk properties of sediments are useful; these are discussed in Section 3.5. In most practical applications, the sediment-water interface is or can be approximated as horizontal. However, surface slopes of bottom sediments are often large enough that they can significantly affect critical stresses and erosion rates. In Section 3.6, a brief presentation of these effects is given.

3.1 DEVICES FOR MEASURING SEDIMENT RESUSPENSION/EROSION

Straight flumes were the earliest devices to quantify sediment transport and generally have been used to measure the bedload of relatively coarse-grained and noncohesive particles. These devices and their applications have been described extensively in the literature (e.g., Van Rijn, 1993; Yang, 1996) and hence their descriptions will not be given here. However, approximate equations to describe bedload are briefly presented in Section 6.2. More recently, many other devices have been developed, primarily to measure sediment resuspension or erosion. Devices of this type are the annular flume, the Shaker, and Sedflume; these are described below.

3.1.1 ANNULAR FLUMES

Annular flumes have often been used to measure the resuspension of fine-grained sediments (Fukuda and Lick, 1980; Mehta et al., 1982; Tsai and Lick, 1988). A typical flume is shown in Figure 3.1. It is 2 m in diameter and has an annular test channel that is 15 cm wide and 21 cm deep. Sediments to be tested are deposited on the bottom of the channel, usually to a depth of about 6 cm. These sediments are usually well-mixed and have relatively uniform properties throughout. Overlying these sediments is a layer of water, typically about 7.6 cm deep. A plexiglass lid, slightly narrower than the channel, fits inside the channel and touches the surface of the water. This lid rotates and causes a flow in the channel that is generally turbulent. This flow causes a shear stress on the bottom.

  For different rates of rotation of the lid, the velocity profiles in the channel, especially near the bottom, have been measured (Fukuda and Lick, 1980;

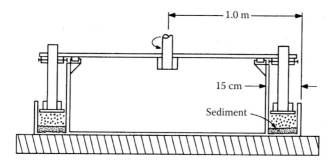

FIGURE 3.1   Schematic of annular flume.

MacIntyre et al., 1990); from this, the bottom shear stress as a function of the rotation rate of the lid can be determined. From the velocity profiles, it can be shown that the shear stress varies gradually in the radial direction by about 10 to 25% at the lower rotation rates but by as much as a factor of two at the higher rotation rates. In most reported results, an average value of the shear stress is used. Although the main flow in the flume is in the azimuthal direction, a secondary flow due to centrifugal forces is also present; it is inward near the sediment-water interface, upward at the inner wall of the flume, outward near the lid-water interface, and downward at the outer wall. Dye measurements and direct velocity measurements show that these secondary currents are relatively small (on the order of a few percent or less) compared to the primary azimuthal current. Because of the annular nature of the flume, the flow and suspended sediment concentration vary only in the radial and vertical directions and not in the azimuthal direction. For fine-grained sediments, bedload is often negligible. When this is true, the erosion and resuspension rates are approximately the same.

For the annular flume, the standard resuspension experiment is as follows. At the beginning of the test (i.e., after the well-mixed sediments are allowed to deposit and consolidate for the desired time), the sediment concentration in the overlying water is generally small, a few milligrams per liter (mg/L). The lid is accelerated slowly and then rotated at a constant rate, a rate that produces the desired shear stress. The sediment concentration in the overlying water is measured as a function of time. A typical result is shown in Figure 3.2. The concentration increases rapidly at first, then more slowly, and eventually reaches a steady state. For each test, the steady-state concentration, C_∞, can be determined directly from the experimental measurements; in addition, the initial rate of resuspension, E_0, can be determined from $E_0 = h(dC/dt)_0$, where h is the depth of the water, C is the sediment concentration, and $(dC/dt)_0$ is the initial slope of the C(t) graph. From a series of tests such as this, the steady-state concentration and initial erosion rate can be determined as a function of shear stress with consolidation time (time after deposition) as a parameter.

There are two conceptually easy, but different and limiting, interpretations as to the appropriate mechanisms that describe the process as shown in Figure 3.2. In the first interpretation, it is assumed that particles are uniform in size and

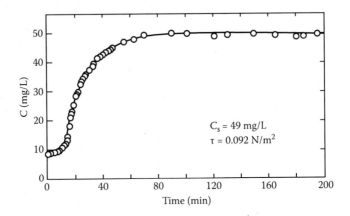

FIGURE 3.2 Suspended sediment concentration in an annular flume as a function of time for a shear stress of 0.09 N/m².

noncohesive, and that the bulk properties of the bottom sediment do not change with depth or time; from this it follows that the resuspension rate does not change with depth or time. The time-dependent variation of the suspended sediment concentration can then be determined from the mass balance equation for the suspended sediment; that is, the increase in suspended sediments in the flume is due to the difference between the resuspension rate and the deposition rate, D, where $D = w_s C$, or

$$h\frac{dC}{dt} = E_o - w_s C \tag{3.1}$$

When the suspended sediment concentration is initially zero, the solution to this equation is

$$C = C_\infty \left(1 - e^{-\frac{w_s t}{h}} \right) \tag{3.2}$$

and the behavior of C(t) is similar to that shown in Figure 3.2. The steady-state concentration, C_∞, is then attributed to a dynamic equilibrium between the resuspension rate and the deposition rate, both of which are occurring more or less simultaneously. From Equation 3.1, $C_\infty = E_o/w_s$.

The second interpretation assumes that the sediments are fine-grained and cohesive, and it takes into account the distribution of sediment particle sizes and the increasing cohesivity of the sediments with depth; however, it also assumes that deposition is negligible. According to this interpretation, as resuspension occurs, (1) the finer particles will be resuspended and will leave the coarser, more-difficult-to-resuspend particles behind, and (2) the less dense and hence less cohesive surficial layers will be resuspended and will expose the denser and more cohesive deeper layers. For a particular shear stress, the resuspension rate will be greatest initially but will then decrease with time as the surficial sediments that are exposed become increasingly more difficult to erode; this will continue until no further sediments can be resuspended. It follows that the suspended solids concentration will increase most rapidly initially but will then approach a constant value as the resuspension rate goes to zero, just as is shown in Figure 3.2. The steady-state concentration, C_∞, is then a measure of the total amount of sediment that can be resuspended at that shear stress.

Because both of the above interpretations indicate the same C(t), it is difficult to decide which of the above interpretations is correct from the experiment as described. However, an experiment that gives additional and discriminatory information is suggested by the following arguments. If the overlying water is continually cleared of sediment (the deposition rate would then be zero), the first interpretation suggests that additional sediment would be resuspended indefinitely with time (or at least until the bottom sediments were all resuspended or changed character). According to the second interpretation, if a steady state is reached and if the overlying water is then continually cleared of sediment, no additional sediment would be resuspended. The experiment needed to distinguish between these interpretations is relatively simple in principle and is: first, a repeat of the experiment shown in Figure 3.2, letting the sediments approach a steady state; and second, a replacement of the turbid water with clear water while measuring the total amount of suspended sediment (in the drained water as well as the small amount still suspended in the flume water at the end of the experiment).

Experiments of this type have been performed (Massion, 1982; Tsai and Lick, 1988) and have shown that the first interpretation is valid for uniform-size, coarse-grained, noncohesive sediments, whereas the second interpretation is valid in the limit of fine-grained, cohesive sediments. These experiments demonstrate the relative significance of the different erosion processes for these two limiting types of sediments. For sediments between these two limits, as the turbid overlying water is removed, additional sediment will be resuspended compared with what was in suspension before the turbidity was reduced; the amount of this additional sediment will decrease with time and also will decrease as the sediment becomes more fine-grained and cohesive.

For the limiting case of fine-grained, cohesive sediments, the experiments described above demonstrate that only a finite amount of sediment, ε, can be resuspended at a particular shear stress. This quantity is referred to as the resuspension potential. From experimental data (Fukuda and Lick, 1980; Lee et al., 1981; Mehta et al., 1982; Lavelle et al., 1984; MacIntyre et al., 1990), the resuspension potential is generally approximated as

$$\varepsilon = \frac{a}{t_d^m}\left[\tau - \tau_c\right]^n \quad \text{for } \tau \geq \tau_c$$
$$= 0 \qquad\qquad \text{for } \tau < \tau_c \qquad\qquad (3.3)$$

where ε is the net amount of resuspended sediment per unit surface area (in g/cm^2); t_d is the time after deposition (in days); τ is the shear stress (N/m^2) produced by wave action and currents; τ_c is an effective critical stress for resuspension; and a, n, and m are constants. Each of the parameters τ_c, a, n, and m depends on the particular sediment (and the effects of benthic organisms) and needs to be determined experimentally.

At shear stresses greater than about 1 N/m^2, bedload in the radial direction may be significant, especially for coarser sediments. In this case, sediments are preferentially eroded near the outer edge because of the higher shear stresses there; the finer sediments are resuspended, but the coarser sediments move radially inward as bedload and are deposited near the inner wall, where they cover previous sediments and coarsen the bed. This reduces the erosion near the inner wall. Over long periods of time, because of this nonuniform erosion, a tilting of the bed surface occurs and further affects the erosion. This nonlinear behavior limits the use of the annular flume to shear stresses less than about 1 N/m^2.

3.1.2 THE SHAKER

Most annular flume experiments are done in the laboratory with reconstructed sediments. For fine-grained, cohesive sediments, these experiments have been very useful and have qualitatively determined the dependence of the resuspension rate and the resuspension potential on various governing parameters such as the applied shear stress; the sediment bulk properties of bulk density, water content, particle size, and mineralogy; time after deposition; and numbers and types of benthic organisms. However, deploying an annular flume in the field for measurements of the resuspension of undisturbed sediments is extremely difficult, and an easier method for measuring resuspension in the field is desirable. For this purpose, a portable device for the rapid measurement of sediment resuspension (called the Shaker) was developed (Tsai and Lick, 1986). The Shaker can be used in the laboratory, but its main use has been in the field for rapid surveys.

The basic Shaker consists of a cylindrical chamber (or core), inside of which a horizontal grid oscillates vertically (Figure 3.3). In a typical laboratory experiment, the sediments whose properties are to be determined are placed at the bottom of the chamber, with water overlying these sediments. In a field test, relatively undisturbed bottom sediments are obtained by inserting the coring tube into the bottom sediments; this core and its contents then are retrieved and inserted into the Shaker frame. The thickness of the sediment in the coring tube is usually about 6 cm. The depth of the water is maintained at 7.6 cm. The amplitude of the grid motion is 2.5 cm, whereas the lowest point of the grid motion is 2.5 cm above the sediment-water interface. The grid oscillates in the water and creates turbulence, which penetrates down to the sediment-water interface and causes the sediment to be resuspended. The turbulence, and hence the amount of sediment resuspended, is proportional to the frequency of the grid oscillation.

The equivalent shear stresses created by the oscillatory grid were determined by comparison of results of resuspension tests in the Shaker with those in an

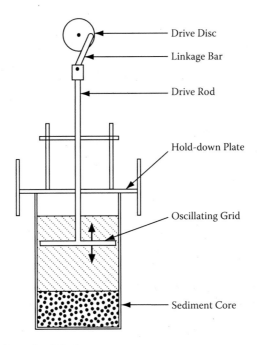

FIGURE 3.3   Schematic of Shaker.

annular flume where the shear stresses had been measured and were known as a function of the rotation rate of the lid of the flume. The basic idea of the calibration is that when the flume and Shaker produce the same concentration of resuspended sediments under the same environmental conditions, the stresses needed to produce these resuspended sediments are equivalent. For calibration purposes, 49 tests of different fine-grained sediments were performed. These tests demonstrated that the results are reproducible and, most importantly, that the equivalent shear stress produced by the Shaker is independent of the sediments and the type of water (fresh or salt) used in the experiments. The Shaker has been extensively used in various aquatic systems.

3.1.3 SEDFLUME

Major limitations of both the annular flume and the Shaker are that they can resuspend only small amounts of sediment (usually only the top few millimeters of the bed) and can measure only net sediment resuspension at shear stresses below about 1 N/m². To measure erosion rates of sediments at high shear stresses and as a function of depth in the sediments, a flume (called Sedflume) was designed, constructed, and tested by McNeil et al. (1996). With this flume, sediment erosion rates have been measured at shear stresses up to 12.8 N/m² and at depths in the sediment up to 2 m. Experiments can be performed either with reconstructed (usually well-mixed) sediments or with relatively undisturbed sediments from field cores. Sedflume is shown in Figure 3.4 and is a straight flume that has a test

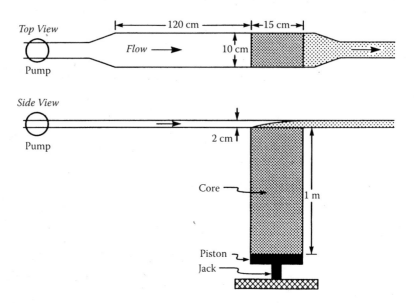

FIGURE 3.4   Schematic of Sedflume.

section with an open bottom through which a coring tube that contains sediment can be inserted. This coring tube has a rectangular cross-section that is 10 by 15 cm and is usually 20 to 100 cm in length. Water is pumped through the flume at varying rates and produces a turbulent shear stress at the sediment-water interface in the test section. This shear stress is known as a function of flow rate from standard turbulent pipe flow theory. As the shear produced by the flow causes the sediments in the core to erode, the sediments are continually moved upward by the operator so that the sediment-water interface remains level with the bottom of the test and inlet sections. The erosion rate (in cm/s) is then recorded as the upward rate of movement of the sediments in the coring tube. The results are reproducible within a ±20% error and are independent of the operator. The erosion rate (in units of $g/cm^2/s$) is then this velocity multiplied by the bulk density of the sediments being eroded.

A quite sophisticated device, SEDCIA, has recently been developed to determine erosion rates by means of multiple laser lines and computer-assisted image analysis (Witt and Westrich, 2003); maximum errors are reported to be 7%, with an average error of 1%. This seems to be more accurate than necessary, because the natural variability of sediments is much greater than this. So far, the device has been developed only for use in the laboratory.

To measure erosion rates at all shear stresses using only one core, the standard procedure with Sedflume is as follows. Starting at a low shear stress, usually about 0.2 N/m^2, the flume is run sequentially at increasingly higher shear stresses. Each shear stress is run until at least 2 mm — but not more than 2 cm — is eroded. The flow then is increased to the next shear stress and so on until the highest desired shear stress is reached. This cycle, starting at the lowest shear stress, is then repeated until all the sediments have eroded from the core. The highest measurable erosion rate is determined by the maximum speed of the hydraulic jack and is about 0.4 cm/s. By means of this device, numerous measurements of erosion rates of relatively undisturbed sediments from the field, as well as of reconstructed sediments in the laboratory, have been made. A few of these results will be illustrated in the following two sections.

A useful parameter in the modeling of sediment transport is a critical shear stress for erosion. This is defined and determined from measurements of erosion rates as follows. As the rate of flow of water over a sediment bed increases, there is a range of velocities (or shear stresses) at which the movement of the smallest and easiest-to-move particles is first noticeable to an observer. These eroded particles then travel a relatively short distance until they come to rest in a new location. This initial motion tends to occur only at a few isolated spots. As the flow velocity and shear stress increase further, more particles participate in this process of erosion, transport, and deposition, and the movement of the particles becomes more sustained. The range of shear stresses over which this transition occurs depends to a great extent on the fluid turbulence and the distributions of particle sizes and cohesivities of the sediments. For uniform-size, noncohesive particles, this range is relatively small and is primarily due to turbulent fluctuations. For fine-grained particles with wide distributions of particle and aggregate sizes as well as cohesivities, this range can be quite large.

Because of this gradual and nonuniform increase in sediment erosion as the shear stress increases, it is often difficult to precisely define a critical velocity or critical stress at which sediment erosion is first initiated, that is, first observed. Much depends on the patience and visual acuity of the observer. More quantitatively and with less ambiguity, a critical shear stress, τ_c, can be defined as the shear stress at which a small, but accurately measurable, rate of erosion occurs. In the use of Sedflume, this rate of erosion has usually been chosen to be 10^{-4} cm/s; this represents 1 mm of erosion in approximately 15 minutes. Because it would be difficult to measure all critical shear stresses at an erosion rate of exactly 10^{-4} cm/s, erosion rates are generally measured above and below 10^{-4} cm/s at shear stresses that differ by a factor of two. The critical shear stress then can be obtained by linear interpolation between the two. This gives results with a 20% accuracy for the critical shear stress. A somewhat easier and more accurate procedure for determining τ_c is to interpolate the measured erosion rates on the basis of an empirical expression for $E(\tau)$ (Section 3.5). Experimental results and quantitative expressions for the critical shear stress for erosion as a function of particle diameter and bulk density are given in Section 3.4.

It should be noted that, as τ_c is defined here, erosion occurs for $\tau < \tau_c$, albeit at a decreasing rate as $\tau \to 0$. This is consistent with experimental observations (where erosion rates have been measured for $\tau < \tau_c$) and is especially evident for fine-grained sediments with wide variations in particle size and cohesivities (e.g., see Section 3.3).

3.1.4 A COMPARISON OF DEVICES

From the previous description of the Shaker, it is clear that the sediment transport process that occurs in the Shaker is net resuspension of the bottom sediment in the absence of any horizontal flow or transport; that is, the amount of sediment in suspension at steady state is a dynamic balance between resuspension and deposition, both of which can and are occurring more or less simultaneously as the turbulence as well as the sediment bulk properties fluctuate in space and time. No bedload is present because there is no horizontal flow. For sediments with a distribution of particle sizes, bed armoring and a subsequent decrease in erosion rates will occur with time due to large particles that are not resuspended, cannot be transported away as bedload, and eventually cover at least part of the sediment surface. In the limit of uniform-size, fine-grained, cohesive sediments (negligible deposition and no bed armoring), the Shaker will give accurate results for the resuspension potential at low suspended concentrations (but only if it is calibrated properly). However, for more general conditions, this is no longer true for the reasons stated above and for additional reasons as described below.

In the annular flume, because of its annular nature, the flow and sediment concentration are independent of the azimuthal direction. Bedload, primarily in the azimuthal direction at low to moderate shear stresses, may be present, but it is the same at all cross-sections. Bed armoring will occur during the experiment. The resuspension processes are essentially the same in both the annular flume and Shaker; that is, the annular flume also measures net resuspension. Once calibrated, these two devices give the same quantitative results.

In contrast to these two devices, Sedflume measures pure erosion, that is, erosion of bed sediment into suspended load and bedload and subsequent transport of these loads downstream with negligible possibility of deposition in the test section. Pure erosion, E, is the quantity that is necessary for the sediment flux equation that generally is used in sediment transport modeling; that is, $q = E - D$.

Because the annular flume/Shaker and Sedflume generally measure two different quantities, a direct comparison of results is not possible. However, the devices should at least be consistent with each other; for example, if erosion rates as measured by Sedflume are used in a sediment dynamics model to predict suspended sediment concentrations under the same conditions as those in the annular flume, then the calculated and measured suspended sediment concentrations in the annular flume should be the same.

For this purpose, experiments were performed with the same fine-grained sediments in both Sedflume and an annular flume at shear stresses of 0.1, 0.2, 0.4, and 0.8 N/m^2 (Lick et al., 1998). The numerical model, SEDZLJ (see Chapter 6), was then used to predict sediment concentrations in the annular flume using Sedflume data. Contrary to expectations, the calculations disagreed with the observations by as much as three orders of magnitude. These differences could not be reduced significantly by fine-tuning the parameters in the model.

One reason for the differences between the calculations and observations was determined to be the following. During experiments in the annular flume, it was observed that eroded sediments in the form of flocs and aggregates tended to collect and subsequently deposit and consolidate in the flow stagnation areas where the sediment-water interface meets the sidewalls. This was more significant at the inner wall; here, many of the flocs that were moved inward by the secondary flow could not be convected upward by the weak flow in the corner and tended to settle there in a volume that had a triangular cross-section. Sediments also collected in the stagnation region near the outer wall, but the amount collected there was generally much smaller than that near the inner wall. The total mass of sediments in the stagnation regions was estimated to be equivalent to a sediment resuspension of 40 mg/cm^2 and varied relatively little with shear stress.

A second reason for the discrepancy between calculations and observations is what can loosely be described as bedload; that is, the consolidated sediments (which were relatively fine and cohesive) tended to erode in aggregates that were then transported horizontally near the sediment-water interface. These aggregates eventually disintegrated with time but generally caused a suspended sediment concentration near the sediment-water interface that was greater than the sediment concentration away from this interface. The sediment concentration in the middle of the water column is what is normally measured in annular flume experiments and, in the experiments described here, did not give an accurate measure of the total amount of sediment in resuspension. More accurate measurements of the vertical distribution in sediment concentration indicated that the mass of sediment in this bedload was negligible at a shear stress of 0.1 N/m^2 but increased with shear stress; at 0.8 N/m^2, it was estimated to be 1 to 3 times the amount of sediment collected in the stagnation regions.

A comparison was made of the net amounts of sediment suspended in the annular flume (1) as determined from measured suspended sediment concentrations in the middle of the water column; (2) corrected as described above, including eroded sediment depositing in stagnation regions and in bedload; and (3) as determined from numerical calculations based on Sedflume data. There were large discrepancies between (1) and both (2) and (3); however, the agreement between the corrected observations (2) and the numerical calculations (3) was quite good. From this it follows that, to obtain accurate results for ε from an annular flume, the standard measurements of concentration must be corrected as indicated above.

In summary, the annular flume and Shaker measure net resuspension in the absence of horizontal transport; due to the small volume of overlying water, bed armoring, and low maximum shear stresses, only small amounts of surficial sediments can be resuspended in these devices. Because of difficulties in accurately determining the net resuspension (as shown above), these devices give only qualitative results. In numerical modeling, the parameter that naturally occurs is the erosion rate, E, and not the net resuspension or resuspension potential, ε. E is what is measured by Sedflume, and this can be done as a function of depth in the sediments and at high shear stresses.

From time to time, other devices have been used to measure sediment resuspension/erosion. Several of these involve rotational flows; these all have similar difficulties to those of the annular flume, that is, rotational flows that cause radially varying erosion rates and centrifugal forces that are different on suspended or surficial bed particles than they are on the fluid. Because of this, nonuniform distributions of flow, erosion/deposition (especially bedload), and sediment (both suspended and deposited) occur. This in turn causes nonuniform bed armoring and even further nonuniform erosion. The result is an inability to accurately interpret and quantify erosion rates for these devices. Some devices determine erosion by measuring the suspended solids concentration after flow through a relatively long flume and then through a vertical pipe (e.g., Ravens, 2007; see critique by Jones and Gailani, 2008); this leads to incorrect concentration measurements because of nonuniform flow and differential settling in the pipe. In addition, bedload will not be included in these measurements but will modify/armor the sediment bed in a nonuniform and non-quantifiable manner. Sedflume accurately reproduces the processes that determine sediment erosion. Other devices do not. For these reasons, only Sedflume or an equivalent flume is recommended for the quantitative determination of erosion rates and for use in numerical modeling.

3.2 RESULTS OF FIELD MEASUREMENTS

By means of Sedflume, erosion rates of relatively undisturbed sediments from field cores have been measured at numerous locations. Examples are the Detroit and Fox Rivers in Michigan (McNeil et al., 1996); Long Beach Harbor in California (Jepsen et al., 1998a); a dump site offshore of New York Harbor (Jepsen et al., 1998b); the Grand River in Michigan (Jepsen et al., 2000); Lake Michigan (McNeil and Lick, 2002a); the Kalamazoo River in Michigan (McNeil and Lick, 2004); the Housatonic River in Massachusetts (Gailani et al., 2006); and Cedar Lake, Indiana (Roberts et al., 2006). Sedflume has now been adopted as a standard device for measuring sediment erosion and is being widely used by the U.S. Environmental Protection Agency, the U.S. Army Corps of Engineers, and consulting companies.

In field tests in shallow waters (less than about 10 m), sediment samples are generally obtained by means of coring tubes attached to aluminum extension poles. For water depths greater than 10 m, this is not possible and other procedures are used. In the dump site in New York Harbor, sediment samples were obtained at depths up to 30 m by means of divers who inserted the coring tubes in the bottom sediments and then retracted them with the sediments retained in the tubes. In Lake Michigan, a series of cores was obtained in water depths from 10 to 45 m as follows. A large box core was first used to sample the sediments on the bottom. This box core was then returned to the surface and subsampled by means of the rectangular cores used in Sedflume. All of the above procedures produced similar and satisfactory results.

Results for the Detroit River and the Kalamazoo River are presented here to illustrate some of the major characteristics as well as the large variability of erosion rates of real sediments in an aquatic system, especially as a function of shear stress and as a function of depth and horizontal location in the sediment.

3.2.1 DETROIT RIVER

Twenty sediment cores were obtained from the Trenton Channel of the Detroit River in October 1993 (McNeil et al., 1996). Three of these cores are discussed here as representatives of a moderately coarse, noncohesive sediment; a finer, more cohesive sediment; and a stratified sediment. For each core, erosion rates were determined as a function of sediment depth and applied shear stress, whereas the sediment properties of bulk density, average particle size, and organic content were determined as a function of depth. The bulk density was obtained by the wet-dry procedure; because of this, the gas fraction could not be determined.

The first core was from the inner edge of a sandbank that separated the shallow water of a lagoon from the deeper water of the channel. The water depth was 2.3 m. The core was 77.5 cm in length and consisted almost entirely of sand and coarse silt. The bulk density was fairly constant at about 1.8 g/cm³. Figure 3.5 shows a plot of the erosion rates as a function of depth with shear stress as a parameter. At the lowest shear stress, the erosion rate is relatively low at the surface and decreases rapidly with depth, whereas at higher shear stresses the erosion rate is higher and relatively constant with depth. This latter behavior of the erosion rate as a function of depth as well as the relatively high bulk density is characteristic of a coarse, noncohesive sediment such as sand.

The second core was 65 cm long and was taken from a deeper location with a water depth of 5.9 m. It was finer-grained than the first and consisted of dark gray

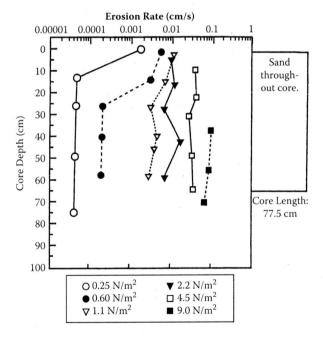

FIGURE 3.5 Trenton Channel, Site 1. Water depth of 2.3 m. Erosion rate as a function of depth with shear stress as a parameter. (*Source:* From McNeil et al., 1996. With permission.)

silt in the upper half and dark silt mixed with gray clay in the lower half. The sediments had a strong petroleum odor and were permeated with small gas bubbles on the order of 1 mm in diameter. The upper surface of the sediments was covered with tubificid oligochaetes and decaying macrophytes. Except for this thin surficial layer, the sediment bulk properties were fairly uniform with depth. The bulk density was approximately 1.4 g/cm^3 throughout the core. For each shear stress, the erosion rate (Figure 3.6) is highest near the surface and decreases rapidly with depth (by as much as two orders of magnitude). This behavior is characteristic of a fine-grained, cohesive (low bulk density) sediment and demonstrates that, for these sediments, only a limited amount of sediment can be resuspended at a particular shear stress (as demonstrated in the previous section by experiments in an annular flume); this is in contrast to the erosive behavior of the more coarse-grained, non-cohesive (high bulk density) sediment illustrated by the previous core.

Figure 3.7 illustrates the erosion rates for the third core, a stratified sediment where the erosion rates varied by an order of magnitude up to as much as three orders of magnitude in distances of a few centimeters. Visual observations determined the layering reasonably accurately. This core was obtained from an area near the mouth of the river at Lake Erie where the water depth was 0.6 m. At this location, large and highly variable shear stresses are often present due to wind waves that propagate into the area from Lake Erie during large storms. The top 2 cm of the sediment were silty and eroded easily; this was followed by a drop in the erosion rate by almost two orders of magnitude due to the presence of

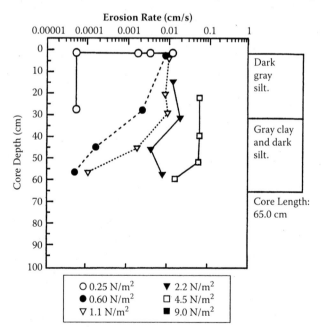

FIGURE 3.6 Trenton Channel, Site 3. Water depth of 5.9 m. Erosion rate as a function of depth with shear stress as a parameter. (*Source:* From McNeil et al., 1996. With permission.)

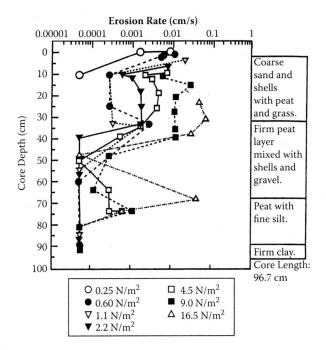

FIGURE 3.7   Trenton Channel, Site 8. Water depth of 0.6 m. Erosion rate as a function of depth with shear stress as a parameter. (*Source:* From McNeil et al., 1996. With permission.)

macrophytes holding the sediment together. Below about 10 cm, a coarser layer composed of sand mixed with peat was present and eroded easily. At 40 cm, firm peat was encountered and the erosion rate dropped to nearly zero for all shear stresses until a layer of peat combined with fine-grained material was exposed at about 75 cm. The erosion rate increased at this point and then dropped back to zero upon reaching a hard-packed clay layer. This type of strong stratification as shown here was not unusual for cores in this area. Although a qualitative correlation between the erosion rates and the sediment bulk properties was indicated, there was insufficient information to determine quantitative relations between the two.

To investigate temporal changes in erosion rates, studies by means of Sedflume were repeated at selected locations in the Trenton Channel in April/May 1994 (Lick et al., 1995). As might be expected, there were significant changes in erosion rates at some locations and very few changes at other locations, with the magnitude of the change depending on the hydrodynamics/sediment transport history during the fall/winter/spring period.

The first site discussed above is an example of a location where there were large temporal changes in sediment properties. In October 1993, the sediment core was 77.5 cm in length and was limited by a very-difficult-to-erode, hard-packed layer below that depth. At this same location in spring 1994, only 20 cm of sand was recovered before hitting the hard-packed clay. This indicates that approximately 60 cm of sand had eroded at this site between fall and spring. In a brief bathymetric survey of the area, it was determined that water depths in this area had increased by up to 1.5 m, again indicating large erosions of sediment on the order of 0.5 to 1.5 m near and at this site. Although the erosion rates of the sands in both cores were approximately the same, erosion rates for the two cores were obviously quite different as a function of depth after 20 cm.

Cores also were taken in fall and spring at a site located near the first site, but toward the inner and more protected part of the lagoon in about 2.5 m of water. This area is primarily a depositional area, with much of the deposited material from a steel plant nearby. The sediments consisted of a black silt deposit approximately 2 m deep, after which there was a sand layer. From fall to spring, there were few changes in the thickness, sediment bulk properties, and erosion rates of the silt deposit.

In this investigation, there were relatively few locations where cores were obtained in both fall and spring. However, the data did indicate that temporal changes in erosion rates were related to temporal changes in the hydrodynamics and sediment transport.

3.2.2 KALAMAZOO RIVER

As an example of a fairly extensive set of measurements of erosion rates and sediment properties, consider the investigation in the Kalamazoo River in Michigan (Figure 3.8) by McNeil and Lick (2002b, 2004). In the study area (extending approximately 36 km from the City of Plainwell to Calkins Dam at Lake Allegan), the river is approximately 65 to 100 m wide and has an average cross-sectional depth of 1.3 to 2.0 m — except for Lake Allegan, which is more than 300 m wide and has an average depth of 3.5 m. Much of the river is characterized by relatively shallow and often fast-moving waters, with numerous meanders and braids formed by small islands; this more or less natural part of the river is interrupted by six dams (Plainwell, Otsego City, Otsego, Trowbridge, Allegan City, and Calkins) that slow the upstream flow and create impoundments for water and sediments. In 1987, the superstructures of three of these dams (Plainwell, Otsego, and Trowbridge) were removed; however, the sills to these dams were retained and still impound water and sediments. Because of the relatively large and rapid changes in the river bathymetry, the hydrodynamics and hence sediment properties and transport also have large spatial variations throughout the river. Because of floods and the lowering of the dams, these quantities also have large temporal variations.

In this area, 35 sediment cores were taken and analyzed. Core locations are shown in Figure 3.8. For each core, erosion rates were determined as a function

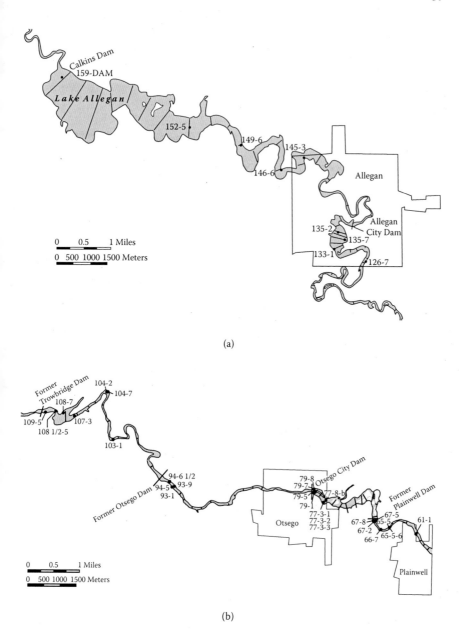

(a)

(b)

FIGURE 3.8   Map of Kalamazoo River. Core locations are numbered. (*Source:* From McNeil and Lick, 2004. With permission.)

of sediment depth and applied shear stress, whereas the sediment properties of bulk density, average particle size, organic carbon content, and gas fraction were determined as a function of depth. In most cores, strong and rapid stratification in one or more of these properties as a function of depth was observed. This layering was clearly delineated using the Density Profiler and further verified by visual observations and by measurements of other bulk properties made at discrete intervals. For many cores, erosion rates often differed by several orders of magnitude between stratified layers. Variations of erosion rates and bulk properties in both the horizontal and vertical directions were large and equivalent in magnitude.

To illustrate these variations, erosion rates and bulk properties of seven cores are shown in Figures 3.9(a) to (g) as a function of depth and are discussed below. For reference purposes, averages of bulk properties over the upper 15 cm for all the cores from the Kalamazoo were determined and are as follows: bulk density = 1.39 g/cm^3, average particle diameter = 134 μm, organic carbon content = 8.0%, and gas fraction = 7.8%. Large deviations from these averages were present and will be evident.

The first core (61-1) was located 2 km upstream from the Plainwell Dam and was in 0.4 m of rapidly flowing water. It was 21 cm in length and consisted of a macrophyte layer at the surface; a distinct 8-cm layer of sand, gravel, and shells; and then a hard-packed silty sand in the remainder of the core (Figure 3.9(a)). In the 8-cm layer, erosion rates were moderately high and reasonably constant with depth; the bulk density was almost constant at about 1.8 to 1.9 g/cm^3. This is typical of a uniform-size, sandy, noncohesive sediment. Below this layer, erosion rates were much less (by three or more orders of magnitude) and decreased rapidly with depth. Near the interface between the layers, the density decreased rapidly in a distance on the order of a centimeter (the resolution of the Density Profiler data as then reported) from about 1.8 g/cm^3 to about 1.2 to 1.5 g/cm^3 (these latter bulk densities are typical of fine- to medium-grained, cohesive sediments) and then stayed reasonably constant as the depth increased. Particle size decreased from about 190 μm in the sandy layer to about 140 μm in the lower layer. In the lower layer, because of its cohesive nature, appreciable erosion occurred only at a shear stress of 12.8 N/m^2. Organic content varied from 4 to 12% in an irregular manner, whereas the gas fraction was moderate (3 to 11%) and generally increased with depth. This core was more or less typical of many of the cores in the Kalamazoo. Of these, many of their vertical profiles of density exhibited even sharper discontinuities than shown here.

Core 67-2 was located just behind the Plainwell Dam and was in 2 m of relatively slow-moving water. This core (Figure 3.9(b)) also was distinctly stratified and consisted of a thin (less than 1 cm) floc layer at the surface; a 4-cm layer of low-density fine sand, gravel, and shells below this; and then a layer of silt

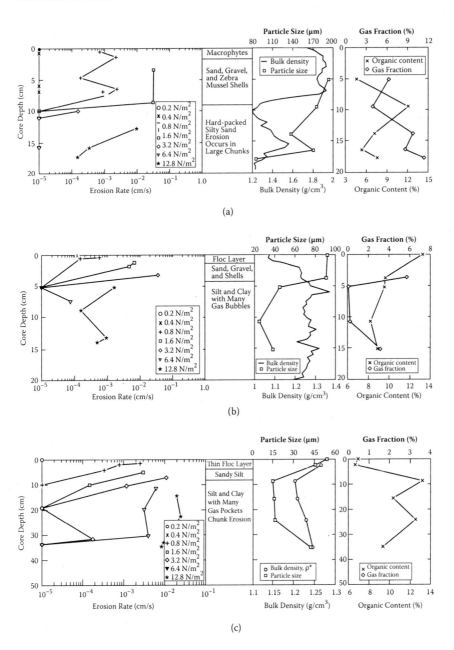

FIGURE 3.9 Erosion rates and bulk properties of sediment cores from the Kalamazoo River as a function of depth: (a) core 61-1, (b) core 67-2, (c) core 77-3-1. (*Source:* From McNeil and Lick, 2004. With permission.)

and clay with many gas bubbles. The bulk density was lowest at the surface and generally low throughout (1.1 to 1.3 g/cm³), indicative of a cohesive sediment. The floc layer eroded rapidly. Compared to the surface layer (average particle size of 90 μm), the lower silty-clay layer was even more fine-grained (20 to 40 μm) and more difficult to erode. Erosion rates above and below the interface between these two layers differed by more than three orders of magnitude.

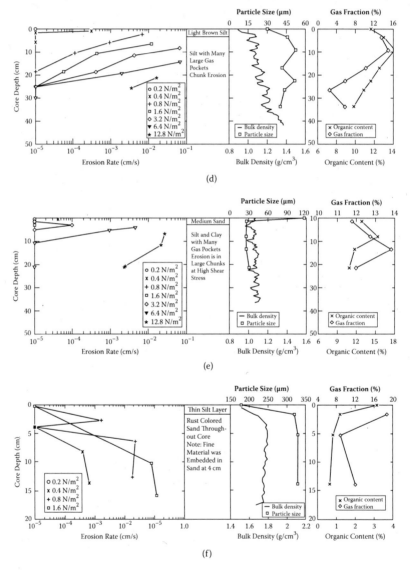

FIGURE 3.9 (CONTINUED) Erosion rates and bulk properties of sediment cores from the Kalamazoo River as a function of depth: (d) core 77-3-2, (e) core 77-3-3, (f) core 77-8. (*Source:* From McNeil and Lick, 2004. With permission.)

Multiple cores were taken at transect 77, which was within a wide, shallow, slow-moving stretch of the river 1.7 km behind the Otsego City Dam. From right to left (looking downstream), the surficial sediments along the transect consisted of fine sand, then coarser sand as the center was approached, followed by fine sand, and then silt toward the left bank of the river. From the local bathymetry, it can be inferred that the flow velocities are higher on the right and center than on the left; this is consistent with the particle sizes. Triplicate cores were obtained from site 77-3 (an area with a low flow rate), which was located about 7 m from the left bank in 1.02 m of water. By means of GPS, the position of the boat was maintained relatively constant while the cores were taken. Properties of the cores are shown in Figures 3.9(c), (d), and (e). Although there are small differences in bulk properties between cores, all three cores show a fine-grained (15 to 45 μm), low-density (approximately 1.0 to 1.25 g/cm^3), cohesive sediment consolidating with depth over the top 20 cm. This consolidation decreases erosion rates by two to three orders of magnitude over this interval as the depth increases. At the surface (the top few centimeters), core 77-3-3 (Figure 3.9(e)) has a thin layer of medium sand (120 μm), which is not present in the other two cores. In this layer, the erosion rates seem to be much lower than those in cores 77-3-1 and 77-3-2. If this layer is subtracted from 77-3-3, erosion rates as a function of depth are then comparable to the other two replicate cores (which are very similar to each other).

Core 77-8 was on the same transect as the 77-3 cores but was in 0.4 m of water in an area with a higher flow rate than the 77-3 cores. At the surface of the core (Figure 3.9(f)), a very thin matted layer of silt and organic matter was present. This was followed by 3 to 4 cm of rust-colored sand with a lower bulk density and particle size but with a higher gas fraction and organic content compared to the sediments below. At 4 cm, a very thin layer of fine sand existed. Below this, the bulk properties were relatively constant with depth. Because of the thin layer of silt and organic matter at the surface, the sediment was very difficult to erode. However, as the sandier sediments below this layer

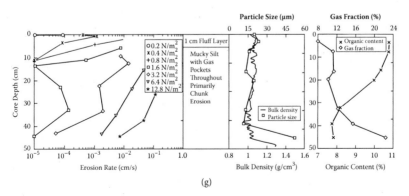

(g)

FIGURE 3.9 (CONTINUED)   Erosion rates and bulk properties of sediment cores from the Kalamazoo River as a function of depth: (g) core 159. (*Source:* From McNeil and Lick, 2004. With permission.)

were exposed, erosion rates increased but only to a depth of 4 cm. At this depth, because of the thin silt layer present there, the sediment was again very difficult to erode and did not erode at 6.4 N/m^2; however, once this layer eroded at 9.0 N/m^2, the sediments below eroded very rapidly. Below 4 cm, the erosion rates were relatively constant and much higher than those at the surface (by more than three orders of magnitude). This increased erosion rate at depth is the reverse of what was found in the previous cores and is the reverse of what is generally found in field cores, where erosion rates usually decrease with depth. The increase in erosion rates is primarily due to the coarser, noncohesive nature of the sediments below the surface layer.

From a comparison of cores 77-3 and 77-8, the core at 77-8 (although in the same transect) shows large differences from 77-3, with coarser sediments (200 to 300 μm), higher bulk densities (about 1.8 g/cm^3), and more easily erodible sediment (by orders of magnitude), with erosion rates increasing with depth. It can be seen that differences between the cores at 77-3 are much less than the differences between the cores at 77-3 and 77-8 (which were in quite different hydraulic regimes).

Core 159 came from behind Calkins Dam, which impounds Lake Allegan; the water depth was 4.4 m. Sediments in this core were typical of most sediments behind the dam. The core (Figure 3.9(g)) had a 1-cm, fluffy, organic layer on top, followed by mucky, fine silt packed with gas. The erosion rates generally decreased rapidly with depth, indicative of a fine-grained, cohesive sediment, and were very low at the bottom of the core. The bulk density was fairly constant, close to 1 g/cm^3, and even less than 1 g/cm^3 at some depths; this was due to the high gas fractions (8 to 23%) and fine-grained nature of the sediments. Organic content was high (8 to 11%). Despite the quiescent nature of the site, the fine-grained sediments, and the high organic content, no evidence of organisms or their activity was observed in the sediment. The sediments were soft and mucky throughout this area and may have been too "soupy" (i.e., low density) to support organisms.

From the data for all 35 cores from the Kalamazoo River as well as the above discussion, a general trend is that, where the river is moving fast, the sediments (when averaged over depth) are coarse, have a higher density, and are easier to erode compared to sediments where the river is moving slow. Although this relation is, in general, qualitatively true, it is too simplistic. Sediment properties are dynamic properties and depend on the spatial and time-dependent variations in the hydrodynamics and sediment transport and not just on the local water depth and average flow rate. Because of the dynamic nature of a river, sediment properties vary greatly with depth in the sediment (quite often with changes by orders of magnitude in a few centimeters) and are not well represented by averages over sediment depth.

The seven cores illustrated here were representative of the other cores obtained in the Kalamazoo River. Many of the cores had a coarse, high-density, easily erodible layer of sand at the surface overlying a finer, low-density, more-difficult-to-erode layer of silt or silt/clay. The reverse of this stratification in one or more parameters as well as multiple stratified layers was also present, but to a lesser extent. A thin floc layer, usually a few millimeters but occasionally as much as a centimeter in thickness, was also often observed at the surface (12 of 35 cores), typically in slower-moving areas of the river. The sandy surface layer often had erosion rates three or more orders of magnitude greater than the finer sediments in the layer below. This sandy layer and the strong and distinct stratification are probably due to a high flow event (greater than 25-year recurrence) that occurred in 1985, possibly modified by the lowering of the three dams in 1987 and the subsequent modification and increase in the flow velocities.

In this river, as shown above in Figures 3.9(c), (d), and (e), as well as in the Housatonic River (Gailani et al., 2006) and Passaic River (Borrowman et al., 2005) where replicate Sedflume cores also were taken, replicate cores (taken near each other but not exactly at the same location) were always similar to each other and distinctly different from those in the same river but in a different hydraulic regime.

In this study, the emphasis was on measuring erosion rates and the properties of bulk density, particle size, organic content, and gas fraction. As the erosion tests were done, general observations were made of the presence of macrophytes and benthic organisms. Only one core (out of 35 cores) had a significant amount of organisms (burrowing worms). Except for this core, no evidence of significant bioturbation or a well-mixed surficial layer due to mixing by organisms was found. As mentioned above, the sediment was often strongly stratified with many cores having 5- to 20-cm thick layers at the surface with constant properties and sharp, distinct interfaces between layers. These layers most certainly were due to erosional/depositional events and not due to bioturbation. Even when present, organisms do not cause sharp vertical changes in bulk density and particle size as were observed here.

3.3 EFFECTS OF BULK PROPERTIES ON EROSION RATES

From field measurements of erosion rates such as those described above, it can be inferred that erosion rates depend on at least the following sediment properties: bulk density, particle size (mean and distribution), mineralogy, organic content, salinity of pore waters, gas volume fraction, and oxidation and other chemical reactions. In addition, benthic organisms, bacteria, macrophytes, and fish also may have significant effects on surficial bulk properties and hence on erosion rates. Nevertheless, despite extensive field measurements, the quantitative dependence of erosion rates on these parameters is difficult to determine from field measurements alone. The reasons are that there are a large number of parameters, each varying more or less independently; in any specific test, all parameters are generally not measured; and measurements are not always as accurate or extensive as desired. As a result, accurately quantifying the effects on erosion rates of each of the above parameters from field tests alone is not practical. By comparison, laboratory tests where only one parameter is varied while the others are kept constant are a more efficient and reliable procedure for determining the effects of each of these parameters. The results of some of these laboratory tests are described below.

3.3.1 BULK DENSITY

As sediments consolidate with time, their bulk densities tend to increase as a function of depth and time. For noncohesive sediments, the increase in bulk density is generally small and erosion rates are minimally affected by these changes. However, for cohesive sediments, the increases in bulk densities are greater and erosion rates are a much more sensitive function of the density. As a consequence, erosion rates for cohesive sediments decrease rapidly as the sediments consolidate with depth and time. To illustrate this, results of laboratory experiments (Jepsen et al., 1997) with reconstructed (well-mixed) but otherwise natural sediments from three locations (the Fox River, the Detroit River, and the Santa Barbara Slough) are discussed here. All sediments were in fresh water. Properties of each of these sediments (along with others to be discussed later) are given in Table 3.1. All three are relatively fine-grained. For each of these sediments and for consolidation times varying from 1 to 60 days, the bulk density as a function of depth and the erosion rate as a function of depth and shear stress were measured.

For reconstructed Detroit River sediments, results for the bulk density as a function of depth were presented and discussed in Section 2.5 (Figure 2.8). Erosion rates as a function of depth with shear stress as a parameter at different consolidation times were also measured. From a cross-plot of this type of data, the erosion rate as a function of bulk density with shear stress as a parameter was determined and is shown in Figure 3.10. For each shear stress, the rapid decrease in the erosion rate as the bulk density increases can be clearly seen. In general, the data is well approximated by an equation of the form

$$E = A \, \tau^n \, \rho^m \tag{3.4}$$

where E is the erosion rate (cm/s), τ is the shear stress (N/m^2), ρ is the bulk density (g/cm^3), and A, n, and m are constants. This equation with n = 2.23 and m = −56 is represented by the solid lines and is clearly a good approximation to the data

---

TABLE 3.1
Properties of Reconstructed Sediments

Sediment	Mean particle diameter (μm)	Organic content (%)	Mineralogy
Quartz	5–1350	0	Quartz
Fox River	20	6.7, 4.1, 2.3	Some clay
Detroit River	12	3.3	Mica, no clay
Santa Barbara	35	1.8	No clay
Grand River	48	4.8	No clay
Long Beach (seawater)	70	0, 0.25	Some clay
Kaolinite	4.5	0	Kaolinite
Bentonite	5	0	Bentonite

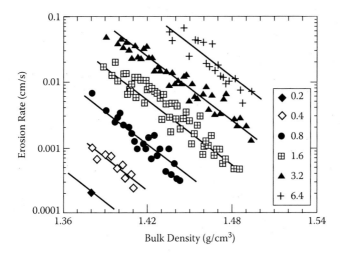

FIGURE 3.10   Erosion rates as a function of bulk density with shear stress (N/m²) as a parameter. Sediments are from the Detroit River. Solid lines are approximations by means of Equation 3.4. (*Source:* From Jepsen et al., 1997. With permission.)

TABLE 3.2
Erosion Parameters for Sediments from the Fox River, Detroit River, and Santa Barbara Slough

	n	m	A
Fox River	1.89	−95	2.69×10^6
Detroit River	2.23	−56	3.65×10^3
Santa Barbara	2.10	−45	4.15×10^5

---

for all erosion rates and densities and for each shear stress. The very sensitive dependence of erosion rates on density is quite evident.

$E(\tau, \rho)$ also was determined for the Fox and Santa Barbara sediments. These data were also well approximated by Equation 3.4. For all three sediments, the parameters n, m, and A are shown in Table 3.2. The parameter n is about 2. Data from other sediments (Section 3.5) also indicate that n is approximately 2 or somewhat greater. From experiments of this type with reconstructed sediments as well as with sediments from field cores, it has been shown that Equation 3.4 is a valid and accurate approximation to almost all existing data for fine-grained (and hence cohesive) sediments with a wide range of bulk properties in both laboratory and field experiments. It shows the effects of hydrodynamics (dependence on τ) and bulk density (where $\rho = \rho(z,t)$, z is depth in the sediments, and t is time after deposition). The parameters A, n, and m are different for each sediment and depend on the other bulk properties (but not density) of the sediment.

3.3.2 PARTICLE SIZE

Extensive laboratory experiments have been performed by Roberts et al. (1998) to understand and quantify the individual and combined effects of bulk density and particle size on the erosion of quartz particles. These experiments were performed with quartz particles with mean sizes ranging from approximately 5 to 1350 µm and were performed and analyzed in a similar manner to those described above. The size distribution for each sediment was fairly narrow. Bulk densities increased with time after deposition and ranged from approximately 1.65 to 1.95 g/cm³.

As representative examples of these data, erosion rates are shown in Figure 3.11 as a function of bulk density with shear stress as a parameter for particles with diameters of (a) 14.8 µm and (b) 1350 µm. For 14.8 µm (a fine-grained sediment), erosion rates are a strongly decreasing function of density, just as in the previous examples. For 1350 µm (a moderately coarse-grained sediment), erosion rates are essentially independent of density. Approximations to the data by means of Equation 3.4 are shown as the solid lines in both figures. The data for 14.8 µm are well approximated by Equation 3.4; however, the data for 1350 µm are not.

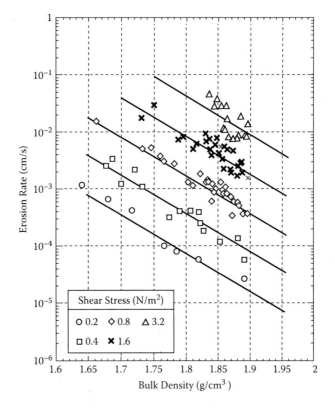

FIGURE 3.11(a)   Erosion rates as a function of bulk density with shear stress as a parameter: 14.8 µm. (*Source:* From Roberts et al., 1998. With permission.)

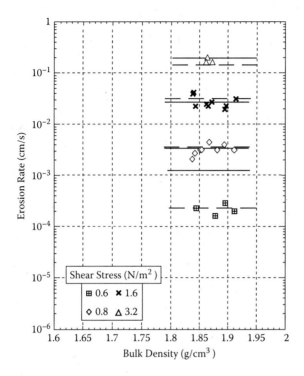

FIGURE 3.11(b)   Erosion rates as a function of bulk density with shear stress as a parameter: 1350 μm. (*Source:* From Roberts et al., 1998. With permission.)

It was shown that Equation 3.4 represents the data well for all the smaller and intermediate-size particles but does not do as well for the largest particles.

For coarse-grained, noncohesive sediments, a generally accepted equation for E(τ) is (e.g., see Van Rijn, 1993):

$$E = A(\tau - \tau_c)^n \qquad (3.5)$$

where $E = 0$ for $\tau < \tau_c$ and A, τ_c, and n are functions of particle diameter but not functions of density. For the 1350-μm particles, E(τ) as given by the above equation is shown in Figure 3.11(b) as the dashed line; it is obviously a better approximation than Equation 3.4, the solid line, and fits the data quite well for all shear stresses. However, for the 14.8-μm particles (as well as for the other small- and intermediate-size particles), Equation 3.5 is not a good approximation because it has no dependence on density; the dependence of E on τ is also incorrect. An approximate equation for erosion rates that is uniformly valid for all particle sizes is given in Section 3.5.

For quartz particles, erosion rates as a function of density for all sizes are compared in Figure 3.12. For simplicity, only erosion rates for a shear stress of 1.6 N/m² are shown. Approximations by means of Equation 3.4 are shown as the solid

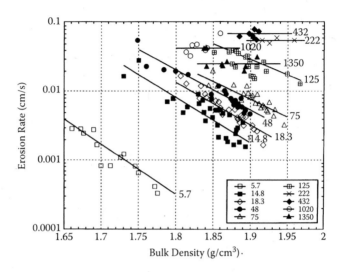

FIGURE 3.12   Erosion rates as a function of bulk density for different uniform-size quartz sediments at a shear stress of 1.6 N/m². Particle diameters are in micrometers. (*Source:* From Roberts et al., 1998. With permission.)

lines. Results for other shear stresses are similar. It can be seen that erosion rates are a very strong decreasing function of density for the finer particles, are less so as the particle size increases, and are essentially independent of density for the coarser particles. In addition, the experimental results generally demonstrated that (1) fine-grained, cohesive sediments tended to be less dense than coarse-grained, noncohesive sediments; (2) for the larger particles, the sediments behaved in a noncohesive manner; that is, they consolidated rapidly, density changes during consolidation were small, and the surface eroded particle by particle; and (3) for the smaller particles, the sediments behaved in a cohesive manner; that is, they consolidated relatively slowly, density changes during consolidation were relatively large, and the surface eroded as aggregates or chunks as well as particles.

3.3.3 MINERALOGY

To illustrate the effects of clay minerals on erosion rates, three related studies are summarized here. In the first (Jin et al., 2002), erosion rates as a function of shear stress were determined experimentally for quartz particles with average diameters of 15, 48, 140, 170, 280, 390, and 1350 μm with no bentonite and with 2% bentonite added. Each of the sediments had a narrow size distribution. Density variations were small. The data are quite extensive and detailed but can be summarized most easily by introducing the quantity R, which is defined as the ratio of the erosion rate of the quartz particles with 2% bentonite to the erosion rate of the same quartz particles with no bentonite, with both rates at the same shear stress and approximately the same bulk density. The quantity R as a function of particle size with shear stress as a parameter is shown in Figure 3.13. It can be seen that

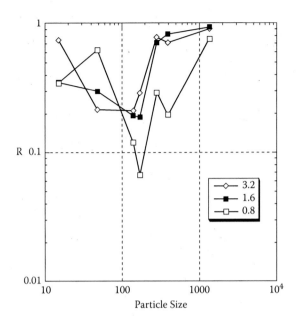

FIGURE 3.13   R as a function of particle size with shear stress (N/m²) as a parameter.

(1) the effect of 2% bentonite is to reduce the erosion rates of quartz particles for all sizes tested (i.e., R < 1); (2) R is least for the particles with average sizes of 140 and 170 μm and approaches unity for both smaller and larger particle sizes; and (3) the effect of the bentonite seems to decrease as the shear stress increases.

In the second study (Jin et al., 2000), small amounts of bentonite (up to 16%) were added to three different sediments (a sand, a topsoil, and a 50/50 mix of the two). For the sand (pure quartz), the mean particle size was 214 μm and the organic content was 0.03%. For the topsoil, the mean particle size was 35 μm, the organic content was 3.3%, and the mineralogy was predominantly quartz, feldspar, and illite with only trace amounts of smectite (bentonite). For the 50/50 mixture, the mean particle size was 125 μm and the organic content was 1.65%. For each sediment and amount of bentonite, erosion rates were measured as a function of shear stress and depth in the sediment. Bulk properties also were determined. Erosion rates decreased rapidly as the amount of bentonite increased. This is illustrated for the sand in Figure 3.14, which shows erosion rates as a function of percent bentonite added with shear stress as a parameter. For all three sediments, the addition of 2% bentonite caused a decrease in erosion rates by one to two orders of magnitude at each shear stress. The addition of larger amounts of bentonite caused further decreases in erosion rates, but the rate of decrease decreased as the amount of bentonite increased. The effect was greatest for the sand (214 μm) and least for the topsoil (35 μm); this is qualitatively consistent with the effect of particle size as described above and summarized for quartz particles in Figure 3.13.

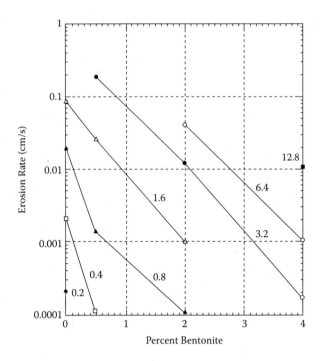

FIGURE 3.14 Erosion rates for a sand (214 μm) as a function of percent bentonite added with shear stress (N/m^2) as a parameter.

In the third study (Lick et al., 2002), small amounts of a clay (bentonite, kaolinite, plainsman clay, or mica) were added to a sediment consisting of 48-μm quartz particles; erosion rates as a function of shear stress were then measured for different amounts of each clay added to the quartz. The clays that were added and their percentages were bentonite (0.5, 2, 4, and 8%); kaolinite (5, 10, and 20%); plainsman clay (10 and 20%); and mica (10, 20, and 40%). Plainsman clay is a commercially available clay with a mineralogy of 37% illite, 25% kaolinite, 3% smectite, and 35% nonclay minerals, primarily quartz. It was used as a surrogate for pure illite, which was not economically available in a pure form. For all four clays, erosion rates decreased as the percentage of each clay increased. The greatest decreases were for bentonite and were comparable to those illustrated in Figure 3.14; smaller but still significant reductions occurred for kaolinite and plainsman clay; only a small effect was observed for mica. The effect of illite was approximately determined from the data on plainsman clay by assuming that the reductions in erosion rates were additive for the different clay minerals. In this way, it was shown that the order of the effects of the different clay minerals on erosion rates were bentonite > illite > kaolinite > mica, a sequence that is

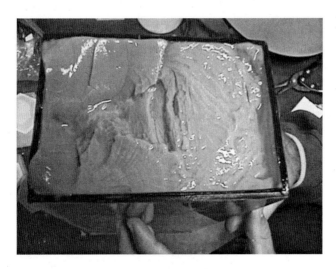

FIGURE 3.15   Fractured bentonite surface. Flow from right to left. 10 parts water to 1 part bentonite by mass. 1 day consolidation. (Photograph courtesy of J. Roberts.)

generally followed in quantifying the effects of clay minerals on soil properties (e.g., Mitchell, 1993).

From the data, the effects of both swelling and gelling were observed. Swelling (or a decrease in density) occurred rapidly (within the first day) and was noticeable for all the clays but was greatest for bentonite. Gelling (thixotropy) caused an increase in cohesivity and a decrease in erosion rates without a change in density or other bulk properties; it became apparent after about 2 days. Of the two processes, gelling had the more significant effect on erosion rates.

These experiments were performed with additions of relatively small amounts of clay, especially for the most cohesive clay (i.e., bentonite). Experiments have also been performed (Roberts, personal communication) on erosion rates of pure bentonite. Pure bentonite has a very low density, behaves as a gel, and is very difficult to erode. For low shear stresses, a small amount of surface erosion occurs by particle-by-particle and small-chunk erosion. For large stresses, the bentonite fails by fracture. Fractured surfaces of this kind are shown in Figure 3.15.

3.3.4 ORGANIC CONTENT

Experiments to investigate the effects of organic matter (measured as organic carbon, or o.c.) on consolidation and erosion rates were performed by heating Fox River (6.7% o.c.) and Long Beach (0.25%) sediments at 500°C for 24 hours to remove the organic matter (Lick and McNeil, 2001). For the Fox River, the organic carbon was reduced to 2.3%. In addition, the Fox River sediments with

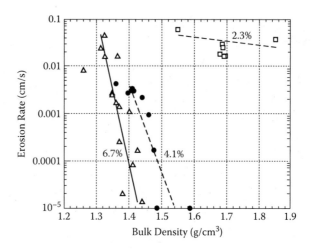

FIGURE 3.16 Erosion rates as a function of bulk density for Fox River sediments with different organic carbon contents at a shear stress of 1.6 N/m². (*Source:* From Lick and McNeil, 2001. With permission.)

high and low organic carbon were mixed in approximately equal amounts so that a sediment with 4.1% o.c. was obtained.

Measurements of erosion rates of these sediments were then made. Results for the Fox River (Figure 3.16) show that organic content has a significant effect on the initial density (the lowest densities tested) of the sediments as well as on $dE/d\rho$; that is, the initial densities of the sediments vary from approximately 1.3 g/cm³ for sediments with 6.7% o.c., to 1.4 g/cm³ for 4.1% o.c., to 1.6 g/cm³ for 2.3% o.c., whereas $dE/d\rho$ decreases so that E becomes almost independent of ρ as the organic content becomes small, similar to that for coarse, almost non-cohesive sediments. The Long Beach sediments behaved in a similar manner. These results demonstrate qualitatively that an increase in organic matter causes sediments to behave in a more cohesive manner.

3.3.5 SALINITY

Several investigators have demonstrated the effects of salinity on sediment erosion. For example, Owen (1975) and Gularti et al. (1980) showed that the critical stress for erosion of cohesive sediments increased as the salinity increased. Owen (1975) also measured erosion rates for a mud and determined that erosion rates decreased by as much as a factor of four as the salinity increased from fresh to sea water. For noncohesive, coarse-grained sediments, salinity did not affect erosion rates. Although the above data are suggestive, additional data are needed to quantify the effects of salinity on erosion rates.

3.3.6 GAS

Gas is often present in sediments and can have a major effect on sediment densities (Section 4.6) and erosion rates (Jepsen et al., 2000; McNeil et al., 2001b). However, because bulk densities are commonly measured by the wet-dry procedure, which ignores the presence of gas, the presence and influence of this gas often are not measured or even recognized. To illustrate the effects of gas on erosion rates, consider first the measurements of erosion rates and bulk properties of relatively undisturbed and reconstructed sediments (both of which often contained considerable amounts of gas) from the Grand River in Michigan (Jepsen et al., 2000). For these sediments, the mean particle sizes ranged from 50 to 100 μm, the organic carbon contents ranged from 1 to 5%, and gas volumes ranged from 1 to 10%. For each site, the bulk density, ρ, as well as the density of the solid-water matrix, ρ^*, were measured as a function of depth. From these two quantities, the fractional gas volume was determined.

As an example, erosion rates as a function of bulk density with shear stress as a parameter for undisturbed and reconstructed sediments from one site in the Grand River are shown in Figure 3.17(a) (sediments with no gas) and Figure 3.17(b) (sediments with 3 to 4% gas). Solid lines are approximations to the data as given by Equation 3.4. By comparing these figures, it can be seen that sediments at a particular density without gas have greater erosion rates than sediments at the same density with gas. This seems somewhat surprising; a possible explanation is as follows. The most obvious effect of gas is to decrease the bulk density. For the present case, the presence of 3 to 4% gas corresponds to a decrease in density of approximately 0.05 g/cm^3 (from Equation 2.19). However, gas also modifies the structure of the sediment-water matrix (1) by forming voids and channels that tend to fill with water and stay open after the passage of a gas bubble and (2) by increasing the density and stiffness of the solid-water matrix near the voids and channels. The presence of numerous voids and channels allows the solid-water matrix to consolidate more rapidly than in the absence of these passageways. The effects of density changes due to the presence of gas can be corrected for by plotting erosion rates as a function of ρ^*; that is, the data in Figure 3.17(b) should be moved to the right by 0.05 g/cm^3. When this is done, the data for sediments with gas then, in general, are equal to or above the analytic approximation for the data in Figure 3.17(a) and indicate an increase in erosion rates (for the same bulk density, ρ^*) as the gas content increases. The implication of this is that the presence of gas bubbles and voids destabilizes the solid sedimentary structure, a result more or less consistent with our conventional ideas of the effects of gas on erosion.

Experiments of this type also have been performed with a proposed capping material that was a 50/50 mixture of topsoil and sand (McNeil et al., 2001a). For this mixture, the mean particle size was 125 μm (significantly coarser than the previous sediment), the organic content was 1.3%, and the gas content varied from

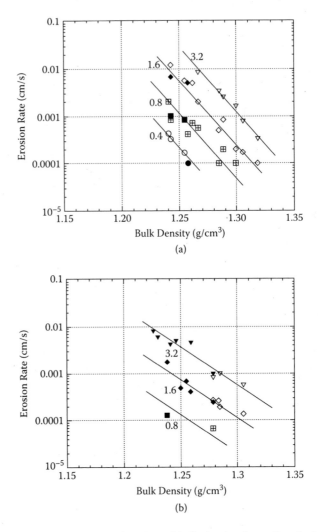

FIGURE 3.17   Erosion rates as a function of bulk density for undisturbed and recon-structed sediments from the Grand River with shear stress as a parameter: (a) sediments with no gas, and (b) sediments with 3 to 4% gas. Solid symbols denote undisturbed sedi-ments, whereas open symbols denote reconstructed sediments. (*Source:* From Jepsen et al., 2000. With permission.)

0 to 5%. Erosion rates for both small and large times (i.e., for negligible and up to 5% gas fraction) are shown as a function of ρ^* in Figure 3.18. All of the data are well approximated by Equation 3.4, with ρ^* substituted for ρ. It follows that E is primarily a function of ρ^*. In experiments where the gas flux was increased but the gas fraction remained constant, this was no longer true and erosion rates were lower. This indicates that a significant effect of gas is to stiffen the solid-water matrix near the voids and channels so as to reduce the erodibility.

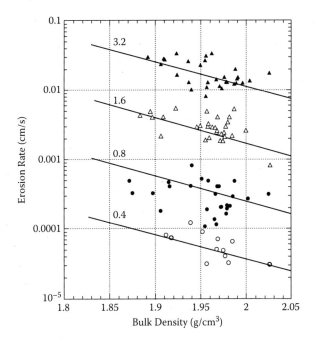

FIGURE 3.18 Erosion rates for a capping material as a function of ρ* with shear stress (N/m²) as a parameter for consolidation times of 1, 2, 4, 8, 64, and 128 days.

Results from both of the above investigations indicated that (1) the erosion rate is primarily a function of ρ*; (2) for the same ρ*, the presence of gas and the associated decrease in the bulk density ρ cause an increase in erodibility; (3) the formation of voids and channels by gas ebullition stiffens the solid-water matrix and reduces erodibility; and (4) this latter effect increases as the cohesivity increases. These are interesting observations but need to be quantified by further experiments and analyses.

3.3.7 COMPARISON OF EROSION RATES

Data for erosion rates of reconstructed sediments are shown as a function of density at a shear stress of 1.6 N/m² in Figure 3.19. All of these data are for sediments in fresh water (except for Long Beach, which was for experiments in seawater) and without gas. Information on mean particle size, organic content, and mineralogy is presented in Table 3.1. In the figure, the dashed lines are for quartz particles of different sizes (as in Figure 3.12). These latter sediments contain no organic carbon, and the mineralogy (pure quartz) is well defined. As such, these data serve as a base against which data from other sediments can be compared. In this way, the effects of other bulk properties on erosion rates can be better understood.

Consider first the reconstructed sediments from the Fox River (mean size of 20 μm). As the organic carbon decreases, the initial bulk density increases from

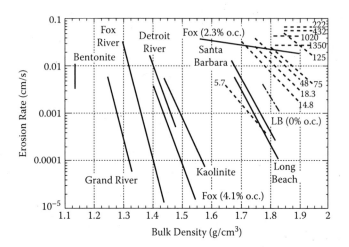

FIGURE 3.19 Erosion rates as a function of bulk density for reconstructed sediments at a shear stress of 1.6 N/m². Dashed lines are for quartz particles of different sizes (diameters in micrometers). (*Source:* From Lick and McNeil, 2001. With permission.)

about 1.3 g/cm³ (6.7% o.c.) to 1.4 g/cm³ (4.1% o.c.) to 1.6 g/cm³ (2.3% o.c.), and the cohesivity decreases (i.e., $dE/d\rho$ decreases). Fox River sediments also contain a small amount of clay (Table 3.1). Eliminating this clay also would increase the bulk density and decrease $dE/d\rho$. The result of this qualitative argument is that the erosion rates for the Fox River sediments, after eliminating the effects of organic carbon and clay content, tend toward those for noncohesive quartz particles.

As another example, sediments from the Detroit River have a mean size of 12 μm and 3.3% o.c.; they contain no clay but do contain mica, which is fine-grained and has a weak cohesive effect. After eliminating the effects of organic carbon and mica, the erosion rates for Detroit sediments tend to those for 12-μm quartz.

In a similar manner, the effects of bulk properties on the erosion rates of other sediments can be qualitatively considered. In general, these results indicate that a decrease in particle size and an increase in clay content, organic content, and salinity each increases the cohesivity of sediments. For deposited sediments, as the cohesivity increases, the bulk density decreases and $dE/d\rho$ (or m in Equation 3.4) increases. To a first approximation, the effects of gas can be corrected for by plotting E as a function of ρ^*. With additional data and understanding, it is hoped that erosion rates for all sediments can be collapsed into a single family of curves such as that for quartz particles, that is, $E(\tau; \rho, d)$, but much additional data is needed for this purpose.

3.3.8 BENTHIC ORGANISMS AND BACTERIA

Benthic organisms and bacteria have significant and diverse effects on sediment bed properties. Both secrete mucus within the sediments, at the sediment-water interface, and on burrow walls. This tends to bind and decrease the erosion rates of the sediments. But benthic organisms also form fecal pellets, which are generally larger and have lower densities than the nonpelletized sediment particles; in addition, benthic organisms burrow, thus stirring the sediments, disrupting the existing sediment fabric, and changing the water content. These factors increase the erosion rates of the sediment. As one example of the effects of benthic organisms, the resuspension properties of sediments from Long Island Sound as modified by Nucula clams have been investigated (Tsai and Lick, 1988). In tests where the sediments were consolidating with time, the resuspension rates of sediments with clams (1) initially decreased for the first 3 or 4 days at a rate comparable to sediments without clams, and (2) then, in the bioturbated zone, remained constant after that time. In contrast, the same sediments without benthic organisms continued to compact with time so that their resuspension rates decreased continuously with time but at a slower and slower rate as time increased. It is believed that other benthic organisms will affect sediment properties in a similar manner.

3.4 INITIATION OF MOTION AND A CRITICAL SHEAR STRESS FOR EROSION

As discussed in Section 3.1, a critical velocity (or critical shear stress) for the initiation of motion (or erosion) is difficult to precisely define and observe (especially for sediments with a wide particle size distribution and/or a wide distribution of cohesive forces between particles); nevertheless, it is a useful parameter, and the general concept of a threshold of particle movement and associated critical velocity and shear stress, τ_c, is generally accepted. Early work on the threshold of particle movement was undertaken by Shields in 1936, and this work has been supplemented by numerous other researchers. The emphasis, in general, has been on the behavior of coarse-grained sediments (for summaries, see texts by Van Rijn (1993) and Chien and Wan (1999)). Some research also has been carried out on the initiation of movement of fine-grained, cohesive sediments. Early experiments were done with sediments consisting of different-size particles and mineralogies (see summaries by Mehta (1994) and Chien and Wan (1999)), but the data were insufficient to distinguish the separate and large effects of changes in bulk density with time (consolidation), particle size, or mineralogy. In the present section, a theoretical description of the initiation of sediment movement is given (Lick et al., 2004a), that includes these effects; the emphasis is on fine-grained, cohesive sediments. The analysis is based on the experiments by Roberts et al. (1998) and Jin et al. (2002) discussed previously in this chapter. A horizontal sediment bed is assumed. Effects of surface slope on critical stresses and erosion rates are considered in Section 3.6.

In the experimental study by Roberts et al. (1998), the effects of particle size and bulk density on the erosion of quartz particles were investigated. In addition to erosion rates, critical stresses for erosion also were determined. For coarse-grained, noncohesive sediments, the usual definition for this quantity is the stress at which sediment movement is first observed. However, because of the difficulty

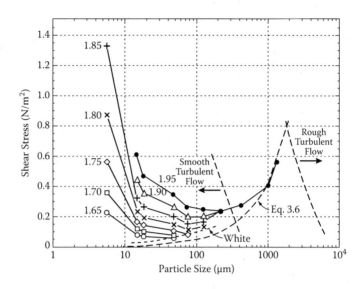

FIGURE 3.20   Critical shear stresses as a function of particle size and bulk density for quartz particles. Data by White (---) and theoretical curve (Equation 3.6) for noncohesive particles (-- --) are shown for comparison. (*Source:* From Lick et al., 2004a. With permission.)

in precisely quantifying this initiation of movement, especially for fine-grained sediments, the critical stress was defined by Roberts et al. as the shear stress at which a small, but accurately measurable, rate of erosion (10^{-4} cm/s) occurred.

Results of this investigation for the critical stress as a function of particle diameter with bulk density as a parameter are shown in Figure 3.20. The critical stress strongly depends on particle size; as d decreases, τ_c first decreases, reaches a minimum, and then increases. For small d, τ_c also strongly depends on the bulk density of the sediments, with the dependence increasing as d decreases. The minimum in τ_c(d) decreases to smaller values of d as the bulk density decreases.

For large d, the sediments behave in a noncohesive manner; that is, changes in bulk density with time after deposition are small and the sediments erode particle by particle. In this region, the results for τ_c agree with the results of others (e.g., see compilation of data by Miller et al. (1977)) and can be expressed as (Chepil, 1959):

$$\tau_c = 0.414 \times 10^3 \, d \qquad (3.6)$$

where τ_c is in units of N/m^2 and d is in m. This result is plotted as a dashed line in Figure 3.20.

As d decreases below about 200 µm, cohesive effects become significant; particle erosion is still dominant, but erosion in small chunks or aggregates of particles begins to occur. The critical stress begins to depend on the bulk density as well as on particle size. As d decreases even further, cohesive effects become more significant; the critical shear stress is an increasingly stronger function of the bulk density and particle size, and erosion as chunks tends to increase relative to erosion as small aggregates or particles. These chunks of sediment tend to increase in size as d decreases and also as the shear stress and bulk density increase. For the most cohesive sediments, erosion occurred by bulk fracture of the sediments, as shown in Figure 3.15.

Also shown in Figure 3.20 are the experimental results of White (1970). His results, within the range of particle sizes that he investigated, agree with the results of Roberts et al. at the lowest density of 1.65 g/cm^3. One reason for this may be that White determined critical shear stresses for the initiation of motion of surficial sediments only and, because surficial, fine-grained sediments often behave as individual, noncohesive particles and generally have low bulk densities, his results presumably correspond to sediments that have low densities. Another reason may be that the surface coatings on his particles may be different from and may cause weaker cohesive forces than those on the particles investigated by Roberts et al. (e.g., see Litton and Olson, 1993; Nairn, 1998). It should be emphasized that all sands (even those with the same average particle size) are not equal. Erosion rates, critical stresses, and angles of repose are all affected by particle size distributions about the mean, by surface coatings, and by grain shapes (e.g., round vs. angular).

3.4.1 THEORETICAL ANALYSIS FOR NONCOHESIVE PARTICLES

A simple theoretical and quite basic description for the initiation of movement of noncohesive particles of uniform size was given by Chepil (1959), and that approach is followed here; it is modified later to include more general conditions and forces between particles. The emphasis is on the dependence of the critical stress as a function of particle diameter. In the analysis, it is assumed that a spherical particle of diameter d rests on other spherical particles of diameter d to form a horizontal bed, as shown in Figure 3.21. A flow with velocity u causes drag

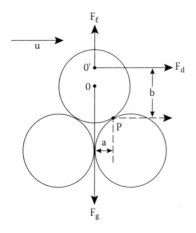

FIGURE 3.21   Forces acting on a particle in a fluid flow on a horizontal bed.

and lift forces on the protruding particle. A gravitational force, F_g, acts vertically through the center of the sphere at point O. The drag force, F_d, and lift force, F_ℓ, are assumed to act through the point O′, a small distance above O. At initiation of movement of the particle, the moment about point P of the forces acting on the particle is zero, or

$$F_d \, b = (F_g - F_\ell) \, a \tag{3.7}$$

where a is the horizontal distance between P and the vertical line connecting O, O′; and b is the vertical distance between O′ and P.

The magnitude of the gravitational force is given by

$$F_g = \frac{\pi}{6}(\rho_s - \rho_w)g d^3 = c_3 \, d^3 \tag{3.8}$$

where $c_3 = \pi(\rho_s - \rho_w)g / 6$. Chepil (1959) has stated that F_ℓ is approximately $0.85 \, F_d$. Saffman (1965) theoretically investigated the lift force on a sphere and found that, for low shear stresses of the magnitude considered here, the lift force was much smaller than that predicted by Chepil. In either case, Equation 3.7 reduces to

$$F_d = c_2 \, F_g \tag{3.9}$$

where c_2 is a constant approximately equal to a/b and therefore of order one.

The drag force is usually written as

$$F_d = \frac{1}{2} \rho_w \, C_d \, A \, u^2 \tag{3.10}$$

where C_d is a drag coefficient and A is the cross-sectional area of the sphere and is given by $\pi d^2/4$. The drag force per unit area of the bed, or shear stress, is generally expressed as $\tau = Cu^2$, where C is another drag coefficient. From these definitions, the above equation can be written as

$$F_d = c_1 \, \tau d^2 \tag{3.11}$$

where $c_1 = \pi \rho_w \, C_d / 8C$.

By substituting Equations 3.11 and 3.8 into Equation 3.9, one obtains

$$\tau_{cn} = \frac{c_2 \, F_g}{c_1 \, d^2} = \frac{c_2 \, c_3}{c_1} d$$

$$= \text{constant d} \tag{3.12}$$

where τ_{cn} is the critical shear stress for the initiation of movement of noncohesive particles. As indicated above, this constant has been determined from experiments and more refined theoretical analyses that have taken into account turbulent fluctuations and the relative protrusion of particles (Grass, 1970; Fenton and Abbott, 1977; Chin and Chiew, 1992); these studies suggested an approximate value of 0.414×10^3 N/m^3. Paphitis (2001) has compiled more recent data on the initiation of noncohesive particles and suggests that $\tau_{cn} = 1.11 \times 10^3 \, d^{1.14}$ for d > 1 mm. You (2000), on the basis of Shields criteria, suggests that $\tau_{cn} = 0.89 \times 10^3$ d for d > 2 mm. For 400 < d < 1000 μm, the flow is in transition from rough turbulent to smooth turbulent so that a linear dependence of τ_{cn} on d may not be adequate. Again, it should be noted that τ_c, as defined in the Roberts et al. and in all Sedflume studies, is based on a critical erosion rate and is somewhat different from that in the above equation (based on the initiation of motion), so that exact comparisons may be difficult. However, the differences in the definitions of τ_c are probably less than the differences in the erosion rates of different, but presumably identical, sands and less than experimental errors in the observations. Additional work needs to be done to determine the dependence of τ_{cn} on d in this transition area. In the present analysis, the emphasis is on fine-grained sediments with d < 200 μm. The specific dependence of τ_{cn} on d for coarse, noncohesive sediments will change the constant in the above equation and possibly the dependence on d but will not change the functional behavior of τ_{cn} (its dependence on particle size, density, or mineralogy) for fine-grained, cohesive quartz particles as described below.

3.4.2 EFFECTS OF COHESIVE FORCES

The curve for $\tau_{cn} = 0.414 \times 10^3$ d is shown in Figure 3.20, agrees with experimental data for d > 1000 μm, but deviates considerably from the data as the particle size decreases below this value. For 400 μm < d < 1000 μm, some of this deviation is due to the fact that the flow is in transition from rough turbulent to smooth turbulent; cohesive effects may also be present. For d < 400 μm, the differences are almost entirely due to cohesive forces, which become dominant as particle size decreases.

It has been shown theoretically and experimentally (Israelachvili, 1992; Ducker et al., 1991, 1992; Raveendran and Amirtharajah, 1995) that the cohesive force between two spherical particles is proportional to the particle diameter, that is,

$$F_c = c_4 \, d \qquad (3.13)$$

where c_4 depends on the cohesive force and is different for particles with different mineralogies and different coatings on the particle surfaces. Measurements of the forces between real, nonspherical quartz particles with approximate diameters of 15 μm have been made by means of an atomic force microscope (R. Jepsen, personal communication) and have shown that these forces were highly variable. These large variations were presumably due to factors such as the nonsphericity

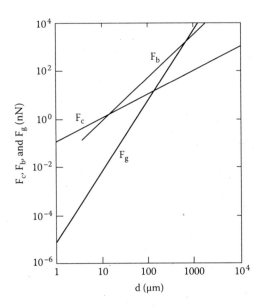

FIGURE 3.22   Gravitational force F_g, cohesive force F_c, and binding force F_b as functions of particle diameter.

and surface roughness of the particles, the relative orientation of the two particles, the nonuniformity of the surface charges, nonuniform surface coatings on the particles, and the presence of water layers trapped between the surfaces.

From these experiments and as a first approximation for F_c, the average attractive force between the 15-μm quartz particles was assumed to be 2 nN. With this value, c_4 in the above equation is then 2 nN/15 μm or 1.33×10^{-4} N/m. F_c and F_g were then calculated from Equations 3.13 and 3.8 and are shown as functions of the particle diameter in Figure 3.22. F_g is dominant for large d, whereas F_c is dominant for small d.

To quantitatively consider the effects of cohesive forces, the analysis presented above for noncohesive particles is now modified so that F_g is replaced by $F_c + F_g$. Implicit in this assumption is the idea that the resultant of the cohesive forces acting between a surficial particle and the surrounding particles can be replaced by an effective cohesive force acting in the vertical direction. Equation 3.12 can now be written as

$$\tau_c = \frac{c_2\,(F_g + F_c)}{c_1\,d^2} \tag{3.14}$$

where τ_c is the critical shear stress when both cohesive and gravitational forces are included. With τ_{cn} from Equation 3.12, F_g from Equation 3.8, and F_c from Equation 3.13, this equation becomes

$$\frac{\tau_c}{\tau_{cn}} = 1 + \frac{F_c}{F_g}$$

$$= 1 + \frac{c_4}{c_3 d^2} \tag{3.15}$$

From this and with c_4 as specified, the critical shear stress can be calculated and is in agreement with the experimental results for a sediment bulk density of 1.85 g/cm^3 (Figure 3.20). In particular, (1) for large d, τ_c will approach τ_{cn}; (2) for small d, τ_c is inversely proportional to d; and (3) the minimum of $\tau_c(d)$ can be shown to be 127 μm, the particle diameter at which F_c and F_g are equal (Figure 3.22).

3.4.3 Effects of Bulk Density

Although the results for τ_c as a function of d agree well with the experimental data for $\rho = 1.85$ g/cm^3, it is also evident from Figure 3.20 that τ_c is a strong function of the sediment bulk density and increases as the bulk density increases. This dependence is due to the increased packing of the particles, collapse of low-density aggregates, and increased contact between particles (and therefore an increased effective cohesive force between particles and an increased value of c_4) as the bulk density increases. In addition, because of the closer packing, individual particles are not greatly exposed to the flow, and the drag and lift forces on the particle must therefore decrease. The result is that the critical shear stresses are relatively low for low bulk densities and increase as the bulk density increases, in qualitative agreement with the observations.

Although the effects of bulk density on c_4 and the critical shear stress cannot be predicted theoretically, the variation of c_4 and hence τ_c as a function of d can be approximately determined for a particular bulk density once τ_c is known as a function of ρ for one value of d. For a constant and small value of d (say 5 μm), it follows from Equation 3.15 that c_4 is proportional to τ_c and is a function of ρ; an effective value for c_4 can then be calculated from

$$c_4(\rho) = c_4^* \frac{\tau_c(\rho,5)}{\tau_c^*} \tag{3.16}$$

where c_4^* is the value of c_4 for $\rho = 1.85$ g/cm^3, $\tau_c(\rho, 5)$ is the value of the critical shear stress at different values of ρ and for d = 5 μm, and τ_c^* is the value of τ_c for $\rho = 1.85$ g/cm^3 and d = 5 μm. For these values of c_4^* and τ_c^*, it follows that $c_4(\rho)$ is numerically approximately equal to $\tau_c(\rho,5) \times 10^{-4}$; it can be approximated as $a_1 e^{b\rho}$, where a_1 and b are constants.

With this approximation, Equation 3.15 can be expressed as

$$\tau_c = \left(1 + \frac{ae^{b\rho}}{d^2} \right) \tau_{cn} \qquad (3.17)$$

where $a = 8.5 \times 10^{-16}$ m^2 and $b = 9.07$ cm^3/g. The critical shear stresses from this equation, along with the experimental results, are shown in Figure 3.23 for specific values of ρ. There is generally good agreement between the theoretical and experimental data. It follows that the above equation is uniformly valid for these quartz particles ranging in size from fine-grained, cohesive particles up to coarse-grained, noncohesive particles.

3.4.4 EFFECTS OF CLAY MINERALS

As discussed in Section 3.3, experiments have been done to determine erosion rates as a function of shear stress for different-size quartz particles with 0 and 2% added bentonite (Jin et al., 2002). From these data, critical shear stresses were estimated and are shown in Figure 3.24 as a function of the diameter of the quartz particles with no bentonite and with 2% added bentonite. With no bentonite, the results are similar to those obtained by Roberts et al. (1998) for similar densities. With 2% bentonite, the critical stresses are significantly larger, with the largest effect occurring for particles on the order of 100 to 200 µm and decreasing as d deviates from this range of values. This effect is similar to the effect on erosion rates as shown in Figure 3.13, where R was least for the particles with average size of 140 and 170 µm and approached one for both smaller and larger particle sizes.

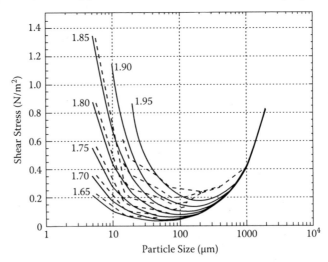

FIGURE 3.23 Comparison of theory and experimental results for critical shear stresses as a function of particle size and bulk density. Solid lines are theory, whereas experimental results are shown as dashed lines. (*Source:* From Lick et al., 2004a. With permission.)

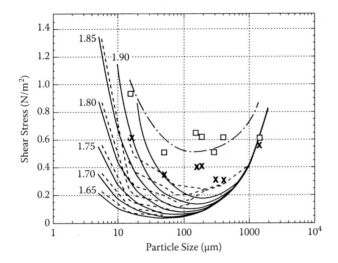

FIGURE 3.24 Critical shear stresses as a function of particle size for quartz particles with 0 (**x**) and 2% (□) added bentonite. The theoretical results for 2% bentonite as expressed by Equation 3.19 are shown as a dot-dash line. (*Source:* From Lick et al., 2004a. With permission.)

The bentonite data for τ_c cannot be well approximated by modifying the value of c_4 as a function of ρ in the cohesive force, that is, in Equation 3.13. If this were done, the theoretical curve could be made to approximate the data for $100 < d < 1000$ µm but would generally be much greater than the experimental data for small d. However, a good approximation for the critical stress can be obtained by assuming an additional binding (cementation) force between the quartz particles that is due to the bentonite and that is proportional to the particle diameter squared. The total force between quartz particles is then given by

$$F = F_g + F_b + F_c \tag{3.18}$$

where $F_b = c_5 d^2$ and c_5 is a constant. From the data, a reasonable approximation is that $c_5 = 7$ N/m². This force as a function of d is shown in Figure 3.22 and can be compared there with F_g and F_c. It is the largest force for 15 µm $< d < 600$ µm but is smaller than F_g for large d and smaller than F_c for small d. The relative effect of F_b on the total force, defined as $(F - F_b)/F$, varies with d in the same manner as R(d), as shown in Figure 3.13.

The critical stress now is given by Equation 3.12, with F substituted for F_g; it can be expressed as

$$\frac{\tau_c}{\tau_{cn}} = 1 + \frac{F_b + F_c}{F_g}$$

$$= 1 + \frac{ae^{b\rho}}{d^2} + \frac{c_5}{c_3 d} \tag{3.19}$$

This approximation (for a sediment bulk density of 1.85 g/cm³) is shown in Figure 3.24 as the dot-dash line and is a reasonable fit to the data.

An explanation for this binding force is as follows. Assume that the amount of added bentonite is relatively small but is sufficient to coat the surfaces of the quartz particles with a uniform layer of thickness δ. These layers overlap over a distance 2ℓ. For fixed amounts of quartz and bentonite and as the quartz particle diameter increases (and as the number of quartz particles decreases), the thickness of the bentonite layer, δ, must increase in proportion to d in order to conserve mass. The length 2ℓ over which the bentonite layers overlap is then proportional to δ and hence proportional to d. The area over which the layers overlap is then proportional to d². The force per unit area required to fracture the bentonite layer is independent of d, and hence the binding force per particle due to the bentonite, F_b, is proportional to d², a result that was shown above to be consistent with the experimental results for $\tau_c(d)$.

For large amounts of bentonite, the bentonite would completely fill the voids between particles. In this case, it can be shown theoretically that the same dependence of F_b on d as above is obtained; that is, F_b is essentially a cementation force due to the bentonite.

From analyses of erosion experiments with other clays (Lick et al., 2002, summarized in Section 3.3), it can be shown that c_5 is different for each clay additive, depends on the cohesivity of the clay, and increases as the amount of clay increases.

3.5 APPROXIMATE EQUATIONS FOR EROSION RATES

In previous sections, data on erosion rates as a function of shear stress and bulk properties were presented. For modeling purposes, it is useful to approximate these data by means of relatively simple algebraic equations. Equations of this type are here illustrated and compared with experimental data (Lick et al., 2006a). Equations are presented that are (1) valid for fine-grained, cohesive sediments but not for coarse-grained, noncohesive sediments; (2) valid for coarse-grained, noncohesive sediments but not for fine-grained, cohesive sediments; and (3) uniformly valid for both fine-grained and coarse-grained sediments. Good agreement between this latter equation, the previous equations (when they are valid), and experimental data on quartz particles is demonstrated. As a preliminary step in quantifying erosion rates for more general sediments and especially to show the effects of clay minerals, the uniformly valid equation also is applied to sediments consisting of quartz particles with the additions of clay minerals.

3.5.1 COHESIVE SEDIMENTS

For fine-grained, cohesive sediments with a wide range of bulk properties, it has been shown that the erosion rate is well approximated by Equation 3.4, which is repeated here for convenience:

$$E = A\tau^n\rho^m \tag{3.20}$$

With the definition of the critical shear stress as the shear stress at which an erosion rate of 10^{-4} cm/s occurs, this can be written as

$$E = 10^{-4}\left(\frac{\tau}{\tau_c}\right)^n \tag{3.21}$$

where τ_c is a function of ρ and A. The above equation is useful because it explicitly shows the dependence of E on τ_c, a quantity that can be measured independently and does not need to be determined from Equation 3.20; that is, the validity of Equation 3.21 does not depend on the validity of Equation 3.20.

A uniformly valid equation for τ_c that includes gravitational and cohesive forces (but not binding forces due to clay minerals) is given by Equation 3.17. Equation 3.21 with τ_c from Equation 3.17 was applied to approximate the erosion data on the quartz particles used by Roberts et al. Nonlinear regression was used to determine the values of n. By this means, it was demonstrated that Equation 3.21 along with Equation 3.17 was a valid approximation to the data for all particle diameters equal to or less than 432 µm but was not a good approximation for noncohesive particles (i.e., for d ≥ 1020 µm). For small and intermediate d (when the above equation is valid), n was approximately 2; for large d (when the above equation is not valid), n increased rapidly to above 5.

3.5.2 Noncohesive Sediments

For coarse-grained, noncohesive sediments, a valid equation for $E(\tau)$ is Equation 3.5, that is,

$$E = A(\tau - \tau_c)^n \tag{3.22}$$

where $E = 0$ for $\tau < \tau_c$ and where A, τ_c, and n are functions of particle diameter but not functions of density. For noncohesive sediments, the parameter $\tau_c = \tau_{cn}$ whereas the parameters A and n can be found from experimental data by means of nonlinear regression. In this way, good agreement between the experimental data and the above equation was obtained. For finer-grained particles, A and n were determined by the same procedure; however, the agreement between the experiment and Equation 3.22 decreased as d decreased. One reason is that, by this formulation, E is not a function of density; this, of course, is required by the experimental results for small d. Even when A and n were allowed to be functions of density, it was shown that the above equation was not valid for small d — that is, the dependence on τ was incorrect. In general, it was demonstrated that Equation 3.22 is a valid approximation for noncohesive sediments but is not a good approximation for cohesive sediments.

3.5.3 A UNIFORMLY VALID EQUATION

To approximate the data for all size ranges, the following equation was proposed:

$$E = 10^{-4} \left(\frac{\tau - \tau_{cn}}{\tau_c - \tau_{cn}} \right)^n \tag{3.23}$$

where $E = 0$ for $\tau < \tau_{cn}$ and where τ_c, τ_{cn}, and n are functions of particle diameter. As $d \rightarrow 0$, $\tau_{cn} \rightarrow 0$ (Equation 3.12) and is much smaller than τ_c or τ; Equation 3.23 then reduces to Equation 3.21, which is valid for fine-grained, cohesive sediments. For large d, $\tau_c - \tau_{cn}$ is then independent of ρ (Equation 3.19), and Equation 3.23 can be written as

$$E = A (\tau - \tau_{cn})^n \tag{3.24}$$

where A is a function of d only. As a result, the above equation has the same form as Equation 3.22 and is therefore valid for coarse-grained, noncohesive particles.

For constant values of a and b as in Equation 3.17, reasonably good agreement was obtained between Equation 3.23 and the experimental data. Improved agreement was obtained by allowing a and b to be functions of d, with their values determined by nonlinear regression. The justification for this was that the quartz particles of different sizes had come from different samples and therefore had somewhat different bulk properties. In general, it was demonstrated that Equation 3.23 was a reasonable approximation to the data and was uniformly valid for all particle sizes tested. In addition, it was demonstrated that Equation 3.23 was as good as or better than Equation 3.21 (valid for fine-grained, cohesive sediments) and as good as or better than Equation 3.22 (valid for coarse-grained, noncohesive sediments). For these quartz particles, n was approximately equal to 2 for all particle diameters. For sediments from field cores, values for n tend to range from 2 to 3 (e.g., see Section 6.4).

3.5.4 EFFECTS OF CLAY MINERALS

As summarized in Section 3.3, erosion rates have been measured as a function of shear stress for (1) sediments consisting of quartz particles of different sizes, without and with 2% added bentonite (Jin et al., 2002), and (2) sediments consisting primarily of single-size (48 µm) quartz particles but with each sediment differing due to the addition of one of four clays, each at several different percentages (Lick et al., 2002). Equations 3.23 and 3.19 were used to approximate these data. The coefficients a and b were assumed to be the same as those for sediments with no clay minerals, whereas c_s and n were coefficients that were determined from the experimental data by nonlinear regression. Values of n ranged from 2 to 4 and increased as the cohesivity of the sediment increased. Generally good agreement between Equation 3.23 and the experimental data was obtained and demonstrated that Equation 3.23 is a valid approximation for these sediments.

3.6 EFFECTS OF SURFACE SLOPE

In most modeling of sediment transport, it is assumed that the sediment-water interface is horizontal. In many cases, this is a reasonable (and often necessary) approximation. However, in many areas of surface waters, such as on the edges of channels in rivers, near river banks and lake shores, and especially in areas of scour near pilings and abutments, the sediment-water interface is not horizontal and may have a large-enough slope that erosion rates can be significantly affected.

Surfaces can slope in the direction of flow (longitudinal direction), perpendicular to the flow (transverse direction), or in an arbitrary direction relative to the flow. Slopes in the longitudinal direction are quantified by the pitch angle, β; uphill flows are here considered positive, whereas downhill flows are negative. Slopes in the transverse direction are quantified by the roll angle, θ, which is considered positive and is measured from the horizontal. In general, the slope at a point is a vector; it has magnitude and direction and is a vector addition of the slopes due to pitch and roll. Erosion rates are influenced significantly by both the magnitude of the slope and the angle between the flow direction and the slope.

3.6.1 NONCOHESIVE SEDIMENTS

Although numerous investigations have been made regarding the effects of slope on critical shear stresses and erosion rates (e.g., Luque, 1972; Luque and Van Beek, 1976; Engelund, 1981; Smart, 1984; Whitehouse, 1991; Damgaard et al., 1997; Whitehouse et al., 2000), almost all research has been done with coarse, noncohesive sediments; of this, most has been concerned with the effects of pitch angle on bedload.

As far as a theoretical determination of the effects of pitch and/or roll on the critical stress is concerned, the same procedure can be used as that used to determine the critical stress on a horizontal bed (Section 3.4). To begin, consider the effects of pitch angle, β. The forces on a particle are as shown in Figure 3.25. The flow is from left to right, and it is assumed for simplicity that all forces act through point O at the center of the sphere lying on the surface. The gravitational force

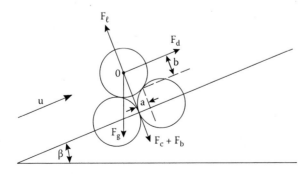

FIGURE 3.25   Forces acting on a particle in a fluid flow on a bed sloped at a pitch angle of β.

acts in the vertical direction; the drag force acts parallel to the sediment surface in the direction of flow; and the lift, cohesive, and binding forces act perpendicular to the sediment surface.

For noncohesive sediments ($F_c = 0$ and $F_b = 0$), the moment of forces about P is

$$(F_d - F_g \sin\beta)b = (F_g \cos\beta - F_\ell)a \qquad (3.25)$$

in contrast to Equation 3.7 for noncohesive sediments on a horizontal bed. Using the same procedure as in Section 3.4, it follows that the critical stress is now given by

$$\frac{\tau_c(\beta)}{\tau_{cn}} = \cos\beta + \frac{b}{a}\sin\beta \qquad (3.26)$$

where τ_{cn} is the critical stress for noncohesive sediments on a horizontal bed, is given by Equation 3.12, and is only a function of particle diameter. Because $a/b = \tan\alpha$, where α is the angle of repose, the above equation also can be written as

$$\frac{\tau_c(\beta)}{\tau_{cn}} = \frac{\sin(\alpha + \beta)}{\sin\alpha} \qquad (3.27)$$

For noncohesive sediments with nonuniform sizes and shapes, experimental results indicate that α varies from approximately 30 to 45°. For $\alpha = 45°$, $\tau_c(\beta)/\tau_{cn}$ is shown in Figure 3.26 and decreases as β decreases, from $2^{1/2}$ at $\beta = +45°$ to zero at $\beta = -45°$. For $|\beta| > \alpha$, the sediment bed will fail by avalanching downhill at zero stress.

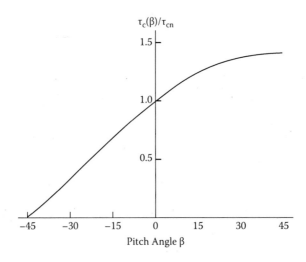

FIGURE 3.26 $\tau_c(\beta)/\tau_{cn}$ as a function of pitch angle for noncohesive sediments.

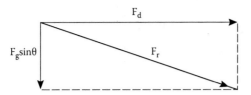

FIGURE 3.27   Forces acting parallel to the bed surface on a particle in a fluid flow on a bed sloping at a roll angle of θ.

Consider now the effects of roll angle. For a particle resting on a bed tilted transverse to the flow at an angle of θ, the forces acting in the plane of the bed are as shown in Figure 3.27. In this plane, the resultant force on the particle, F_r, is given by

$$F_r = \left(F_d^2 + F_g^2 \sin^2 \theta\right)^{1/2} \tag{3.28}$$

and acts at an angle to the flow. Because F_ℓ acts perpendicular to the plane of the bed, the moment of forces on the particle is now

$$F_r b = (F_g \cos \theta - F_\ell)a \tag{3.29}$$

Using the same procedure as above and in Section 3.4, the critical stress is now determined to be

$$\frac{\tau_c(\theta)}{\tau_{cn}} = \left[1 - \frac{\tan^2 \theta}{\tan^2 \alpha}\right]^{1/2} \cos \theta \tag{3.30}$$

For α = 45°, this ratio is shown in Figure 3.28 as a function of θ, is symmetric with respect to θ, is a maximum at θ = 0, and decreases to zero as θ increases to α.

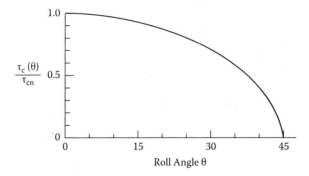

FIGURE 3.28 $\tau_c(\theta)/\tau_{cn}$ as a function of roll angle for noncohesive sediments.

For a slope with pitch and roll relative to the flow direction, the shear stress can be determined by the same procedure as above and is given by (Soulsby, 1997; Soulsby and Whitehouse, 1997):

$$\frac{\tau_c(\beta,\theta)}{\tau_{cn}} = \frac{b}{a}\sin\beta + \cos\beta\cos\theta \left[1 - \frac{\tan^2\theta}{\tan^2\alpha}\right]^{1/2} \tag{3.31}$$

In the limit as $\theta \rightarrow 0$ or $\beta \rightarrow 0$, this reduces to Equation 3.26 or 3.30, respectively.

For the effect of slope on bedload transport, Damgaard et al. (1997) have concluded, on the basis of extensive experiments of their own as well as those of others, that the bedload transport as predicted by the Meyer-Peter and Muller formula can be approximated for positive and small negative pitch angles by simply modifying the critical Shields parameter for the effects of slope, as indicated by Equation 3.26; however, for large negative slopes, additional corrections are needed. Their experiments were done with a fine sand with a median diameter of 208 μm. Roberts et al. (1998) and Lodge (2006), for their sands of this size (which may be different from those of Damgaard et al., 1997), have demonstrated that cohesive effects were present and tended to modify erosion rates and critical shear stresses as compared with those for completely noncohesive sediments.

3.6.2 CRITICAL STRESSES FOR COHESIVE SEDIMENTS

For cohesive sediments, the moment of forces now must include the cohesive force, F_c, and the binding force, F_b. For a pitch angle of β, the moment of forces about P is now

$$(F_d - F_g\sin\beta)b = (F_g\cos\beta - F_\ell + F_c + F_b)a \tag{3.32}$$

By the same procedure as before, the critical shear stress is now

$$\frac{\tau_c(\beta)}{\tau_{cn}} = \cos\beta + \frac{b}{a}\sin\beta + \frac{F_c + F_b}{F_g} \tag{3.33}$$

and reduces to Equation 3.26 for noncohesive sediments. The quantity $a/b = \tan\alpha$, where α retains its definition as the angle of repose for noncohesive sediments of the same size.

For a roll angle of θ, the moment of forces can now be written as

$$F_r b = (F_g\cos\theta - F_\ell + F_c + F_b)a \tag{3.34}$$

and the critical shear stress is

$$\frac{\tau_c(\theta)}{\tau_{cn}} = \cos\theta \left[-\frac{\tan^2\theta}{\tan^2\alpha} + \left(1 + \frac{F_c + F_b}{F_g\cos\theta}\right)^2\right]^{1/2} \tag{3.35}$$

For a slope with pitch and roll, the critical stress is given by

$$\frac{\tau(\beta,\theta)}{\tau_{cn}} = \frac{b}{a}\sin\beta + \cos\beta\cos\theta \left[-\frac{\tan^2\theta}{\tan^2\alpha} + \left(1 + \frac{F_c + F_b}{F_g\cos\beta\cos\theta}\right)^2 \right]^{1/2} \quad (3.36)$$

In the limit as $\theta \to 0$ or $\beta \to 0$, this reduces to Equation 3.33 or 3.35, respectively.

3.6.3 EXPERIMENTAL RESULTS FOR COHESIVE SEDIMENTS

Relatively little work has been done on the effects of slope on critical stresses and erosion rates for cohesive sediments. A preliminary experimental investigation of this problem has been made by Lodge (2006), who measured erosion rates of different size quartz particles from noncohesive (1350 μm) to fine-grained cohesive (5 μm) as a function of shear stress (0.4 to 3.2 N/m^2), pitch angle (−30 to +35°), and roll angle (0 to as much as 75°), and for consolidation times of 3 and 10 days. Some illustrative results are presented here.

Experimental results for erosion rates as a function of pitch angle and shear stress for three different cases are shown in Figure 3.29 (1350 μm, 3-day consolidation time), Figure 3.30 (48 μm, 3-day consolidation time), and Figure 3.31 (48 μm, 10-day consolidation time). In general, for a pitch angle of zero, results agree with those of Roberts et al. (1998) described in the previous sections. Erosion rates increase nonlinearly as the shear stress increases and as the absolute value of the pitch angle increases. For 1350-μm quartz particles (which are noncohesive) and a

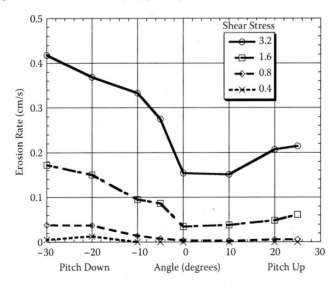

FIGURE 3.29   Erosion rates as a function of pitch angle with shear stress (N/m^2) as a parameter for 1350-μm quartz. Consolidation time is 3 days. (*Source:* From Lodge, 2006. With permission.)

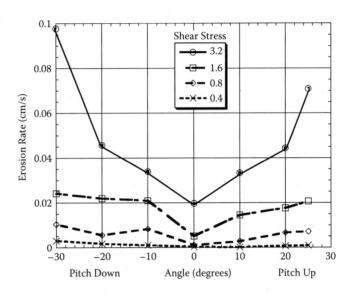

FIGURE 3.30 Erosion rates as a function of pitch angle with shear stress (N/m^2) as a parameter for 48-μm quartz. Consolidation time is 3 days. (*Source:* From Lodge, 2006. With permission.)

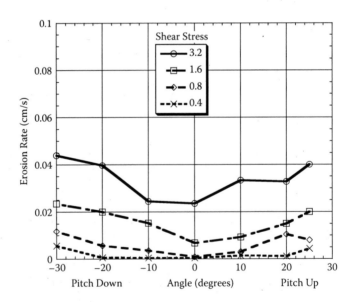

FIGURE 3.31 Erosion rates as a function of pitch angle with shear stress (N/m^2) as a parameter for 48-μm quartz. Consolidation time is 10 days. (*Source:* From Lodge, 2006. With permission.)

consolidation time of 3 days, erosion rates are a minimum at zero angle (or possibly at a small positive angle) and then increase for both positive (uphill) and negative (downhill) angles. However, as would be expected from a qualitative consideration of gravity, the increase is more rapid for negative angles than for positive angles. The increases in erosion rates as the absolute value of β increases are significant. For example, for a shear stress of 3.2 N/m^2, the ratio of E(β)/E(0) is approximately 2.6 for β = –30° and 1.5 for β = +25°.

These experiments were meant to quantify erosion rates as a function of shear stress and did not emphasize the delineation of critical stresses as a function of pitch angle. Nevertheless, it can be seen from Figure 3.29 (1350 μm) that the critical stresses are in qualitative agreement with Equation 3.26 and Figure 3.26, and generally increase as the pitch angle increases. For large β, the data indicate that τ_c may decrease as β increases and approaches the angle of repose. This may be due to turbulent fluctuations and/or failure of the sediment bed by avalanching or slumping, all of which become more significant as β approaches α. For a consolidation time of 10 days, the results are essentially the same as those for 3 days because coarse-grained quartz particles do not consolidate significantly.

Erosion rates for 48-μm particles as a function of pitch angle with shear stress as a parameter and at a consolidation time of 3 days are shown in Figure 3.30. The rates have the same qualitative dependence on angle and shear stress as those for 1350 μm. However, they are considerably less in magnitude and are more symmetric with respect to β than those for 1350-μm particles. This is because of the cohesive nature of the finer particles and the associated consolidation. For a consolidation time of 10 days, results are shown in Figure 3.31. Because of the longer consolidation time and greater cohesive effects, erosion rates are lower than those for 3 days and increase less rapidly as a function of pitch angle. In general, the experimental results show that as particle size decreases and consolidation time increases, erosion rates decrease and also become less dependent on pitch angle. The reason for this is that, in this limit, cohesive forces become more important relative to gravitational forces. As with the 1350 μm particles, the critical stresses are in agreement with Equation 3.26 for negative and small positive β but, for large β, decrease as β increases.

The experimental results also demonstrate that critical stresses for cohesive sediments are less dependent on pitch angle than those for noncohesive sediments, consistent with Equation 3.33. This can be shown more clearly by rewriting Equation 3.33 as

$$\frac{\tau_c(\beta)}{\tau_c(0)} = 1 + \frac{\cos\beta + \sin\beta - 1}{\tau_c(0) / \tau_{cn}} \tag{3.37}$$

For quartz particles, $\tau_c(0)/\tau_{cn}$ can be determined from Figures 3.20 and/or 3.23. As particle size decreases below 200 μm, this parameter increases rapidly (with the rate depending on density); the above equation then demonstrates the decreased dependence of τ_c on β as the sediments become more cohesive and $\tau_c(0)/\tau_{cn}$ increases.

Typical experimental results for erosion rates as a function of roll angle with shear stress as a parameter for three different particle sizes are shown in Figure 3.32 (1350 μm, 3-day consolidation time), Figure 3.33 (48 μm, 10-day consolidation time), and Figure 3.34 (5 μm, 10-day consolidation time). Erosion rates increase nonlinearly with shear stress and with roll angle. For 48-μm particles, because of the increased cohesive forces between particles, erosion rates are lower and increase less rapidly as a function of roll angle as compared with the 1350-μm particles. For 5-μm particles, erosion rates are only a function of shear stress and are independent of roll angle for all θ up to at least 75°.

For 1350-μm particles, the critical stress decreases as the roll angle increases, in general agreement with Equation 3.30. For 48-μm particles, $\tau_c(\theta)$ is almost independent of θ for small θ, whereas for 5-μm particles, τ_c is independent of θ. These results for 48- and 5-μm particles are consistent with Equation 3.35 for large cohesive forces.

Preliminary experimental results for the angle of repose of different-size quartz particles also were reported by Lodge (2006) and are shown in Table 3.3. As particle size decreases and consolidation time increases, the angle of repose (when $\tau_c(\beta) = 0$) increases from 44° to 90°. Equation 3.33 indicates that this

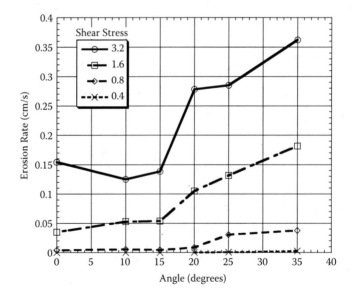

FIGURE 3.32 Erosion rates as a function of roll angle with shear stress (N/m²) as a parameter for 1350-μm quartz. Consolidation time is 3 days. (*Source:* From Lodge, 2006. With permission.)

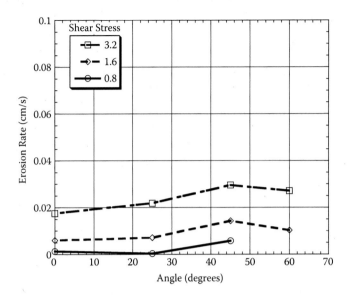

FIGURE 3.33 Erosion rates as a function of roll angle with shear stress (N/m²) as a parameter for 48-μm quartz. Consolidation time is 10 days. (*Source:* From Lodge, 2006. With permission.)

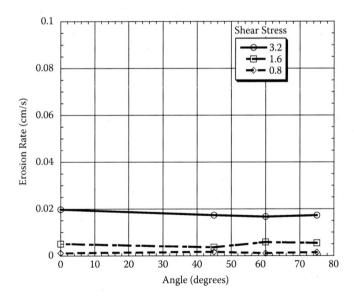

FIGURE 3.34 Erosion rates as a function of roll angle with shear stress (N/m²) as a parameter for 5-μm quartz. Consolidation time is 10 days. (*Source:* From Lodge, 2006. With permission.)

TABLE 3.3
Angle of Repose

Particle Size	Angle of Repose at 3 Days	Angle of Repose at 10 Days
1350	44	N/A
400	50	45
160	50	50
100	50	90
75	60	90
48	50	90
20	65	90

increase in the angle of repose depends on the ratio of cohesive to gravitational forces.

For different-size particles, the data presented in Sections 3.3 and 3.4 and above are sufficient to make preliminary estimates of the effects of surface slope on the critical shear stress for erosion. The data on erosion rates as a function of β and θ are quite sparse and have significant variability. However, with this caveat, the data for all β and θ are consistent with Equations 3.22 and 3.24 in the appropriate limits for these equations as long as τ_c is determined from the experiments.

4 Flocculation, Settling, Deposition, and Consolidation

Much of the suspended particulate matter in rivers, lakes, and oceans exists in the form of flocs. These flocs are aggregates of smaller, solid particles that may be inorganic or organic. The emphasis here is on the dynamics of aggregates that are primarily inorganic — that is, fine-grained sediments. These are especially prevalent in rivers, in the near-shore areas of lakes and oceans, and throughout estuaries. Because they are fine-grained, these sediments have large surface-to-mass ratios and hence adsorb and transport many contaminants and other substances with them as they move through an aquatic system. Flocculation is a dynamic process; that is, flocs both aggregate and disaggregate with time, and their sizes and properties change accordingly. The net rate of change of floc properties depends on the relative magnitudes of the rates of aggregation and disaggregation, with the steady state being a dynamic balance between the two. It is this dynamic nature of a floc that is particularly interesting and one of the major concerns in the present chapter.

Section 4.1 considers the basic theory of the aggregation of suspended particles. This is useful for a preliminary understanding of the experimental results on the flocculation of suspended particles presented in Section 4.2. As flocs are formed, their sizes and densities change with time and are quite different from the sizes and densities of the individual particles that make up a floc; this greatly modifies the settling speed of the floc. Experimental measurements on the settling speeds of flocs are presented in Section 4.3. Based on the elementary theory of aggregation presented in Section 4.1 and especially the experimental data presented in Sections 4.2 and 4.3, numerical models of the aggregation and disaggregation of suspended particles have been developed; these are discussed in Section 4.4.

As sediments (both individual particles and flocs) are transported through surface waters, they tend to settle to the bottom. The rate at which they deposit on the bottom depends on the sediment concentration, the settling speeds of the particles and flocs, and the dynamics of the fluid. This is discussed in Section 4.5. Finally, as particles and flocs are deposited on the bottom, the bottom sediments consolidate; that is, the bulk density, water content, gas content, and erosive strength of the bottom sediments change with depth and time. Information on these processes is given in Section 4.6.

4.1 BASIC THEORY OF AGGREGATION

4.1.1 COLLISION FREQUENCY

A general formula for the time rate of change of a suspended particle/floc size distribution will be developed later in Section 4.4. However, basic to this general formula and to a preliminary understanding of the experimental work is a knowledge of the collision frequency, N_{ij} (the number of collisions occurring between particles in size class i and particles in size class j per unit volume and per unit time). If binary collisions are assumed, N_{ij} can be written as

$$N_{ij} = \beta_{ij} \, n_i n_j \qquad (4.1)$$

where β_{ij} is the collision frequency function (volume per unit time) for collisions between particles i and j, and n_k is the number of particles per unit volume in size range k.

The quantities β_{ij} depend on the collision mechanisms of Brownian motion, fluid shear, and differential settling. The original collision rate theories are due to Smoluchowski (1917), and additional work has been performed by Camp and Stein (1943). Ives (1978) presents the expressions for the different collision functions as follows. For Brownian motion,

$$\beta_{ij} = \beta_b = \frac{2}{3} \frac{kT}{\mu} \frac{(d_i + d_j)^2}{d_i d_j} \qquad (4.2)$$

where k is the Boltzmann constant (1.38×10^{-23} Nm/°K), T is the absolute temperature (in Kelvin), μ is the dynamic viscosity of the fluid, and d_i and d_j are the diameters of the colliding particles. For fluid shear,

$$\beta_{ij} = \beta_f = \frac{G}{6}(d_i + d_j)^3 \qquad (4.3)$$

where G (s^{-1}) is the mean velocity gradient in the fluid. For a turbulent fluid, G can be approximated by $(\varepsilon/\nu)^{1/2}$ where ε is the energy dissipation and ν is the kinematic viscosity (Saffman and Turner, 1956). For differential settling,

$$\beta_{ij} = \beta_d = \frac{\pi}{4}(d_i + d_j)^2 \left| w_{si} - w_{sj} \right|$$

$$= \frac{\pi g}{72\mu}(\rho_f - \rho_w)(d_i + d_j)^2 \left| d_i^2 - d_j^2 \right| \qquad (4.4)$$

where w_{si} is the settling speed of the i-th particle/floc, ρ_f is the density of a particle/floc, ρ_w is the density of the water, and g is the acceleration due to gravity. The first equality is generally true, whereas the second equality is only valid when the densities of all flocs are approximately the same and when Stokes law, Equation 2.4, is valid.

A comparison of collision functions for collisions of an arbitrary-size particle with a particle of 1 μm diameter is shown in Figure 4.1(a). For this comparison, the following data used were: T = 20°C = 293K, G = 200/s, ρ_f = 2.65 g/cm³, and ρ_w = 1.0 g/cm³. As can be seen, Brownian motion is only important for collisions of a 1-μm particle with particles less than 0.1 μm. For collisions of a 1-μm particle with particles between 0.1 and 50 μm, collisions are caused primarily by fluid shear, whereas collisions of a 1-μm particle with particles greater than 50 μm are caused primarily by differential settling. As the applied shear decreases, β_f will decrease and the range of diameters over which fluid shear is dominant in causing collisions will decrease. As the floc density decreases, β_d and the effects of differential settling will decrease. Figure 4.1(b) shows the collision functions for collisions of an arbitrary-size particle with a particle of 25 μm. Again, fluid shear is the dominant mechanism for collisions of a 25-μm particle with other particles up to 50 μm, whereas differential settling is more important for larger particles. The effects of Brownian motion are negligible.

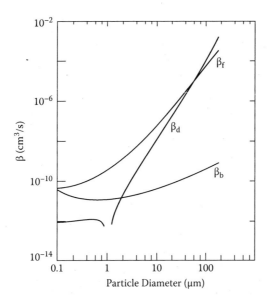

FIGURE 4.1(a) Collision function β as a function of particle size: collisions with a 1-μm particle. β_f is the collision function due to fluid shear, β_d is the collision function due to differential settling, and β_b is the collision function due to Brownian motion. For these calculations, T = 298K, G = 200/s, and $\rho_p - \rho_w$ = 1.65 g/cm³.

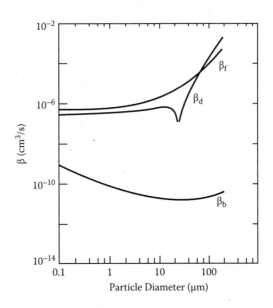

FIGURE 4.1(b)   Collision function β as a function of particle size: collisions with a 25-μm particle. β_f is the collision function due to fluid shear, β_d is the collision function due to differential settling, and β_b is the collision function due to Brownian motion. For these calculations, T = 298K, G = 200/s, and $\rho_p - \rho_w$ = 1.65 g/cm³.

4.1.2 PARTICLE INTERACTIONS

In the derivation of Equations 4.2 through 4.4 for the collision frequency functions, it was assumed that forces between particles were negligible. For predicting the rate of collisions between particles, this approximation is quite accurate. For a more accurate analysis of the entire collision process, these forces (which can be both attractive and repulsive) must be considered. It can be shown that these forces somewhat modify the rates of collisions (Friedlander, 1977), but, more importantly, they affect the probability of cohesion of particles during the collision process. The VODL theory by Verwey and Overbeek (1948) and Darjagin and Landau has been developed that includes these forces and assists in understanding the cohesion of colliding particles. This theory is only briefly summarized here; for more details, extensions, and additional theories, see Hiemenz (1986) and Stumm and Morgan (1996).

Figure 4.2 is a schematic of the potential energies between two interacting particles. The forces between the two particles are proportional to the slopes of the potential energy curves. The figure shows repulsive and attractive potential energies whose magnitudes decrease as the distance between particles increases. For small separations, on the order of twice the radius of the particles, more complex interactions occur; these forces are generally strongly repulsive.

The attractive forces are of the van der Waals type. The resultant attractive force is the summation of the pairwise interactions of the molecules making up

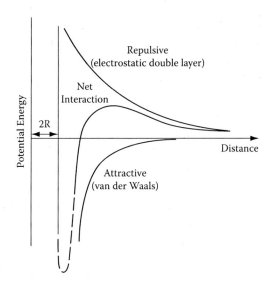

FIGURE 4.2 Schematic of potential energies between two interacting particles.

the individual particles. It depends on the molecules making up the individual particles but does not depend significantly on external parameters.

Clay particles (as well as coated, nonclay particles) in solution are normally charged due to unbalanced cations at their surfaces; this leads to a repulsive force between particles. When ions are present in a system that contains an interface, there will be a variation in the ion density near that interface; an electrical double layer results. The net effect of this double layer is to reduce the repulsive force. This force is a function of the charge or potential of the particle as well as the ionic strength (salinity) of the solution. In particular, as the ionic strength of the solution increases, this repulsive force decreases.

The net potential energy of two colliding particles is the sum of the repulsive and attractive potentials. In Figure 4.2, the net potential energy is positive at large distances and negative at small distances. A dashed curve has been added to show schematically the bottom of the potential well. Because of the variations in the repulsive and attractive potentials in magnitude and with distance, other forms of potential energy curves also can occur. Possible potential energies for different ionic strengths (low, medium, high) of the solution are shown in Figure 4.3. For low ionic strength, the maximum potential energy is relatively large and decreases slowly with increasing distance between particles. Because of the large potential energy, the speed of particle aggregation is relatively low because the particles cannot readily overcome this barrier. As the ionic strength increases, the maximum potential energy decreases and the speed of aggregation of particles should therefore increase. In some cases, there may be a secondary minimum (as shown); this is seldom very deep but could possibly explain some weak forms of flocculation. For high ionic strengths, the potential energy is negative everywhere and the rate of aggregation should therefore be quite high. As constructed, the

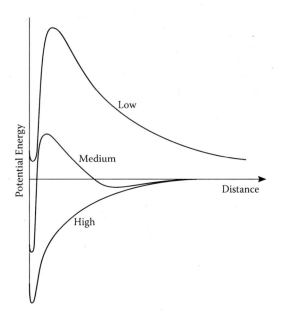

FIGURE 4.3   Potential energies between two interacting particles for different ionic strengths.

depths of the potential wells (the difference between the maximum and minimum potential energies) are approximately the same.

4.2 RESULTS OF FLOCCULATION EXPERIMENTS

For aggregation to occur, particles must collide and, after collision, they must then stick together. As described above, the rates at which particles collide are reasonably well understood. However, as particles collide, usually only a small fraction of the collisions actually results in cohesion of the particles; this process is not well understood or well quantified. Because of this, experiments are needed to determine the probability of cohesion and the parameters on which this probability depends. In addition, during transport, the disaggregation of flocs also can occur. This disaggregation is due to fluid shear and, most importantly for many conditions, is due to collisions between flocs. The probability of disaggregation and the parameters on which it depends cannot be predicted from present-day theory and thus must be determined from experiments.

To quantify the rates of aggregation and disaggregation and determine the parameters on which these processes depend, two types of flocculation experiments have been performed. In the first type of experiment, a Couette flocculator was used to determine the effects of an applied fluid shear on flocculation; fluid shear, sediment concentration, and salinity were varied as parameters. In these experiments, differential settling of particles/flocs was inherently present. In the second type of experiment, a disk flocculator was used to isolate and hence determine the effects of differential settling on flocculation in the absence of an applied fluid shear; sediment concentration and salinity were varied as parameters. The emphasis in both sets of experiments was to characterize flocculation as a function of time and to determine the effects of fluid shear, sediment concentration, and salinity on this flocculation.

4.2.1 FLOCCULATION DUE TO FLUID SHEAR

A typical Couette flocculator is shown in Figure 4.4 (Tsai et al., 1987). It consists of two concentric cylinders, with the outer cylinder rotating and the inner cylinder stationary. In this way, a reasonably uniform velocity gradient can be generated in the fluid in the annular gap between the cylinders. The width of the gap is 2 mm. The flow is uniform and laminar for shears up to 900/s, after which the flow is no longer spatially uniform and eventually becomes turbulent for higher shears.

For the experiments reported here, the sediments were natural bottom sediments from the Detroit River inlet of Lake Erie. These were prepared by filtering and settling so that the initial (disaggregated) size distribution had approximately 90% of its mass in particles less than 10 μm in diameter. The median diameter of the disaggregated particles was about 4 μm. The experiments were performed at different sediment concentrations; at shears of 100, 200, and 400/s; and in waters of different salinities. Because $\tau = \mu G$ and μ is approximately 10^{-3} Ns/m^2, these shears correspond to shear stresses of 0.1, 0.2, and 0.4 N/m^2. These shear stresses are typical of conditions in the nearshores of surface waters. In open waters, they would be less than this. Particle size distributions were periodically measured during the tests using a Malvern particle size analyzer.

As shown by Camp and Stein (1943) and Saffman and Turner (1956), the flocculation rate in a laminar flow can be related to the rate in a turbulent flow by replacing the laminar shear by an effective turbulent shear equal to $(\varepsilon/\nu)^{1/2}$, where ε is the average turbulent energy dissipation rate per unit mass and ν is the kinematic viscosity of water. This relation is valid as long as the size of the flocs is less than the size of the turbulent eddies. In isotropic turbulence in open water,

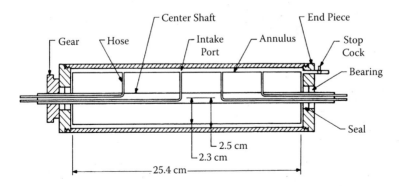

FIGURE 4.4   Schematic of Couette flocculator. (*Source:* From Tsai et al., 1987. With permission.)

this size is given by the Kolmogorov microscale of turbulence. For estuaries and coastal seas, the eddies are on the order of a few millimeters or larger (Eisma, 1986). In the bottom benthic layer, the turbulence is no longer isotropic, but eddy sizes are generally 1 mm or larger. Because the sizes of the flocs in the present experiments are less than 400 μm, and generally much less, the above relation is generally valid and the conditions in the flocculator can be related to turbulent flows in rivers, lakes, estuaries, and oceans.

In flocculation experiments, some investigators have used a tank with some sort of agitator or blade. Flows in this type of apparatus are turbulent but far from uniform, with very high shears produced near the agitator and generally quite low shears elsewhere. The high shears will dominate the aggregation–disaggregation processes. Because of this, the effective shear will be much higher (by an undetermined amount) than the average shear, and therefore the use of the average shear in correlating the experimental results is not accurate.

In the first set of experiments described here (Tsai et al., 1987; Burban et al., 1989), tests were conducted with identical sediments in fresh water, in sea water, and in an equal mixture of fresh and sea water so as to mimic estuarine waters. In fresh water, the first tests were double shear stress tests and were done as follows. The sediments in the flocculator were initially disaggregated. The flocculator was then operated at a constant shear stress for about 2 hr. During this time, the sediments flocculated, with the median particle size initially increasing with time and then approaching a steady state in which the median particle size remained approximately constant with time. For the conditions in these experiments, this generally occurred in times of less than 2 hr. After this, the shear stress of the flocculator was changed to a new value and kept there for another 2 hr. Again, after an initial transient of less than 2 hr, a new steady state was reached. The initial shears were 100, 200, and 400/s and these were changed to 200, 400, and 100/s, respectively. For each shear, the experiments were run at sediment concentrations of 50, 100, 400, and 800 mg/L.

For fresh water and a sediment concentration of 100 mg/L, the median particle diameters as a function of time are shown in Figure 4.5. The initial diameters were about 4 μm. For each shear, the particle size initially increased relatively slowly. After about 15 minutes, the size increased more rapidly but then approached a steady state in about an hour. After the change in shear stress, a transient occurred, after which a new steady state was reached. For a particular shear stress, the median particle size for the second steady state is approximately the same as the median particle size after the first steady state. These results, along with results of other experiments of this type, indicate that the steady-state median particle size for a particular shear is independent of the manner in which the steady state is approached.

From Figure 4.5 it can be seen that both the steady-state floc size and the time to steady state decrease as the shear increases. From observations of the flocs as well as measurements of settling speeds (see Section 4.3), it can be shown that the flocs formed at the lower shears are fluffier, more fragile, and have lower effective densities than do the flocs formed at the higher shears. The steady-state size distribution for each shear is shown in Figure 4.6.

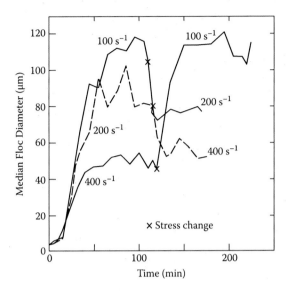

FIGURE 4.5 Median floc diameter as a function of time for a sediment concentration of 100 mg/L and different fluid shears. Detroit River sediments in fresh water. Fluid shears were changed at times marked by an ×. (*Source*: From Tsai et al., 1987. With permission.)

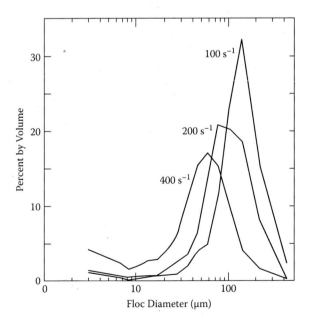

FIGURE 4.6 Steady-state floc size distribution for a sediment concentration of 100 mg/L and different fluid shears. Fresh water. (*Source*: From Tsai et al., 1987. With permission.)

In most experiments, single shear stress tests were done rather than double shear stress tests. To illustrate the effects of sediment concentration on
flocculation, results for single shear tests in fresh water, a fluid shear of 200/s,
and sediment concentrations of 50, 100, 400, and 800 mg/L are shown in Figure 4.7. For other shears and sediment concentrations, the results are similar and
demonstrate that, as the sediment concentration increases, both the steady-state
floc diameter and the time to steady state decrease. This dependence on sediment concentration is qualitatively the same as the dependence on fluid shear
(compare Figures 4.5 and 4.7).

Tests similar to those described above were done with the same sediments in
sea water. The sediment concentrations ranged from 10 to 800 mg/L and the shears
varied from 100 to 600/s. Results of one set of experiments at an applied shear of
200/s and different sediment concentrations are shown in Figure 4.8. In general,
as for fresh water, an increase in fluid shear and sediment concentration causes a
decrease in the steady-state floc size and a decrease in the time to steady state.

For tests with a 50/50 mixture of fresh water and sea water (Burban et al.,
1989), results for the steady-state floc size and the time to steady state were
approximately the average of those for fresh water and sea water. This indicates
that these quantities in estuarine waters of arbitrary salinity between that of fresh
water and sea water are approximately weighted averages of these same quantities
for fresh and sea waters.

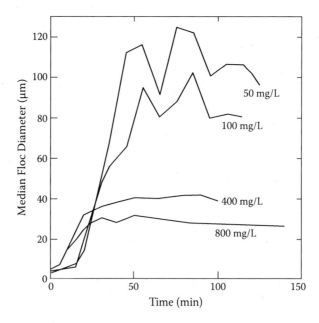

FIGURE 4.7   Median floc diameter as a function of time for a fluid shear of 200/s and different sediment concentrations. Fresh water. (*Source*: From Tsai et al., 1987. With permission.)

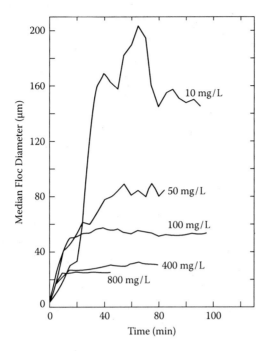

FIGURE 4.8   Median floc diameter as a function of time for a fluid shear of 200/s and different sediment concentrations. Sea water. (*Source*: From Burban et al., 1989. With permission.)

For a more general investigation of the effects of salinity (ionic strength), flocculation tests also were performed for the same sediment in sea water, fresh water, and de-ionized water (Lick and Huang, 1993). Results are shown in Figure 4.9. For all tests, the applied shear and concentration were identical. The ionic strength is greatest for the sea water, less for fresh water, and least for de-ionized water. As the ionic strength increases, the particles aggregate more rapidly, whereas the steady-state floc size decreases. The more rapid aggregation as the ionic strength increases is consistent with the VODL theory described in Section 4.1 (see Figure 4.3). However, VODL theory does not assist in describing steady-state conditions or the approach to steady state. From Figure 4.9 as well as from a comparison of Figures 4.7 and 4.8, it can be seen that the steady-state floc sizes are somewhat greater for fresh water than for sea water, but the differences are not large. All these results are consistent with the experimental results described above.

From these figures as well as other data, it can be demonstrated that (1) changes (e.g., increases) in sediment concentration, fluid shear, or salinity have similar qualitative effects on the steady-state median floc size and the time to steady state, and (2) the effects of sediment concentration and fluid shear can be approximately quantified in terms of the product of the sediment concentration and fluid shear.

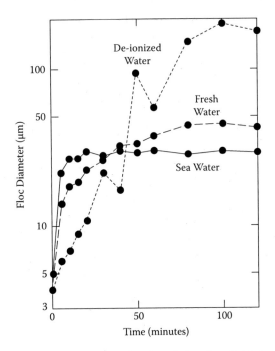

FIGURE 4.9 Median floc diameter as a function of time for a fluid shear of 400/s and a sediment concentration of 100 mg/L. Experiments in sea water, fresh water, and de-ionized water. (*Source*: From Lick and Huang, 1993. With permission.)

As an example of this last statement, for Detroit River sediments in fresh water, the data for the steady-state median diameter can be approximated as

$$d_s = 9.0(CG)^{-0.56} \tag{4.5}$$

where d_s is in micrometers, C is the sediment concentration (g/cm^3), and G is the fluid shear (s^{-1}). For these same sediments in sea water, the steady-state median diameter can be approximated as

$$d_s = 10.5(CG)^{-0.40} \tag{4.6}$$

These results are shown in Figure 4.10.

A time to steady state, T_s, can be defined as the time for the median floc diameter to reach 90% of its steady-state value. Consistent with the idea that d_s is a function of the product of CG, it can also be conjectured that T_s is a function of CG. This is demonstrated in Figure 4.11, where T_s (in minutes) is plotted as a function of CG for both fresh and sea waters. In this case, the experimental data can be approximated by

$$T_s = 12.2(CG)^{-0.36} \tag{4.7}$$

$$T_s = 4.95(CG)^{-0.44} \tag{4.8}$$

for fresh water and sea water, respectively.

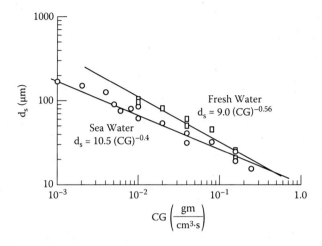

FIGURE 4.10 Steady-state median floc diameter as a function of the product of sediment concentration and fluid shear, CG, for both fresh water and sea water. (*Source*: From Lick et al., 1993. With permission.)

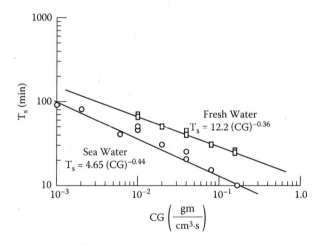

FIGURE 4.11 Time to steady state as a function of the product of sediment concentration and fluid shear, CG, for both fresh water and sea water. (*Source*: From Lick et al., 1993. With permission.)

Figures 4.7 and 4.8 show that the curves for $d_s(t)$ are similar in shape. This fact and the above relations suggest that the median diameter as a function of time can be described by the self-similarity relation,

$$\frac{d}{d_s} = f\left(\frac{t}{T_s}\right) \tag{4.9}$$

From the experimental data, this relation can be shown to be approximately valid for all values of fluid shear and sediment concentration for this sediment.

4.2.2 FLOCCULATION DUE TO DIFFERENTIAL SETTLING

As the fluid shear applied to flocs decreases, the frequency of collisions between flocs due to fluid shear also decreases. As this occurs, the sizes of the flocs, the settling speeds of the flocs, and hence the frequency of collisions due to differential settling will all increase. As a result, the primary mechanism for particles to collide then becomes differential settling rather than fluid shear.

To study the effects of differential settling on the flocculation of fine-grained sediments in the absence of fluid shear, disk flocculators have been designed, constructed, and used (Lick et al., 1993). Two of these flocculators are shown in Figure 4.12. The diameters of the small and large disks are 0.3 and 1.0 m, respectively, whereas the widths are 2.8 cm and 3.5 cm, respectively. The small disk is rotated by rollers, whereas the large disk is rotated by gears that connect the axis of the disk to a motor. Most experiments were done in the small disk; the large disk was primarily used for experiments at low sediment concentrations. For the same conditions, tests in both disks were consistent with each other.

To initiate the experiments, the sediments were disaggregated and then, at a specified concentration, introduced into a disk. This disk was then slowly accelerated until it rotated at a constant rate. After an initial transient of less than 2 min (during which time fluid shear is present but is very small), the fluid in the disk thereafter rotates as a solid body, and hence no applied fluid shear is present. The particles settle in the water, aggregate with time due to differential settling, and form flocs. Because of the rotation, the flocs stay in suspension. The rotation rate (approximately 2 rpm) (1) is fast enough that only a very small fraction of the flocs settle onto the walls during any rotation period, and (2) is slow enough that centrifugal forces are negligible.

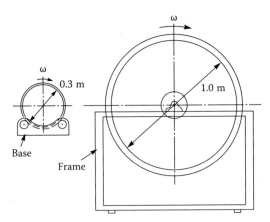

FIGURE 4.12   Schematic of disk flocculators. (*Source*: From Lick et al., 1993. With permission.)

Floc sizes were measured at various time intervals by withdrawing a sample from the disk and then inserting this sample into a Malvern particle sizer. This procedure was only possible for floc median diameters approximately equal to or less than 200 μm. The reasons for this are that (1) the Malvern particle sizer used at that time could only measure floc diameters up to 500 μm and, when an appreciable fraction of the floc diameters became greater than that, this usually corresponded to a median diameter of about 200 μm; (2) despite great care, floc breakage during sampling became severe for the large fragile flocs; and (3) the large flocs settled very rapidly and were difficult to capture. For these reasons, the size measurements using the Malvern particle sizer were generally limited to median diameters approximately equal to or less than 200 μm.

As the median floc size increased beyond this, measurements of floc size were made from photographs of the sediment suspension. The floc size distributions were determined from these photographs, and floc median sizes were then calculated. For floc diameters approximately equal to 200 μm, good agreement between the Malvern and the photographic technique was obtained. At low concentrations, the determination of the floc median size became problematic because of the small number of large flocs present in a disk (sometimes only one or two in the entire disk).

Tests were done for Detroit River sediments in both fresh and sea waters at solids concentrations of 1, 2, 5, 10, 25, 50, 100, and 200 mg/L. For fresh water and for concentrations of 50, 100, and 200 mg/L, the results for the median diameters as a function of time are shown in Figure 4.13. The general character of the variation of median diameter with time is the same as for the Couette flocculator tests where fluid shear is dominant. That is, there is an initial time period during which the median diameter of the flocs is small and changes relatively slowly. As the flocs increase in size, the collision rate increases, and the median size changes

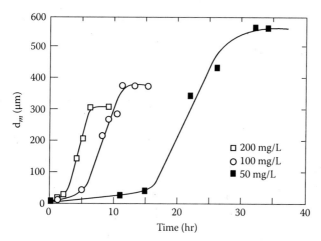

FIGURE 4.13   Median floc diameter as a function of time in fresh water during settling at different concentrations. (*Source*: From Lick et al., 1993. With permission.)

more rapidly. Still later, as the concentration of large flocs increases, disaggregation becomes more significant. A steady state is then approached where the rate of disaggregation is equal to the rate of aggregation. For other sediment concentrations, the variations of median diameter with time were similar in character.

In the experiments, a steady state was reached in all cases. The steady-state median floc size and the time to steady state depend on the concentration and are shown as functions of concentration in Figures 4.14 and 4.15. Both d_s and T_s decrease as the concentration increases. For fresh water, at the lowest concentration of 2 mg/L, the median diameter is about 2 cm, and the time to reach steady state has increased to about 25 days. From Figures 4.14 and 4.15, it can be seen that both $\log d_s$ and $\log T_s$ are linear functions of $\log C$; that is, both d_s and T_s are proportional to a power of C. These relations are shown in the figures.

From the above data, it can be seen that d_s and T_s are much greater in the differential settling tests than in the fluid shear tests. The reason for the slower rate of flocculation is that, in the absence of an applied shear, the collision rate between small particles is relatively small.

The transition in effects between situations when fluid shear is dominant and when differential settling is dominant is of significance. To investigate this, the fluid shear experiments of Burban et al. (1989) at a concentration of 100 mg/L and fluid shears of 100, 200, and 400/s were extended to lower shears of 50, 25, and 10/s (Lick et al., 1993). These results, along with the differential settling results (G = 0) and the previous results of Burban et al. (all at 100 mg/L), are shown in Figure 4.16. As the applied fluid shear approaches zero, the floc size increases

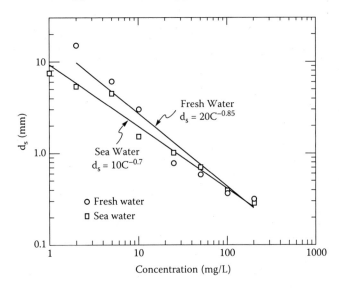

FIGURE 4.14 Steady-state median floc diameter during settling as a function of sediment concentration for fresh water and sea water. (*Source*: From Lick et al., 1993. With permission.)

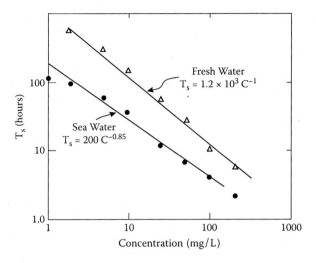

FIGURE 4.15 Time to steady state for floc during settling as a function of sediment concentration for fresh water and sea water. (*Source*: From Lick et al., 1993. With permission.)

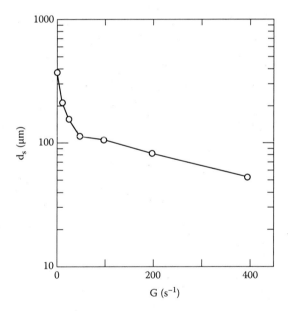

FIGURE 4.16 Steady-state median floc diameter at a concentration of 100 mg/L as a function of shear for fresh water. (*Source:* From Lick et al., 1993. With permission.)

rapidly. However, there is a smooth transition in d_s from the fluid-shear-dominated region to the region where differential settling is dominant.

To investigate the effects on flocculation of organic matter contained in or on the particles, experiments were done in the disk flocculators using Detroit River sediments with the organic matter removed and then comparing these results with the above results where organic matter (approximately 2%) was naturally present (Lick et al., 1993). It was demonstrated that the removal of organic matter causes the steady-state median diameters and the times to steady state to decrease. The effect is not large at the higher sediment concentrations but becomes more significant as the concentration decreases. As with erosion rates (Section 3.3), this effect is consistent with the general concept that the presence of organic matter causes sediments to behave in a more cohesive manner.

4.3 SETTLING SPEEDS OF FLOCS

For small spherical particles with known density, the settling speed and diameter are related by Stokes law. However, most aggregated particles are not spherical; more than that, the average density of a floc is less, often much less, than the solid particles in the floc and depends on parameters such as the fluid shear, sediment concentration, salinity of the water, and properties of the particle (e.g., particle size, mineralogy, and organic content). In addition, the flow field is modified due to flow through the floc as well as around the floc. For these reasons, the settling speed of a flocculated particle is not related to floc diameter as in Stokes law and, in fact, may not even have a unique relation to the diameter, as will be shown below.

Settling speeds of flocs have been measured for both fresh water and sea water as a function of fluid shear and sediment concentration (Burban et al., 1990; Lick et al., 1993); some of these results are shown here. These speeds were often quite small, on the order of 10^{-2} cm/s or less. Because of this, great care was taken in these experiments to eliminate thermally driven currents and vibrations and, hence, to measure the settling speeds accurately. Measurements of settling speeds were made in a carefully insulated square tube approximately 1 m long and 10 cm wide. The tube was insulated with styrofoam on all sides, top, and bottom and kept away from windows and drafts. Experiments were usually performed in the late afternoon and evening when the air temperature was about 20°C and relatively constant. To reduce vibrations and the associated convective instabilities, the settling tube and camera/microscope assembly used to observe the flocs were both rigidly mounted on the concrete floor and kept separate from each other. A pipette was used to introduce the sample from the flocculator into the settling tube. The flocs then settled in the tube. After a short initial transient, their speeds were essentially constant. The settling speed of a floc was determined by measuring the distance between two successively observed positions of the floc and then dividing by the time interval between observations.

4.3.1 FLOCS PRODUCED IN A COUETTE FLOCCULATOR

For sediments from the Detroit River, the settling speeds of flocs produced in a Couette flocculator at different applied fluid shears, sediment concentrations, and salinities were measured by Burban et al. (1990). Typical results are shown in Figure 4.17 (fresh water, a sediment concentration of 100 mg/L, and applied shears

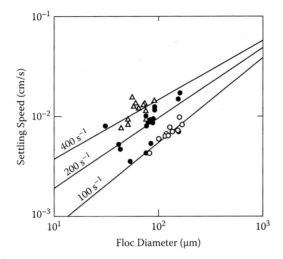

FIGURE 4.17   Floc settling speeds as a function of diameter for a sediment concentration of 100 mg/L and different fluid shears. (*Source*: From Burban et al., 1990. With permission.)

of 100, 200, and 400/s). For each concentration and fluid shear, the data can be approximated by the equation (solid line)

$$w_s = a \, d^m \qquad (4.10)$$

where a and m are parameters that depend on the fluid shear and sediment concentration. From measurements such as this, it was shown that the settling speeds of flocs depended on the conditions in which they were produced. For the same diameter, flocs produced at higher fluid shears and sediment concentrations generally have higher settling speeds (and hence floc densities) than do flocs produced at lower fluid shears and sediment concentrations.

The dependencies of the settling speeds of flocs on fluid shear and sediment concentration are qualitatively the same in both fresh and sea waters. For the same fluid shear, sediment concentration, and floc diameter, the settling speeds of flocs in sea water are somewhat greater than the settling speeds of flocs in fresh water, but generally by no more than 50%. However, the median diameter of a floc is a function of CG; for the same CG, the median diameter for flocs in fresh water is greater than that for flocs in sea water (Figure 4.10). Because of this effect, the settling speeds of the median-diameter flocs in fresh water are somewhat greater than those of the median-diameter flocs in sea water. As a first crude approximation, the settling speeds of flocs formed under the same conditions are about the same in both fresh and sea waters.

The fact that settling speeds of flocs in both fresh and sea waters are very similar, together with the fact that flocs are only a little smaller in sea water than in fresh water (Burban et al., 1989), is further evidence that the dependence of flocculation on salinity is unlikely as a cause in the formation of a turbidity maximum in estuaries; see Eisma (1986) for a similar conclusion. Of much more importance in causing changes in flocculation in an estuary are the changes in shear stress (turbulence) and sediment concentration, both generally decreasing as the flocs are transported from the river out into the estuary. However, the net effect of the changes in these parameters on settling speeds is difficult to determine *a priori* without knowing specific values of these parameters. To be specific and to illustrate the maximum change in settling speed within the present range of parameters, consider the changes in floc size and settling speed as conditions change in an estuary from (1) fresh water, a fluid shear of 400/s, and a sediment concentration of 400 mg/L, to (2) sea water, a fluid shear of 100/s, and a sediment concentration of 10 mg/L. From the above data and from Burban et al. (1989, 1990), it follows that for (1), the median floc diameter is 20 μm and the corresponding settling speed is 9×10^{-3} cm/s, whereas for (2), the median floc diameter is 172 μm and the settling speed is 5×10^{-3} cm/s. It can be seen that although the floc size changes considerably, the change in settling speed is relatively small.

This example indicates that changes in the settling speeds of flocs as they are modified and transported through an estuary are rather small. Of more importance than flocculation in causing the observed turbidity maximum in estuaries are the hydrodynamics of the stratified flow caused by the intrusion of sea water and the fresh water flowing over the resulting sea water wedge. This is discussed in Section 6.6.

4.3.2 FLOCS PRODUCED IN A DISK FLOCCULATOR

The settling speeds of flocs produced in disk flocculators also have been measured (Lick et al., 1993). The sediments were from the Detroit River. In the experiments, sediment concentration and salinity were varied as parameters. For floc diameters less than 1 mm, the same measurement techniques as described above were used. Flocs larger than 1 mm were quite fragile and were very difficult to capture and measure. These large flocs were only present at small sediment concentrations, and only a few of them were present at any one time. They generally moved independently of one another and seldom collided with each other or with the wall of the disk. For these flocs, an alternate procedure for measuring their settling speeds was used, and was as follows. A floc in a rotating disk can be observed to move in an almost circular orbit whose radius varies slowly with time but whose center is displaced by a constant distance r_0 from the center of the disk. In this situation, it has been demonstrated (Tooby et al., 1977) that the settling speed w_s of an isolated particle in a rotating disk is given by

$$w_s = \omega r_0 \qquad (4.11)$$

where ω is the rotation rate of the disk. For flocs less than 1 mm in diameter, comparisons between w_s from this relation and measurements of w_s in a settling tube were made; good agreement between the two was obtained. For flocs greater than 1 mm, settling speeds were therefore calculated from Equation 4.11. For these flocs, their orbiting motion could be readily observed, and the center of the orbit and hence r_0 could then be readily determined.

For fresh water, the results are shown in Figure 4.18(a). For flocs less than 1 mm in diameter (all of which were produced at 50, 100, and 200 mg/L), the data can be reasonably approximated by

$$w_s = ad^m = 0.268\ d^{1.56} \qquad\qquad (4.12)$$

For these flocs, there was no significant effect of sediment concentration on the settling speeds. The effect is probably small and, for this narrow range of concentrations, is probably masked by the experimental scatter. For flocs larger than 1 mm (which were produced at 2, 5, and 10 mg/L), the settling speeds fall below the line given by Equation 4.12. Reasons for this may be that (1) the flocs have not yet reached their steady-state density; and (2) for these flocs, the Reynolds numbers are from 10 to 100 and therefore significantly greater than the limiting Reynolds number for Stokes flow of about 0.5; this decreases the settling speed below that for laminar flow around the floc (Figure 2.5).

By comparison of Figure 4.18(a) with Figure 4.17, it can be seen that the settling speeds of flocs produced by differential settling are significantly greater

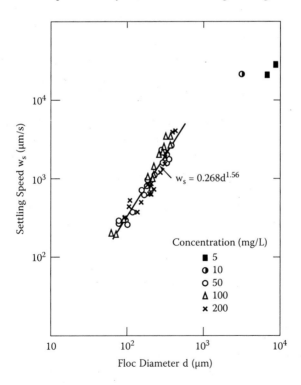

FIGURE 4.18(a)   Settling speeds of flocs formed by settling at different concentrations: fresh water. (*Source*: From Lick et al., 1993. With permission.)

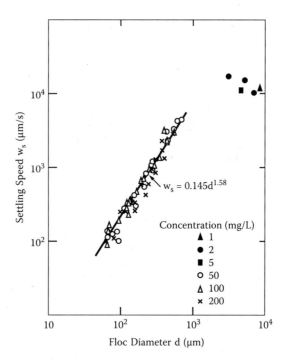

FIGURE 4.18(b)   Settling speeds of flocs formed by settling at different concentrations: sea water. (*Source*: From Lick et al., 1993. With permission.)

than the settling speeds of flocs produced by fluid shear. This is primarily due to the larger sizes of the flocs produced by settling. However, it also can be seen that the dependence of w_s on d is quite different, with the slope of log w_s for the settling experiments (m = 1.56) being significantly greater than the slopes of log w_s for the shear experiments (the values of m were between 0.25 and 1.0, depended on sediment concentration and fluid shear, but had an average value of about 0.6). For the same diameter, the settling speed of a floc produced by differential settling is greater than that for a floc produced by fluid shear.

For sea water, the results for settling speeds are shown in Figure 4.18(b). The same general trends as for fresh water are evident. For flocs less than 1 mm in diameter, the data can be approximated by

$$w_s = 0.145 \; d^{1.58} \qquad\qquad (4.13)$$

For flocs greater than 1 mm (which were produced at 1, 2, and 5 mg/L), the settling speeds fall below this line, as for fresh water. From Figures 4.18(a) and (b), it can be seen that, for the same diameter, the settling speeds of flocs in fresh water are somewhat greater than those of flocs in sea water, but not by much.

An effective density for settling can be defined from Stokes law as

$$\Delta\rho = \rho_f - \rho_w = \frac{18\mu}{gd^2}\, w_s$$

$$= 18\frac{\mu a}{g}\, d^{m-2} \qquad (4.14)$$

Because m is less than 2, this demonstrates the decrease in density of a floc as the diameter of the floc increases.

4.3.3 AN APPROXIMATE AND UNIFORMLY VALID EQUATION FOR THE SETTLING SPEED OF A FLOC

The above results were for the settling speed as a function of diameter for all flocs produced (not just median-size flocs) and measured in steady-state conditions at a specified shear and sediment concentration. In Section 4.2 (Figure 4.10 as an example), data for the steady-state median sizes of flocs as a function of CG were presented. By combining these results, one can then determine settling speeds of median steady-state flocs at specified shears and sediment concentrations. As an example, for Detroit River sediments in fresh water and a sediment concentration of 100 mg/L, the settling speed of the median steady-state floc is shown as a function of shear in Figure 4.19. As the fluid shear increases from

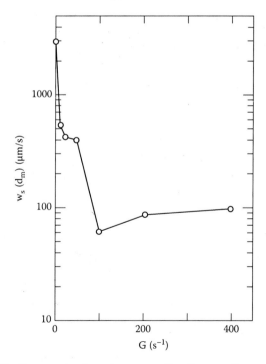

FIGURE 4.19 Settling speeds of steady-state median diameter floc produced at a concentration of 100 mg/L as a function of shear.

zero, the settling speed first decreases rapidly and then increases slowly. Similar curves are obtained for different sediment concentrations. For different sediment concentrations, the curves of $w_s(d)$ are similar in shape, and $w_s(d)$ decreases in magnitude as the sediment concentration decreases.

In general, the settling speed of the steady-state median size floc is a function of the diameter of the floc but also depends on the fluid shear and sediment concentration, with the dependence on concentration being least. Because the diameter of the floc is also a function of C and G, all three parameters (d, C, and G) are not independent; only two of the three are independent. As a result, the settling speed can be written as $w_s(d, C)$, $w_s(d, G)$, or $w_s(C, G)$. For convenience, the settling speed will here be approximated as $w_s(d, G)$.

The available data on settling speeds, and especially the conditions under which they were produced, are rather meager. On the basis of the data by Burban et al. (1990) and Lick et al. (1993) and as a first approximation, the following equation for w_s is suggested:

$$w_s = w_\infty + (w_0 - w_\infty)e^{-cG} \qquad (4.15)$$

$$w_0 = \lim_{G \to 0} w_s = .268\,d^{1.56} \qquad (4.16)$$

where $w_\infty = 80\ \mu m/s$ and $c = 0.04$ s. The above equations are approximate but are consistent with the data shown in Figures 4.17, 4.18(a), and 4.19 for flocs in fresh water. An analogous formula can be determined for flocs in sea water; the appropriate parameters are now $w_0 = 0.145d^{1.58}$, $w_\infty = 60\ \mu m/s$, and $c = 0.04$ s.

4.4 MODELS OF FLOCCULATION

4.4.1 GENERAL FORMULATION AND MODEL

Experimental results demonstrate that the formation of flocs is a dynamic process, with the floc size distribution at any particular time determined by the rates at which individual flocs aggregate and disaggregate. In general, these rates are not equal so that the median size of the flocs either increases or decreases, depending on whether aggregation or disaggregation is dominant. The steady state is determined as a dynamic equilibrium between aggregation and disaggregation.

The rate of aggregation depends on the rate at which collisions occur and on the probability of cohesion after collision; the rate of disaggregation depends on collisions between particles, on fluid shear, and on the probability of disaggregation due to collisions. A quite general formula for the time rate of change of the particle size distribution that includes the above mechanisms can be written as follows. Denote the number of particles per unit volume in size range k by n_k. The time rate of change of n_k is then given by (Lick and Lick, 1988)

$$\frac{dn_k}{dt} = \frac{1}{2} \sum_{i+j=k} A_{ij}\beta_{ij}n_i n_j - n_k \sum_{i=1}^{\infty} A_{ik}\beta_{ik}n_i$$

$$- B_k n_k + \sum_{j=k+1}^{\infty} \gamma_{jk} B_j n_j$$

$$- n_k \sum_{i=1}^{\infty} C_{ik}\beta_{ik}n_i + \sum_{j=k+1}^{\infty} \gamma_{jk} n_j \sum_{i=1}^{\infty} C_{ij}\beta_{ij}n_i \qquad (4.17)$$

The first term on the right-hand side of the above equation is the rate of formation of flocs of size k by cohesive collisions between particles of size i and j. The second term represents the loss of flocs of size k due to cohesive collisions with all other particles. Binary collisions have been assumed. The quantities A_{ij} are the probabilities of cohesion of particles i and j after collision. They depend on the properties of the particles in a floc, cannot be determined at present on the basis of theoretical arguments, but can be approximated from experimental results. The quantities β_{ij} are the collision frequency functions for collisions between particles i and j and depend on the collision mechanisms of Brownian motion, fluid shear, and differential settling (Section 4.1).

The third term on the right-hand size of Equation 4.17 represents the loss of flocs of size k due to disaggregation by fluid shear, and the fourth term represents the gain of flocs of size k due to the disaggregation of flocs of size j > k due to shear. In general, the coefficient B_k should be a function of the fluid shear, floc diameter, and effective density of the floc, as well as depend on the particular sediment. Numerous analyses (Matsuo and Unno, 1981; Parker, 1982; Clark, 1982; Hunt, 1984) have attempted to determine this quantity from basic theoretical considerations. However, this quantity is still not well understood or quantified. The quantity γ_{jk} is the probability that a particle of size k will be formed after the disaggregation of a particle of size j. Its value depends on the manner of breakup of the particle of size j, many of which are possible. If it is assumed that each floc can break up into two smaller flocs of size i and j such that i + j = k and that there is no preferred mode of breakup, then

$$\gamma_{jk} = \frac{2}{j-1} \qquad (4.18)$$

The second-to-last term on the right-hand side of Equation 4.17 represents the disaggregation of flocs of size k due to collisions with all other particles. The last term represents the gain of flocs of size k after disaggregation due to collisions between all particles i and j, where j is greater than k. Binary collisions have been assumed. The quantity C_{ik} is the probability of disaggregation of a particle of size k after collision with a particle of size i.

It has been shown that the above analysis as written is sufficient to describe many experimental results but is still inadequate in general (Lick and Lick, 1988). In particular, for the theoretical analysis to approximate the experimental result that the median diameter decreases as the sediment concentration increases, it is necessary that three-body collisions be significant in the disaggregation process. This can be demonstrated as follows. Note that aggregation is proportional to the square of the concentration, disaggregation due to fluid shear is proportional to the concentration, and disaggregation due to binary collisions is proportional to the square of the concentration. If it is assumed that fluid shear is the dominant mechanism for disaggregation, then the median size would increase as the concentration increases until a balance between aggregation and disaggregation occurs. If it is assumed that binary collisions are the dominant mechanism for disaggregation, then the median size would remain approximately constant as the concentration increases. However, if it is assumed that three-body collisions are the dominant mechanism for disaggregation, then the median size should decrease as the sediment concentration increases, in agreement with the experiments.

The probability of three-body collisions increases as the concentration increases and becomes significant when a nonnegligible fraction of the total volume is occupied by flocs. In the experiments described in Section 4.2, the sediment concentrations ranged from 10 to 800 mg/L so that, in their disaggregated state, the particles generally occupied less than 0.1% of the total volume. In this situation, three-body collisions would be negligible. However, the largest flocs are generally on the order of 99% water or more. In this case, the volume occupied by the flocs is then on the order of 100 times the volume occupied by the disaggregated particles and can amount to several percent of the total volume. Both of these latter statements were suggested and have been verified by direct observations of the flocculated mixture. In this situation, three-body collisions are quite probable and must be considered.

To show this more quantitatively, the approach was to use Equation 4.17 and determine the coefficients in that equation so as to obtain the best agreement between the calculated results for the median diameter and the experimental results. Details of the procedure used to determine C_{ik} and A_{ik} are given by Lick and Lick (1988). The coefficient B_{ik} was found not to be necessary, as will be demonstrated below, and was set to zero. If only binary collisions were important, the C_{ik} value would be independent of concentration. However, if three-body collisions are important, the C_{ik} value should increase with concentration. For the range of parameters for the experiments described in Section 4.2, it was found that the C_{ik} values did increase with concentration; they were also proportional to the shear, as would be expected if three-body collisions were important.

By means of Equation 4.17, calculations were made for the median diameter as a function of time for some of the experiments described in Section 4.2. The calculated results for d are compared with measurements in Figure 4.20(a) for sea water, a sediment concentration of 100 mg/L, and a shear of 200/s. For these same parameters, the measured and calculated steady-state floc size distributions are shown in Figure 4.20(b). It can be seen that the agreement between the calculated and

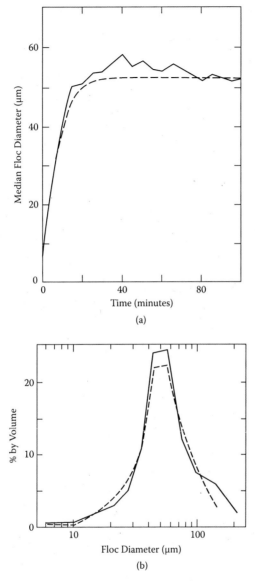

FIGURE 4.20 Measurements and calculations for a sediment concentration of 400 mg/L and a shear of 200/s: (a) median floc diameter as a function of time and (b) floc size distribution. (*Source*: From Lick and Lick, 1988. With permission.)

experimental results is quite good. The parameters used in the calculations were determined so that there was reasonable agreement between the calculated and experimental results for the median diameter as a function of time. The calculations of the floc size distribution did not involve any additional parameters so that this agreement between the calculated and experimental results further substantiates the analysis. For other values of shear and concentration, the agreement between the calculated and experimental results was equally good.

Calculations such as this were done only for the experiments done in the Couette flocculator, that is, when fluid shear was the dominant mechanism for collisions; at that time, experiments when differential settling was dominant had not yet been done.

4.4.2 A SIMPLE MODEL

In the theoretical analyses of flocculation as described above, the numerical computations, although relatively simple, can be extremely lengthy and often require a great deal of computer time. In the model as described, the state of each floc is defined by the number of particles in the floc; that is, a floc consisting of k particles would be in the k-th state. Because the mass and the number of particles in a floc increase rapidly with the diameter of the floc (as the cube of the diameter if the density of the floc remains constant), the number of states needed to describe the evolution of large flocs also increases rapidly. As an example, if it is assumed that the basic particles are 1 μm in diameter and the flocs are densely packed, then flocs 10 μm in diameter would consist of a thousand particles and k would equal 1000; flocs 100 μm in diameter would consist of 10^6 particles and k would equal 10^6. Of course, flocs are not densely packed so that the above is an upper limit. Nevertheless, it can be seen that as flocs increase in size, the number of particles in a floc, and hence the number of states needed to define all the flocs up to that size, increases dramatically. Because the rate equation (Equation 4.17) must be solved for each state, k, and in addition several summations over all states are present, the number of computations required increases even more rapidly than the number of states and the procedure as described becomes impossible for large ranges in floc sizes, even for the most advanced computers. For a flocculation model to be included in a transport calculation, where flocculation calculations must be made at each of up to several thousand grid points and possibly for long time periods, the flocculation model must be simplified and made more efficient so that computation times are reduced by several orders of magnitude!

One solution to this problem is to reduce the number of states by grouping flocs of similar size together (Lick et al., 1992). Each group of flocs then defines a new state with its own average floc diameter and density and, hence, mass. These states are defined to be independent of time. In defining new averaged states, an almost unlimited number of possibilities exists. In the approach by Lick et al. (1992), states were defined by nonuniform increments in diameter, with the increments increasing in size as the diameter increased. In an example calculation, results were shown for models with the number of states equal to 16, 10, 5, and 3 and were compared with experimental results and a model with several thousand states (which agreed well with the experimental results). Models with 16 and 10 states agreed well with the experimental results. Models with 5 and 3 states agreed less so but were probably satisfactory for many engineering calculations. Additional calculations have been made with the model to investigate the effects of flocculation on the vertical transport of particles (Lick et al., 1992, summarized in Section 6.3) and on the surface and submerged discharges of particles into a flow in a channel (Lick et al., 1994b).

Despite the relative simplicity of this model, there were difficulties in defining states, especially for problems with a large range of floc sizes; the calculations, although much simpler than previous, still required considerable time. For these reasons, a new approach was devised and is as follows.

For transport calculations, the quantities of most interest are the average properties of the flocs, for example, the average diameter, average density, and average settling speed. In reality, there are distributions of these quantities around their averages. However, these distributions may be more information than is necessary or even desirable. Rather than trying to determine these distributions, the present approach is to describe the dynamics of flocs by means of average or representative properties of flocs.

Two quantities that are useful in this modeling are the total number of flocs per unit volume, N, and the average diameter of the flocs, d. Both quantities are generally functions of time. Equations for their time rate of change can be derived by summation of the terms in Equation 4.17 over all states. Unquantified parameters appearing in these equations can be determined from the experimental results. The main simplifying assumption in the derivation is that the floc size distribution is narrow (not bi-modal, for example) and qualitatively similar to a log-normal Gaussian distribution (as in Figure 4.5). To illustrate the procedure, consider collisions due to Brownian motion alone. If all flocs are similar in size ($d_i \cong d_j \cong d$), then the collision frequency function for Brownian motion (from Equation 4.2) becomes

$$\beta_b = \frac{8}{3} \frac{kT}{\mu} \qquad (4.19)$$

Similarly, it is assumed that $A_{ij} = A_d(d)$ and is only a function of d. The rate of aggregation (the first two terms on the right-hand side of Equation 4.17) is then given by

$$\frac{dn_k}{dt} = \frac{1}{2} A_d \beta_b \sum_{i+j=k} n_i n_j - n_k A_d \beta_b \sum_{i=1}^{\infty} n_i \qquad (4.20)$$

By summing the terms in this equation over all states, one obtains

$$\sum_k n_k = N \qquad (4.21)$$

$$\frac{dN}{dt} = \frac{1}{2} A_d \beta_b \sum_{k=1}^{\infty} \sum_{i+j=k} n_i n_j - N A_d \beta_b \sum_{i=1}^{\infty} n_i \qquad (4.22)$$

It can be shown that

$$\sum_{k=1}^{\infty} \sum_{i+j=k} n_i n_j = \sum_i \sum_j n_i n_j \qquad (4.23)$$

By means of this equality, Equation 4.22 becomes

$$\frac{dN}{dt} = \frac{1}{2} A_d \beta_b N^2 - A_d \beta_b N^2$$

$$= -\frac{1}{2} A_d \beta_b N^2 \qquad (4.24)$$

The fourth term on the right-hand side of Equation 4.17 can be simplified by noting that

$$\sum_{k=1}^{\infty} \sum_{j=k+1}^{\infty} \gamma_{jk} n_j = \sum_{j=1}^{\infty} \sum_{k=1}^{j} \gamma_{jk} n_j$$

$$= 2 \sum_{j=1}^{\infty} n_j$$

$$= 2 N \qquad (4.25)$$

where it is assumed that γ_{jk} is given by Equation 4.18; that is, each floc breaks into two parts. By means of this relation and summation over all states, the third and fourth terms on the right-hand side of Equation 4.17 reduce to

$$\frac{dN}{dt} = B_d (G, d) N \qquad (4.26)$$

where $B_k = B_d$ and B_d is a function of G and d only. By essentially the same procedures, the fifth and sixth terms of Equation 4.17 reduce to

$$\frac{dN}{dt} = C_d \beta_b N^2 \qquad (4.27)$$

where $C_{ij} = C_d(d)$.

As demonstrated by experiments and the previous analysis, three-body collisions must be included for a more complete and predictive model of flocculation. However, little quantitative information is available for the description of three-body collisions. In the kinetic theory of gases (e.g., Clarke and McChesney, 1964), three-body collisions are analyzed using the assumption that two particles collide to form a complex; a third particle then collides with this complex. A major unknown is the time of duration for the complex. Although this parameter is uncertain, it follows from the above argument that three-body collisions should be proportional to the square of the rate of collisions due to binary collisions. By means of this assumption and summing over all states, the time rate of change of N due to three-body collisions is then given by

$$\frac{dN}{dt} = D_d \beta_b^2 N^3 \qquad (4.28)$$

where D_d is a function of the floc diameter but is otherwise unknown.

The resulting equation for the time rate of change of N due to Brownian motion is the sum of Equations 4.24, 4.26, 4.27, and 4.28, or

$$\frac{dN}{dt} = -\frac{1}{2} A_d \beta_b N^2 + B_d N + C_d \beta_b N^2 + D_d \beta_b^2 N^3 \qquad (4.29)$$

where β_b is given by Equation 4.19; A_d, B_d, C_d, and D_d are functions of the floc diameter; and B_d is also a function of the fluid shear. In order, terms on the right-hand side of this equation denote aggregation due to binary collisions, disaggregation due to fluid shear, disaggregation due to binary collisions, and disaggregation due to three-body collisions.

For collisions due to fluid shear, the collision frequency function (Equation 4.3) for a narrow size distribution reduces to

$$\beta_f = \frac{4}{3} G d^3 \qquad (4.30)$$

In a similar manner to that for Brownian motion, an equation for the time rate of change of N due to collisions caused by fluid shear can be derived by summation over all states; the resulting equation is the same as Equation 4.29, with β_f substituted for β_b.

For collisions due to differential settling, Equation 4.4 gives

$$\beta_d = \frac{\pi}{4} (d_i + d_j)^2 |w_{si} - w_{sj}| \qquad (4.31)$$

where the settling speed can be approximated by Equation 4.10 and is

$$w_{si} = a d_i^m \qquad (4.32)$$

If all particles are exactly the same size and density and hence have the same settling speed, then $\beta_d = 0$ and there is no contribution of differential settling to dN/dt in this limit. As the difference in particle sizes increases, the difference in settling speeds also increases, and Equation 4.31 shows that β_d increases.

However, because of the assumed narrow size distribution, there are relatively few pairs of flocs with large differences in settling speeds. Hence, there is no contribution of differential settling to dN/dt in this limit, also.

However, in between these two limits, there is an intermediate range of floc sizes that can significantly contribute to dN/dt. This may be quantified as follows. As an approximation, assume that the floc sizes have a log-normal Gaussian distribution so that

$$n(d) = cNe^{-\alpha(\ell n\, d - \ell n\, d_i)^2} \qquad (4.33)$$

where c is a normalization constant. For convenience, summation over all states will be replaced by integration. The summation

$$\sum_j \beta_{ij} n_j,$$

for example, then leads to

$$\pi d^2 \int_{-\infty}^{+\infty} n(u) |w_s - w_{si}|\, du = cd^2 N \int_{-\infty}^{+\infty} \pi e^{-\alpha(\ell n\, d - \ell n\, d_i)^2} a |d^m - d_i^m|\, du$$

$$= ac\pi d^{2+m} N \int_0^{\infty} e^{-\alpha u^2} (e^u - 1)\, du$$

$$= d^{2+m}\, N\, I \qquad (4.34)$$

where $u = \ell n\, d - \ell n\, d_i$ and I is the product of the integral and the other constants appearing in the above equation. The value of the integral is a number of 0(1); its exact value is irrelevant because it will be absorbed into another constant below, which will then be determined from the experimental results.

With this relation, an equation for the time rate of change of N due to differential settling can now be derived in the same manner as above. The result is Equation 4.29 but with β_b now replaced by

$$\beta_d = \alpha_d w_s d^2 \qquad (4.35)$$

where α_d is a constant.

When collisions due to Brownian motion, fluid shear, and differential settling are all included (and additivity is presumed), the resulting time-dependent equation for N is

$$\frac{dN}{dt} = -\frac{1}{2} A_d \beta N^2 + B_d N + C_d \beta N^2 + D_d \beta^2 N^3 \qquad (4.36)$$

where β is given by

$$\beta = \frac{8}{3}\frac{kT}{\mu} + \frac{4}{3}Gd^3 + \alpha_d w_s d^2 \tag{4.37}$$

For the solution of these equations, a relation between N and d is needed; this relation follows from consideration of the conservation of the mass of solid particles as flocculation occurs in a closed volume. Because the volume of a floc, v_f, is $\pi\, d^3/6$ and the mass of solids in a floc is $\rho_p v_f\, (\rho_f - \rho_w)/(\rho_p - \rho_w)$, the total mass of solids per unit volume in flocs is given by

$$C = \rho_p \left(\frac{\rho_f - \rho_w}{\rho_p - \rho_w} \right) \frac{\pi}{6} d^3 N \tag{4.38}$$

where C is the solids mass concentration and ρ_p is the density of a sedimentary particle and is approximately 2.65 g/cm³. Because C is conserved, it follows that

$$\Delta\rho\, d^3 N = 1.65\, d_o^3 N_o \tag{4.39}$$

where $\Delta\rho = \rho_f - \rho_w$. The subscript "o" denotes the disaggregated state. The effective density, $\Delta\rho$, can be determined from settling speed or equivalent measurements (Section 4.3) and is given by

$$\Delta\rho = \frac{18\mu}{g} ad^{m-2} \tag{4.40}$$

Equations 4.36, 4.37, 4.39, and 4.40 constitute a complete set of equations from which N(t) can be determined.

The quantity N was convenient to use in the above derivation; however, a more directly useful quantity in flocculation and transport calculations is the average floc diameter, d. An equation for the time rate of change of d can be determined by substituting Equation 4.40 into Equation 4.39 to eliminate $\Delta\rho$ and then substituting the resulting equation into Equation 4.36 to eliminate N. The result is

$$\left[\frac{f + f'd}{f} \right] \frac{d(d)}{dt} = \frac{A_1 A_d \beta C}{2f} - B_d d - \frac{A_1 C_d \beta C}{f} - \frac{A_1^2 D_d \beta^2 C^2}{f^2 d} \tag{4.41}$$

where $\beta = \beta(d)$ from Equation 4.37, A_1 is a constant, $w_s = f(d,G)$, and $f' = (\partial f/\partial d)_G$. The quantity $(f + f'd)/f = 1 + f'd/f$ is a slowly varying function of d.

The above equation is quite general and includes aggregation due to binary collisions, disaggregation due to fluid shear, disaggregation due to binary collisions, and disaggregation due to three-body collisions. It clearly illustrates the dependence of each term on the sediment concentration. From this, it is also clear (as stated previously) that three-body collisions (proportional to C^2) are necessary to obtain the experimental result that the steady-state average diameter decreases as C increases (see Figures 4.7 and 4.14).

From the numerical calculations described previously in this section, it was inferred that the two processes of aggregation due to binary collisions and disaggregation due to three-body collisions are necessary to describe the experimental results described and referenced above; however, the processes of disaggregation due to fluid shear and disaggregation due to binary collisions are not necessary. This will be further substantiated in the present analysis. If only the necessary terms are retained, the above equation reduces to

$$\left[\frac{f + f'd}{f}\right]\frac{d(d)}{dt} = \frac{A_1 A_d C\beta}{2f} - \frac{A_1^2 D_d C^2 \beta^2}{f^2 d} \tag{4.42}$$

This can be rewritten as

$$\frac{d(d)}{dt} = A^* \frac{C\beta}{f}\left[1 - \frac{D^* C\beta}{A^* fd}\right] \tag{4.43}$$

where $A^* = A_1 A_d f/(f + f'd)$, $D^* = A_1 D_d f/(f + f'd)$, and $D^*/A^* = D_d/A_d$.

In these equations, A^* and D^* are parameters that depend on d as well as other particle properties; however, they cannot be determined theoretically and must be determined from experiments. In the following, it will be assumed that there is a simple polynomial dependence of A^* and D^* on d, that is,

$$A^* = a_1 d^{a_2} \tag{4.44}$$

$$D^* = b_1 d^{b_2} \tag{4.45}$$

where a_1, a_2, b_1, and b_2 are constants. With this assumption and for each of the collision mechanisms of Brownian motion, fluid shear, and differential settling, it is shown in Appendix A at the end of this chapter that Equation 4.43 can be transformed into the general form

$$\frac{d\eta}{d\tau} = \eta^\alpha (1 - \eta) \tag{4.46}$$

where η and τ are defined by Equations A.2 and A.3, respectively, and where α depends on a_2 and b_2. When A^* and D^* are constants, $\alpha = 2$. Exact analytic solutions to Equation 4.46 are possible when α is an integer. For $\alpha = 1$, 2, and 3, the solutions are given in Appendix A and shown in Figure 4.21. It can be seen that (1) the general forms of the solutions are quite similar to the experimental results (Figures 4.5, 4.7, and 4.13), and (2) the time to steady state is primarily dependent

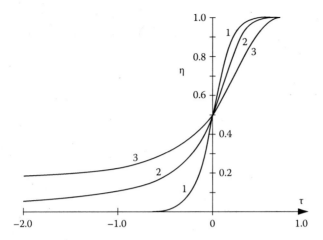

FIGURE 4.21   Normalized floc diameter as a function of a normalized time for different values of α.

on the parameter α and increases as α increases. It is expected that the solution to Equation 4.43 for realistic values of A_d^* and D_d^* will be qualitatively similar to the solution of the above equation when α = 2, with small deviations from this solution because of the dependence of A* and D* on d.

Consider first the case when fluid shear is negligible and differential settling is dominant, CG = 0, and $\beta = \beta_d = \alpha_d \, w_s d^2$. In the steady state, Figure 4.14 shows that d = 1.6 × 10⁻⁵ C⁻⁰·⁸⁵. The parameter D*/A* in Equation 4.43 is chosen so that the steady-state solution reproduces this result. The bracketed term can then be written as

$$1 - \frac{D^* C \beta}{A^* f d} = 1 - B^+ C d^{1.18} \qquad (4.47)$$

where B⁺ = 4.6 × 10⁵. Once this is done, A* can then be determined so as to give the correct time to steady state. Equation 4.43 can then be written as

$$\frac{d(d)}{dt} = A^+ C d^{1.2} [1 - B^+ C d^{1.18}] \qquad (4.48)$$

where A⁺ = 3.0 × 10². Integration of this equation gives essentially the same results as those shown for the steady-state diameter in Figure 4.14 and for the time to steady state shown in Figure 4.15.

When fluid shear is dominant, the terms in Equation 4.43 can be evaluated in a similar fashion, that is, by first evaluating D*/A* from the steady-state experimental results and then evaluating A*(d) from the time-dependent experimental results. However, it should be realized that although fluid shear may be dominant in some experiments but not present in others, differential settling is inherently present in all experiments. Because of this, the approach is to include differential settling as described by the above equation and then include fluid shear by adding the effective β values due to differential settling and fluid shear. In this way, the governing equation for the time rate of change of d, including differential settling and fluid shear, becomes

$$\frac{d(d)}{dt} = A^+ \left[3.0(CG)^{0.5} d^{2.2} + Cd^{1.2} \right]\left[1 - B^+ (3.5CGd^{2.18} + Cd^{1.18}) \right] \quad (4.49)$$

where d is in centimeters (cm), G is in reciprocal seconds (s^{-1}), and C is in grams per cubic centimeter (g/cm^3).

The above equation is uniformly valid for all values of C and CG. It accurately reproduces all the experimental results (for fresh water) for d and T_s shown in Figures 4.10 and 4.11 (when fluid shear is dominant) and in Figures 4.14 and 4.15 (when differential settling is dominant). It also accurately describes the transition in experimental results when G $\rightarrow$ 0 (Figure 4.16). In all cases (except for one outlying experimental result), the agreement between experiment and calculated result is within 25%.

For different sediments and/or waters with different salinities, the coefficients A^+ and B^+ and the powers of d may need to be modified, but the general form of the above equation should be valid for a wide variety of sediments and waters. By comparison with the original formulation, Equation 4.17, the above equation is quite simple and, in particular, the computation times are less by several orders of magnitude. The present approach is therefore quite efficient and suitable for use in calculating sediment and contaminant transport in surface waters.

4.4.3 A Very Simple Model

For exploratory purposes and as a first approximation in transport problems where flocculation may be significant, a very simple and approximate model can be determined from the experimental results and is as follows. When aggregation is relatively fast so that times to steady state are small and/or when conditions (fluid shear, sediment concentration, and salinity) are changing slowly, it may be reasonable to approximate floc diameters that are slowly changing with time by their quasi-steady-state values. For the experiments described in Section 4.2, analytic results for d(CG) and d(C) when fluid shear is dominant and when differential settling is dominant can be obtained from the experimental results and are shown in Figures 4.10 and 4.14, respectively. An approximate analytic relation for d that is more general and includes these relations can be obtained from the steady-state results implied by Equation 4.49, that is, d(d)/dt = 0, so that

$$1 - B^+(3.5CGd^{2.18} + Cd^{1.18}) = 0 \quad (4.50)$$

Although this is an implicit equation for d, the solution for d(C, CG) can be found rather simply by iteration. The solution is uniformly valid for all values of C and

CG and reduces to those shown in Figures 4.10 (when shear is dominant) and 4.14 (when differential settling is dominant).

An approximate expression for T_s can be determined by a linearization of Equation 4.49 for small d. The result is a uniformly valid expression for T_s for all values of CG and C and includes the relations shown in Figures 4.11 and 4.15. The reason that this linearization is a valid approximation is that flocculation time is primarily determined by the collision rate when d is small (e.g., see Figures 4.7 and 4.13).

4.4.3.1 An Alternate Derivation

It is interesting (and useful) to consider an alternate and more physically based derivation of an equation equivalent to Equation 4.50. When fluid shear is dominant, experimental results for the median steady-state floc diameter in fresh water (Figure 4.10) can be summarized as

$$d = 9(CG)^{-0.56} \qquad (4.51)$$

When the applied fluid shear is absent (G = 0), flocculation still occurs due to differential settling, but the above relation is not meaningful and cannot be used. However, even in the absence of an applied fluid shear, a particle/floc does experience a local fluid shear due to its settling and the associated drag on the particle/floc. This shear can be determined theoretically for slow, viscous flow and is on the order of w_s/d (e.g., Schlichting, 1955). In addition, at steady state when flocs are large, the effective shear between particles due to differential settling is of this order. This suggests the equivalence of G and w_s/d and, in the simple model described above, this equivalence (except for a constant of order one) is indeed demonstrated by the fact that Equation 4.30 for β_f has the same form as Equation 4.35 for β_d when G is replaced by w_s/d.

This equivalence suggests that Equation 4.51 and a similar equation for sea water are also valid when G = 0, except that the applied shear, G, must be replaced by the local shear, w_s/d. As an example to assist in the verification of this equivalence hypothesis, replace G by w_s/d in Equation 4.51, where w_s can be determined from Figure 4.18(a) or calculated from

$$w_s = 0.268d^{1.56} \qquad (4.52)$$

With this substitution, Equation 4.51 becomes

$$d = 3C^{-0.43} \qquad (4.53)$$

where d is in millimeters (mm) and C is in milligrams per liter (mg/L). This should be compared with the results for d(C) for differential settling, as shown in Figure 4.14. Somewhat surprisingly, the analytic dependence of d shown there $(d = 20\ C^{-0.85})$ is not the same as in the above equation. Closer examination shows that results for d calculated from Equation 4.53 agree well with the experimental results for d for C > 10 mg/L but do not agree for C < 10 mg/L. The reason for this is that for C < 10 mg/L, w_s is not given accurately by Equation 4.52 for large flocs, as shown in Figure 4.18(a). When w_s for large flocs is determined directly from Figure 4.18(a) and the resulting w_s/d is substituted for G in Equation 4.51, then d as calculated from Equation 4.51 compares well with d from the experiments from Figure 4.14.

The equivalence of G and w_s/d also suggests that for a uniformly valid result for all values of C and CG, the substitution of $G + \alpha w_s/d$ (where α is a constant of order one) for G in Equation 4.51 may be made. When this is done, Equation 4.51 becomes an implicit equation for d and its solution requires iteration. The solution procedure is then to assume an initial value for d, determine w_s from Equation 4.52 or from Figure 4.18(a), and then calculate a new value for d from Equation 4.51. The procedure continues until the results for d converge, which generally occurs in a few iterations.

With this approach, uniformly valid results for d for all values of C and CG can be obtained from Equation 4.51. The results agree with experiments and are approximately the same as (actually somewhat better than) those given by Equation 4.50. For other sediments that have different bulk properties and hence different coefficients appearing in the above equations, this more general view of flocculation will greatly reduce the number of experiments required to determine these coefficients.

With the equivalence of G and $\alpha w_s/d$, all the above results for flocculation due to fluid shear and/or differential settling now can be approximately reduced to a self-similar expression for the median diameter as a function of time in the same manner as described by Equation 4.9 when fluid shear was dominant, that is, $d/d_s = f(t/T_s)$.

4.4.4 FRACTAL THEORY

To describe experimental results on particle flocculation, some investigators have invoked the concept of fractal theory (Mandelbrot, 1977). In its most basic form, the main assumption of fractal theory as applied to sediment flocculation is that all flocs are similar in appearance, independent of floc size and density. From this statement, it follows that the number of particles, n_p, in a floc increases with the floc diameter such that, for large enough n_p,

$$n_p \sim d^D \tag{4.54}$$

where D is defined as the fractal dimension.

When this proportionality is true, the dependence of w_s on d may be different from that given by Stokes law and can be determined as follows. For a floc with effective density ρ_f and diameter d, Stokes law gives

$$w_s = \frac{g}{18\,\mu}(\rho_f - \rho_w)d^2 \tag{4.55}$$

The solids volume fraction in a floc, x_s, is given by (see the derivation of Equation 4.38)

$$x_s = \frac{\rho_f - \rho_w}{\rho_s - \rho_w} \qquad (4.56)$$

where x_s is generally a function of d. For regular packing as described for bottom sediments in Section 2.5, x_s is a constant. For fractal geometry, x_s is proportional to a power of d, as is shown below. By combining these last two equations, one obtains

$$w_s = \frac{g}{18\mu}(\rho_s - \rho_w)x_s d^2 \qquad (4.57)$$

In general, the number of particles in a floc is given by

$$n_p = x_s \frac{d^3}{d_p^3} \qquad (4.58)$$

where d_p is the diameter of a particle in the floc. The quantity x_s is then proportional to n_p/d^3, or d^{D-3} from Equation 4.54. By substituting x_s from this equation into Equation 4.57 and using Equation 4.54, one obtains

$$w_s \sim d^{D-1} \qquad (4.59)$$

For a solid particle or regular packing of particles, $D = 3$, and the above equation reduces to Stokes law. Experiments with flocs show that D is less than 3.

More recently, instead of similarity in appearance, the above equation (or Equation 4.54) has been assumed to be the fundamental basis of fractal theory; i.e., if Equation 4.59 is valid, then a fractal description of the process is valid. However, it should be noted that although similarity in appearance leads to the above equation, the reverse is not true, and the approximate satisfaction of Equation 4.59 by experimental data does not lead to self-similar appearance of flocs in all cases.

Over a large range of data, Equation 4.59 with a constant D may not fit the data well. Because of this, the concept of piecewise fractal theory was introduced. In this case, it was assumed that the above equation with the appropriate value of D was a good fit to the data over a certain range of particle diameters but that for another range of diameters, Equation 4.59 was still valid but with a different value of D.

As described previously in this chapter, experimental results for the settling speeds of flocs can certainly be described by Equation 4.59. For example, w_s was given as a function of the floc diameter for fresh water by Equation 4.10 when fluid shear was dominant and by Equation 4.12 when differential settling was dominant. Analogous formulas are valid for flocs in sea water. The different dependencies of w_s on d for different processes, and hence different ranges, of d also demonstrates the necessity for a piecewise description of the data or, better yet, a uniformly valid description of w_s such as that in Equation 4.15. It follows that fractal theory can be used to approximately describe both the results of flocculation experiments and a theory based on physical processes. But the real question is whether fractal theory is useful; for example, can it give additional information, or even replace the more conventional approach of experiment and theory based on experiment as described previously in this section?

For this purpose, the following facts should be considered:

1. The appearance of different flocs is not independent of floc size, as can be seen in Figures 2.7(a) through (f), so that the simple concept of self-similarity in appearance is not valid. Experimental results, not fractal theory, demonstrate that Equations 4.54 and 4.59 are valid.
2. The fractal dimension D in the above equations is not determined by fractal theory, but must be determined from experiments.
3. For piecewise fractal theory, the ranges of particle diameters for which Equation 4.59 with constant D is valid are not given by fractal theory, but must be determined from experiments.
4. When all is said and done, the general assumption of fractal theory that $y \sim x^m$ (where y is the dependent variable and x is the independent variable) is a very elementary assumption, almost always made by experimenters and theorists when describing and approximating experimental data, y(x), so that it is not surprising that it is valid in the present case.

In summary, experimental results and theory based on physical processes lead to equations such as those above and also determine the constants in these equations and the ranges of validity of these equations. Fractal theory neither proves that flocs are self-similar in appearance nor determines the constants in the above equations or the ranges of validity of piecewise fractal approximations. Fractal theory as applied in its simplest formulation (self-similarity in appearance) is incorrect, whereas its more general statement as $w_s \sim d^{D-1}$ does not contribute to our understanding or ability to quantify the flocculation process; that is, fractal theory as applied to flocculation is either incorrect or uninformative.

4.5 DEPOSITION

The rate of sediment deposition is proportional to the sediment concentration and settling speed and is usually described by

$$D_i = p_i \, w_{si} \, C_i \qquad (4.60)$$

where the subscript i denotes the size class i, p_i is the probability of deposition of particles in size class i, and $0 < p_i < 1$. There is general agreement on the form of this equation, but there is considerable disagreement on the value of p_i and the parameters on which it depends. Much of the difficulty in the use of the above equation is due to conceptual disagreement as to what is nondeposition of some fraction of the suspended sediments ($p_i < 1$) and what is deposition ($p_i = 1$) followed by erosion of the recently deposited sediments. Part of this may be semantics, but a clearer understanding of deposition, erosion, and their interactions will allow a better understanding and quantification of p_i.

It should be emphasized that, on the average over space and time and at a particular average shear stress, erosion and deposition can occur more or less simultaneously. This was demonstrated quite clearly by the results of the annular flume experiments described in Section 3.1. A major cause of this simultaneity is spatial and temporal variations in fluid motions and sediment properties. For example, in turbulent flow, there are spatial and temporal variations in fluid velocities and pressures where relatively quiet flows when deposition can occur are interrupted on an intermittent basis by large turbulent eddies where erosion occurs. In addition, there are spatial and temporal variations (1) in particle/floc sizes and their distributions, especially those differences between eroding and depositing particles/flocs, (2) in binding forces between particles due to changes in orientation of impacting particles and also due to variable electrostatic and chemical forces on particle surfaces caused by variable surface coatings, and (3) on a larger scale, in flow conditions and sediment properties within a modeling segment.

In the absence of spatial and temporal fluctuations, erosion and deposition still occur simultaneously, but their definitions become somewhat ambiguous. To understand the problem, consider the one-dimensional, time-dependent sediment-water flux in a system with the following properties: no spatial and temporal fluctuations, for example, laminar flow (a constant shear stress) over uniform-size, noncohesive spherical particles with uniform properties and uniform packing; flow and sediment conditions that are uniform in the horizontal; and the basin depth that is h and is shallow enough that the suspended sediments are uniformly mixed. In the absence of deposition (e.g., when the suspended sediment concentration, C, is small), the erosion rate is uniquely defined and measurable as

$$E = h \frac{dC}{dt} \qquad (4.61)$$

When suspended sediments are present, the suspended sediments will settle and impinge on the particles previously deposited on the bottom at a rate

$$D = w_s C \qquad (4.62)$$

After collision, the particles can (1) immediately and completely rebound into the overlying water ($p = 0$), (2) possibly continue along the sediment surface by skips and jumps until some fraction will eventually settle while the rest will rebound ($p < 1$), or (3) be immediately captured by the bottom sediments ($p = 1$); any captured sediments may be eroded later. In any case, the net result is that some of the suspended sediments are deposited more or less permanently on the bottom and some are not. In this situation, the definition of what is nondeposition (p < 1) and what is deposition (p = 1) followed by erosion is somewhat ambiguous and hence difficult to quantify.

To avoid this difficulty, operationally well-defined and unambiguous definitions of erosion and deposition are proposed as follows. In erosion experiments such as those in Sedflume where the test section is short, erosion occurs and erosion rates can be readily measured; because of the short test section, deposition of suspended sediments is negligible. From this operational point of view, the erosion rate is uniquely defined and measurable.

However, in contrast, it is difficult to devise experiments where deposition occurs in the absence of erosion. This is illustrated in the following by the analyses of several types of deposition experiments. As indicated above, the general problem is that, as deposition occurs, the deposited sediments may erode and this erosion can modify the observed deposition rate. In deposition experiments, deposition and erosion inherently occur simultaneously and their effects are difficult to separate.

As suggested above, the erosion rate, E, can be defined as the sediment-water flux in the absence of deposition (readily measurable as in Sedflume). Consistent with this, the deposition rate then can be operationally well-defined as the erosion rate minus the sediment-water flux (also readily measurable). It should be emphasized that the erosion rate is that of the surficial sediments. When net erosion is occurring, these are the newly exposed sediments; their erosion rates can be determined as a function of depth from Sedflume measurements. When net deposition is occurring, the erosion rate must be determined from erosion experiments with recently deposited sediments; this can also be done by means of Sedflume.

For the simple and idealized case of noncohesive sediments described above, the sediment-water flux is given as

$$q = h \frac{dC}{dt} = E - D$$

$$= E - pw_s C \qquad (4.63)$$

The deposition rate is then $E - hdC/dt$. If the initial suspended sediment concentration is C_i, the solution to this equation is

$$C = (C_i - C_\infty) e^{-t/t_1} + C_\infty \qquad (4.64)$$

where t_1 is a decay time given by h/pw_s. The steady state is a balance between erosion and deposition (i.e., $E - pw_s C = 0$), and hence $C_\infty = E/pw_s$. If E and w_s were constants and E, w_s, and C_∞ were measured, then the determination of p would be straightforward, that is, $p = E/w_s C_\infty$. However, as discussed in Chapter 3 and in the previous sections of this chapter, E and w_s are not constants. This complicates but does not negate the determination of p by means of deposition experiments.

In the following, various processes and parameters that affect deposition rates and p_i (fluid turbulence, particle dynamics, particle size distributions, flocculation, bed armoring/consolidation, and partial coverage of previously deposited sediments by recently deposited sediments) are discussed. Based on this discussion, results of several deposition experiments are presented and analyzed. A consistent approach that is valid for describing all experiments and for modeling purposes is then proposed.

4.5.1 PROCESSES AND PARAMETERS THAT AFFECT DEPOSITION

4.5.1.1 Fluid Turbulence

Because of turbulent fluctuations, the shear stress at the sediment-water interface fluctuates in space and time. In theoretical analyses (e.g., see Gessler, 1967), the statistical distribution of this shear stress often is assumed to be Gaussian (Figure 4.22(a)). As the average shear stress, τ_{avg}, increases, an increasing fraction of the instantaneous turbulent fluctuations, τ, becomes greater than a fixed critical

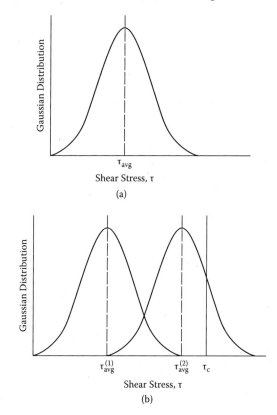

FIGURE 4.22 Shear stress due to turbulent fluctuations: (a) Gaussian distribution of instantaneous fluctuations about average shear stress, τ_{avg}, and (b) shear stress for two different distributions relative to a critical shear stress, τ_c.

stress, τ_c (Figure 4.22(b)). The probability, P, or the fraction of all instantaneous shear stresses that is greater than τ_c is then given by the area under the curve for $\tau > \tau_c$, or

$$P = \frac{1}{\sqrt{\pi}} \int_Y^\infty e^{-x^2} dx \qquad (4.65)$$

The parameter Y is proportional to the turbulent fluctuations and is

$$Y = \frac{\tau_c - \tau_{avg}}{\sigma \tau_{avg} \sqrt{2}} \qquad (4.66)$$

where σ is the standard deviation of the Gaussian distribution. Einstein (1950) states that σ is a constant, equal to 0.57. More generally, it is known that the statistical distribution of the shear stress depends on the Reynolds number and is not symmetric as in the above equations but is skewed toward higher shear stresses because of the occurrence of turbulent bursts (Gessler, 1967; Obi et al., 1996; Gust and Muller, 1997; Miyagi et al., 2000). A schematic diagram (with a nonsymmetric distribution of turbulent fluctuations) of $P(\tau)$ is shown in Figure 4.23(a).

In general, the actual erosion rate as influenced by turbulent fluctuations is then

$$E = \int_\tau^\infty \frac{dP(t)}{dt} E_0(t) dt \qquad (4.67)$$

where $E_0(\tau)$ is the (theoretical and somewhat idealized) erosion rate in the absence of turbulence fluctuations, usually represented as

$$E_0 = f(\tau - \tau_c) S(\tau_c) \qquad (4.68)$$

where $f(\tau - \tau_c)$ is some function of $\tau - \tau_c$ and $S(\tau_c)$ is the step function. That is, $S = 1$ for $\tau > \tau_c$ and $S = 0$ for $\tau < \tau_c$. The result of this is that as the shear stress increases from zero, the erosion rate will be non-zero for all shear stresses and increase slowly as a function of shear stress; that is, there is no exact value at which erosion begins, which is consistent with experiments. Qualitatively, the erosion rate as a function of shear stress therefore appears as in Figure 4.23(b) for small shear stresses and is compared there with $E_0(\tau)$, with f approximated as a linear function of $\tau - \tau_c$. If E_0 is a slowly varying function of τ compared to the changes in P, then Equation 4.67 can be approximated as

$$E = P(\tau) E_0(\tau) \qquad (4.69)$$

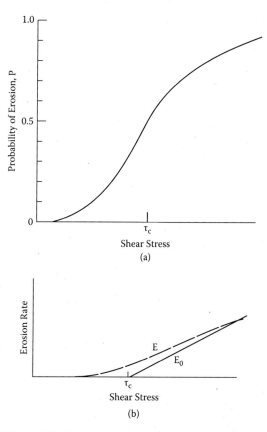

FIGURE 4.23 Effects of fluid turbulence on sediment erosion: (a) probability distribution, and (b) erosion rates as measured in turbulent flow and in laminar flow.

A critical shear stress for deposition, τ_d, also can be defined. For noncohesive sediments, this often is stated to be the same as the critical shear stress for erosion. In this case, the probability of deposition is given by

$$p = 1 - P \tag{4.70}$$

and is the inverse of P shown in Figure 4.23(a). Other expressions for $p(\tau)$ also have been proposed; for example, a linear dependence of p on τ has been suggested by Krone (1962) and is discussed below.

However, in the absence of other processes, the effects of turbulence on deposition can best be described as a deposition at the rate w_sC offset by an erosion of the recently deposited sediment. The advantage of this is that the erosion rate $E(\tau)$ is generally determined from experiments, effects due to turbulent fluctuations are inherently included in the erosion measurements, and hence no independent determination of p is necessary. As far as the effects of turbulence on deposition are concerned, it follows from this that $p = 1$.

4.5.1.2 Particle Dynamics

In a quiescent fluid, particles settle and deposit at a rate given by Equation 4.60, with $p_i = 1$. However, as the fluid velocity increases, settling particles impact the sediment surface with a certain horizontal momentum; they then may be captured, they may rebound, or they may continue along the surface by skips and jumps before eventually depositing or rebounding. This effect causes the deposition rate and hence the probability of deposition to decrease as the shear stress and particle size increase. For fine-grained particles, because of their low settling velocity, low Reynolds number, and low momentum relative to the fluid, this effect is probably small and hence $p = 1$. For coarse-grained particles, effects of particle dynamics and the associated bed load are significant and must be considered. This latter problem is discussed in Section 6.2.

4.5.1.3 Particle Size Distribution

Particle sizes (diameters) for bottom sediments in a typical system vary over two to three orders of magnitude (Section 2.1). Because settling speeds for particles of constant density are proportional to the diameter squared (Stokes law), this implies that settling speeds (and hence deposition rates) vary by as much as four to six orders of magnitude! This is a major influence on deposition rates and must be considered in the interpretation of experimental results and in the modeling of deposition rates. Because of this, an assumption of an average particle size and hence an average settling speed may not be an adequate approximation for actual settling speeds that vary over several orders of magnitude. For example, for a flow in a river, large particles may settle out in longitudinal distances of a few meters, average-size particles in distances of a few kilometers, and the finest particles may not settle out for hundreds of kilometers, essentially staying suspended throughout the length of the river. This distributed process cannot be described adequately using a single, constant settling speed. Multiple size classes, each with a separate settling speed, must be used.

4.5.1.4 Flocculation

As discussed in the previous sections, flocculation is a dynamic process that causes changes with time in the sizes and densities of flocs as well as their settling speeds and deposition rates. Flocculation becomes increasingly significant as particles become smaller and more cohesive. Floc sizes range from that for a single particle (micrometers) to millimeters and even centimeters in some cases. In addition, because of the large variations (usually increases) in fluid shear and sediment concentration near the sediment-water interface and the strong dependence of floc size and density on these same two parameters, the effects of flocculation on settling speeds and hence deposition are often significant. In particular, recently deposited, fine-grained sediments are more flocculent, have lower densities, and are much easier to erode than these same sediments after even short periods of consolidation. Because of this, the spatial and temporal variations in floc size, density, and settling speed must be considered.

4.5.1.5 Bed Armoring/Consolidation

As described in Section 3.3, experimental results demonstrate that erosion rates decrease as cohesive sediments consolidate and the density increases. Bed armoring (described in Section 6.2) similarly causes a decrease in erosion rates as finer sediments are eroded while coarser particles remain behind and coarsen (armor) the bed. Both of these processes can be significant in reducing the erosion rates of recently deposited sediments. In most depositional experiments, their effects are similar and difficult to separate without further measurements of deposited sediment properties.

4.5.1.6 Partial Coverage of Previously Deposited Sediments by Recently Deposited Sediments

As sediments are deposited, they consolidate with time. As a result, recently deposited sediments can be resuspended more readily than previously deposited sediments. In many laboratory experiments, the amount of suspended sediments and the amount of recently deposited sediments are relatively small. For example, consider an experiment in an annular flume with a water depth of 10 cm and a suspended sediment concentration of 100 mg/L. If all the suspended sediment were deposited, this would amount to an average deposited sediment layer thickness of about 10^{-3} cm, or only 10 µm. The thickness of the layer of recently deposited sediments actually on the bottom during the experiment would be less than this. For most sediments, a thickness of even 10 µm is less than the diameters of most particles. This implies that the deposited layer is not uniform but is patchy, with some areas having no deposited sediments while other areas have layers that are only one or several particles thick. This result is consistent with observations.

In the deposition experiments described below, erosion and deposition generally occur simultaneously. However, erosion will be limited by the amount of sediment available for erosion. If the bottom is only partially covered by recently deposited sediments, this will decrease the erosion rate as compared with that which is present when the bottom is completely covered. This limitation is significant in the interpretation of laboratory results but is probably not of importance in the interpretation and modeling of most cases of sediment transport in the field, where much more sediment is active in the transport process.

4.5.2 EXPERIMENTAL RESULTS AND ANALYSES

Various experiments have been performed in an effort to understand and quantify sediment deposition. Several illustrative and informative sets of these experiments are discussed here. It will be seen that results of all experiments are "explainable." The problem is that some experimental results will have more than one set of explanations; which one is correct, or at least which is the most significant, depends on various measurements, which generally were not made at the time of the experiments. A set of explanations that are uniformly valid for all experiments is desirable.

Consider first a set of experiments in an annular flume as reported by Lick and Kang (1987). These were similar to those by Partheniades (1965) and Mehta (1973) and were performed as follows. While the flume was operating at various prescribed stresses, sediments were introduced into the water (which had a depth of 7.6 cm and was initially clear) and allowed to deposit. The suspended sediment concentrations were monitored as a function of time. The sediments were pure sands with a fairly narrow size distribution. They had a median size of 23 μm, with 80% of the sediments between 13 and 40 μm and were relatively noncohesive.

The suspended sediment concentrations as a function of time with shear stress as a parameter are shown in Figure 4.24. Several significant results and possible explanations for these results are as follows.

1. The long-time (steady-state) concentration increases as the shear stress increases. If it is assumed that $p = 1$ and because $C_\infty = E/pw_s$ (Equation 4.63), this is primarily due to the erosion rate increasing with shear stress.

2. At a specific shear stress, the sediment concentration decreases approximately exponentially with time to C_∞. The exponential decay time as given by Equation 4.64 is $t_1 = h/pw_s$. For the lowest shear stresses and with p approximately equal to 1, the calculated value for t_1 of approximately 6 to 8 min for the smaller particles ($d = 13$ to 20 μm) agrees reasonably well with the experimental results.

3. The decay time increases from a few minutes to about 20 to 30 min as the shear stress increases. This is not due to particle size variations (wrong direction of effect) or flocculation (minimal flocculation occurring in water column and wrong direction of effect). Possible reasons include:
   a. The effect of particle dynamics
   b. The effect of radial quasi-bed load, the deposition of this bed load in the stagnation regions of the annular flume (see Section 3.1), and the subsequent bed-armoring and/or consolidation of the sediment bed that slowly reduces the erosion rate
   Both of these processes would cause the decay time to increase as the shear stress increases. For large time, rather than an exactly steady state, the concentration seems to decrease slowly with time. This also may be due to bed armoring/consolidation of the bed with time.

Similar experiments were performed with sediments from the Detroit River (Lick and Kang, 1987). These sediments were finer and had a much wider distribution of particle sizes, with approximately 50% of the particles in the clay-size range (<4 μm) and with 20% greater than 10 μm. Measured sediment concentrations as a function of time were similar to those in the above experiments. For these experiments, concentrations as a function of time with shear stress as a parameter are shown in Figure 4.25. In contrast to the above experiments, the decay time decreased with increasing shear stress. This is primarily due to the wide distribution in particle sizes. For small stresses, almost all particles are deposited. The decay time is then determined by those particles that have the longest decay time

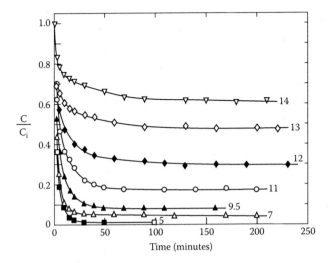

FIGURE 4.24 Concentration as a function of time with shear stress (dynes/cm^2) as a parameter during deposition experiments with quartz particles. (*Source*: From Lick and Kang, 1987. With permission.)

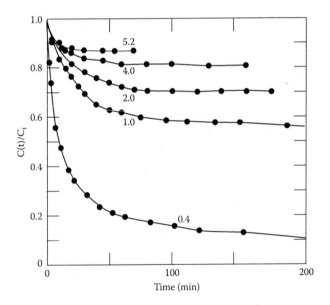

FIGURE 4.25 Concentration as a function of time with shear stress (dynes/cm^2) as a parameter during deposition experiments with Detroit River sediments. (*Source*: From Lick and Kang, 1987. With permission.)

(i.e., the finest particles). For the disaggregated and fine particles, calculated settling times are on the order of 1000 min. Flocculation would decrease this time. For large stresses, only the heaviest particles settle out and the lighter particles stay in suspension. The decay time then is determined by the settling speed of the heavy particles (for d = 10 μm, this is about 10 min) and is therefore smaller than the decay time for the smaller stresses, in agreement with the experiments. Bed armoring/consolidation may be significant in causing, or contributing to, the slow decrease in concentration for a long time.

Somewhat different experiments, erosion followed by deposition, have been performed by Lee et al. (1981) and demonstrate an interesting hysteresis effect. The experiments were done in an annular flume. The sediments were from the Western Basin of Lake Erie and had a median size of about 60 μm with a fairly wide size distribution. They were allowed to consolidate for various periods of time before the erosion/deposition experiments were begun. At the beginning of each set of experiments (each with the same consolidation time), the shear stress was relatively low and was kept constant; the sediments subsequently eroded, and the suspended sediment concentration reached a steady-state value in a period on the order of an hour. After this, the stress was increased to a new constant value, and the concentration again reached a new steady state as time increased. This step-up procedure was repeated four to five times. The shear stress then was decreased to a lower and constant value; the sediment concentration then decreased with time until a new steady state was reached. This step-down procedure was repeated four to five times.

Results of these experiments for the steady-state concentrations as a function of shear stress with time of consolidation (water content) as a parameter are shown in Figure 4.26. The interesting result here is that a hysteresis phenomenon is present; that is, at a certain stress, the value of C_∞/C_i when the stress is being increased (lower branch of the curve) is less than the value of C_∞/C_i when the stress is being decreased (upper branch of the curve). This behavior can be readily explained as follows. Initially, the sediment is consolidated to some degree and has an erosion rate that decreases as the consolidation increases. When sediment is freshly deposited, it has a high water content, is not strongly consolidated, and has a relatively high erosion rate. On the lower branch of each hysteresis curve, the only sediment available for deposition is that which was eroded at that particular shear stress. This is probably insufficient to completely cover the surface of the bottom sediment. In contrast, on the upper branch of the hysteresis curve, the sediment available for deposition is that which was eroded at the previous (and higher) shear stress. This will cover more of the bottom sediments and cause a higher erosion rate and, hence, higher concentrations on the upper branch of the hysteresis curve as compared with the lower branch. This accounts for the observed hysteresis. For different consolidation times, the hysteresis curves for longer consolidation times are below those for shorter consolidation times.

Another set of interesting deposition experiments was performed by Partheniades (1986). These experiments were performed in an annular flume and were similar to the deposition experiments described above except that, for a particular shear stress and all other conditions constant, the initial concentrations were varied in six increments from about 50 to 1500 mg/L. The interesting result was that C_∞/C_i was the same for all experiments; that is, the steady-state concentration

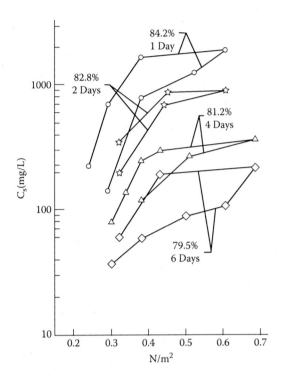

FIGURE 4.26   Hysteresis experiment: steady-state concentration as a function of shear stress with water content (time after deposition) as a parameter. (*Source*: From Lee et al., 1981. With permission.)

was proportional to the initial concentration. As indicated above and as described more quantitatively in Appendix B at the end of this chapter, this again follows if it is assumed that the sediment surface is only partially covered by recently deposited sediments. Then, the steady-state suspended sediment concentration is proportional to the sediment available for erosion (i.e., the initial concentration), and therefore C_∞/C_i is constant.

Krone (1962) has performed experiments to investigate deposition in a straight, recirculating flume rather than in an annular flume as in the experiments described above. The sediments were a mud from San Francisco Bay. They were mixed in fresh water with sodium chloride (to mimic brackish water) at an initial sediment concentration of about 1 g/L. This mixture was then circulated through the flume at high velocity so that all sediments were in suspension. The flow velocity was decreased and maintained at a lower value while the suspended sediments deposited on the bottom. Run times were quite long, up to 180 hr. Suspended sediment concentrations were measured as a function of time. This deposition procedure was repeated for different flow velocities. The suspended sediment concentrations decreased as a function of time, with the decrease being most rapid at the lower velocities.

Krone described the results of the experiments in terms of a probability of erosion, $P = 1 - p$, and assumed p was given by

$$p = 1 - \frac{\tau}{\tau_d} \qquad (4.71)$$

where $0 < p < 1$ and τ_d is the critical shear stress for deposition. The predicted results agreed well with the experimental results. This probability of deposition has been widely used by modelers. Winterwerp (2007) argues that the results can be better described by $p = 1$ and that the decrease in sediment concentration with time is due to bed consolidation and the associated decrease in erosion rate of the deposited sediments. This decrease also could be attributed to bed coarsening, as indicated above for other deposition experiments. The assumption of $p = 1$ and a decrease in erosion rate due to bed armoring/consolidation is consistent with the previous descriptions.

4.5.3 IMPLICATIONS FOR MODELING DEPOSITION

Various processes and parameters that affect deposition and the determination of p_i (fluid turbulence, particle dynamics, particle size distributions, flocculation, bed armoring/consolidation, and partial coverage of previously deposited sediments by recently deposited sediments) have been discussed. These discussions were then used to analyze the results of several different experiments meant to quantify deposition. In these analyses, the following rules were used throughout to describe the results of all experiments in a uniform manner.

1. The effects of fluid turbulent fluctuations on erosion are significant but the effects are inherent in measurements of erosion rates and do not need to be considered separately.
2. The deposition rate can be defined operationally and unambiguously as the erosion rate minus the sediment-water flux. Both of these latter quantities are readily measurable. With this definition, the effects of turbulence on p_i are minimized.
3. The effects of particle dynamics on deposition are significant for coarse-grained particles. However, for finer-grained particles, because of their low momentum relative to the fluid, their effects are probably not significant, i.e., $p = 1$ in this limit.
4. Effects of particle size distribution are significant and must be included in analyses and modeling.
5. For cohesive sediments, effects of flocculation are significant and must be included in analyses and modeling.
6. Effects of bed coarsening/consolidation are significant and must be included in analyses and modeling.
7. Effects of partial coverage of previously deposited sediments by recently deposited sediments are important in the interpretation of laboratory

experiments but probably not important in the modeling of sediment transport in the field.

If the above rules are followed, a uniformly valid description of deposition is obtained. For the finer-grained sediments, $p_i = 1$ and the rate of deposition is given by $w_{si}C_i$.

At the present time, numerical models generally do not accurately include all of the processes described above and/or do not have accurate values of the parameters necessary to describe these processes. Examples are (1) a quantitative description of flocculation near the sediment-water interface and (2) accurate measurements of erosion rates of recently deposited sediments. Because of these limitations, modelers necessarily assume values for p_i that are less than 1.

4.6 CONSOLIDATION

At sufficiently low water velocities, suspended sediments in the form of particles and flocs will begin to deposit at the surface of the bottom sediments. As these particles/flocs settle on the surface, their weight will force the underlying particles/flocs closer together and also will compress the flocs; this movement will be modified by fluid friction and cohesive forces between particles/flocs. As this process occurs, water and gas trapped within the pores of the solid matrix will be expelled and transported toward the surface. In addition to any gas that is initially present, gas may be generated within the sediments due to the decay of organic matter. Differential settling and differential movement of different-size particles also may occur, especially during the early stages of consolidation.

In general, this overall process of settling, deposition, and consolidation depends on the distribution of particle and floc sizes, their settling speeds and rates of deposition, the dynamics of aggregation, the hydrodynamics (especially the shear stress near the sediment-water interface), and the bulk properties of the bottom sediments. Several investigations of this settling/consolidation process have been made (Been, 1981; Been and Sills, 1981; Fitch, 1983; Toorman, 1996, 1999). However, the effects of all the above variables, and especially the interactions between these variables, have not been considered, so that the description of the settling/consolidation process is still incomplete. The presence and effects of gas have generally been ignored. Because of this complexity, the emphasis here will be on the consolidation process alone and will not include descriptions in detail of the settling/consolidation processes in the sediment-water interfacial region.

Results of several consolidation experiments are shown here to illustrate some of the major characteristics and the general behavior of different types of sediments as they consolidate. A theory of consolidation is then described and applied to further quantify these experimental results.

4.6.1 Experimental Results

The first example is one referred to in Section 2.5, that is, a reconstructed sediment from the Detroit River (Jepsen et al., 1997). These sediments had an average particle size of 12 μm and an organic carbon content of 3.3%. The sediments were mixed thoroughly with water to form a slurry; they were then allowed to consolidate in cores of different lengths: 20, 40, and 80 cm. For each core length, replicate cores were prepared. At different times (usually 1, 2, 5, 12, 21, 32, and 60 days), a core was sacrificed and the bulk density of the core was measured as a function of depth by means of the wet-dry procedure. Erosion tests (see Section 3.3) also were performed. During this time period, no gas was present in any of the cores.

Results of the density measurements for core lengths of 20, 40, and 80 cm are shown in Figures 4.27(a), (b), and (c), respectively. For all core lengths, the densities increase with time and usually increase with depth. As noted above, the reason for this is the expulsion of water from the solid matrix due to the weight of the overlying solids and the subsequent transport of that water toward the surface. The increase in the density with time is most rapid initially and then decreases as time increases; a steady state seems to be approached for a long time but is not attained during the 60-day experimental time period. From a comparison of the results for all three core lengths, it can be seen that the increase in density (and hence expulsion of water) at depth is more rapid initially for the longer cores; this is due to the greater weight of the overlying sediment at depth as the core length increases.

An interesting feature is that there are deviations from a monotonic increase of bulk density with depth. For example, consider the 40-cm core (Figure 4.27(b)).

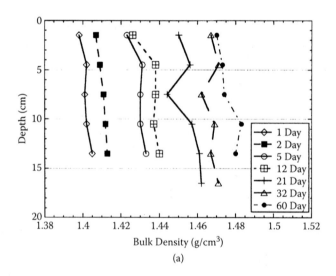

(a)

FIGURE 4.27   Bulk density as a function of depth for Detroit River sediments for consolidation times of 1, 2, 5, 12, 21, 32, and 60 days: (a) 20-cm cores. (*Source*: From Jepsen et al., 1997. With permission.)

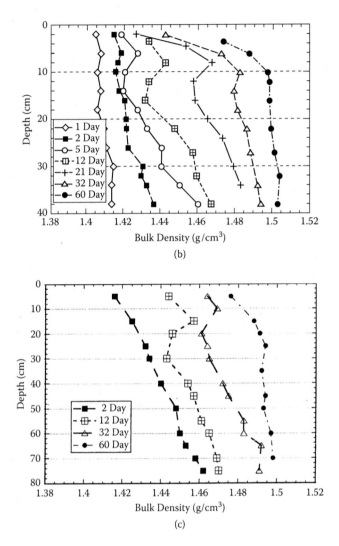

FIGURE 4.27 (CONTINUED)   Bulk density as a function of depth for Detroit River sediments for consolidation times of 1, 2, 5, 12, 21, 32, and 60 days: (b) 40-cm cores, and (c) 80-cm cores. (*Source*: From Jepsen et al., 1997. With permission.)

At about 5 days, the increase in bulk density with depth becomes negative between 5 and 15 cm; that is, there is a greater water content (porosity) at these depths relative to the sediments above and below this layer. This indicates that water above this layer is moving upward more slowly than the waters within and below this layer. The reason for this is that very fine-grained sediments move upward with the pore water through the solid matrix and then collect and hence reduce the permeability of the sediments near the surface. This effect increases for the next 7 days but then decreases with time and disappears by 60 days. For the 20-cm core

(Figure 4.27(a)), the effect is less obvious because the amount of water and very fine-grained sediments in the core and hence being transported to the surface is less. For the 80-cm core, the effect is more evident, with the low-density layer occurring at the 15 to 30 cm depth and lasting somewhat longer.

Similar consolidation tests were performed with sediments from the Fox River and Santa Barbara Slough (Jepsen et al., 1997). For these sediments, average particle sizes were 20 and 35 μm, whereas organic carbon contents were 4.1 and 1.8% respectively. Cores were 20 cm in length. Results of these tests are shown in Figure 4.28 and Figure 4.29. For the Fox River, the low-density layer due to the trapping of the water temporarily appears and is quite evident; however, it is almost nonexistent for sediments from the Santa Barbara Slough, which were relatively coarse and porous. The ranges of bulk densities for these sediments during consolidation are quite different: 1.27 to 1.32 g/cm³ for the Fox River, 1.4 to 1.5 g/cm³ for the Detroit River, and 1.7 to 1.9 g/cm³ for the Santa Barbara Slough. This indicates that the cohesivities of the sediments are in the order: Fox River > Detroit River > Santa Barbara Slough.

During these experiments, no gas was observed in any of these cores, although gas was usually generated in these cores at later times. However, in many sediments of interest, especially contaminated sediments, gas produced by the decay of organic matter is often present in sufficient amounts that it can have significant effects on both the density and erosion rates. In fact, it often can be the dominant effect. As an illustration, consider the consolidation of a thin layer of a capping material placed over a reconstructed sediment from the Grasse River (McNeil et al., 2001b). Consolidation experiments were performed with (1) the Grasse River sediments alone, (2) the capping material alone, and (3) the capping material overlaying the Grasse River sediments. In

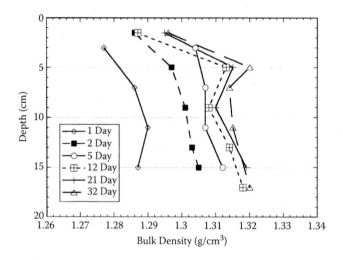

FIGURE 4.28   Bulk density as a function of depth with consolidation time as a parameter for 20-cm cores of Fox River sediments. (*Source*: From Jepsen et al., 1997. With permission.)

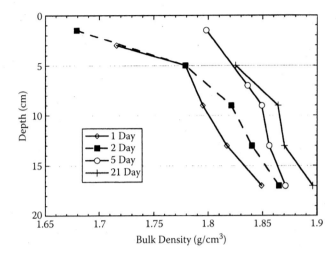

FIGURE 4.29   Bulk density as a function of depth with consolidation time as a parameter for 20-cm cores of Santa Barbara Slough sediments. (*Source*: From Jepsen et al., 1997. With permission.)

these tests, both the actual bulk density, ρ, and the density with no gas, ρ^*, were measured.

For the sediments from the Grasse River, the mean particle size was 45 μm, the organic content was 2.0%, and the mineralogy was dominated by quartz, feldspar, and illite. For a 20-cm core of this sediment alone (with no cap), bulk densities as a function of depth at different times after deposition from 1 to 128 days were measured by means of the Density Profiler and are shown in Figure 4.30(a). The sediments were initially well mixed. After 1 day of consolidation, the densities are fairly uniform with depth and range from 1.53 g/cm³ near the surface to 1.57 g/cm³ at the bottom of the core. For times up to 4 days, the density increases with time and with depth in the same qualitative manner as when no gas is present, except at a slower rate.

However, after 4 days, the densities begin to decrease with time. This is due to the generation of gas caused by the degradation of organic matter in the sediments. The densities continue to decrease until about 64 to 128 days, when they seem to approach a steady state. Throughout the core, local fluctuations with depth and time occur due to the time-dependent formation and upward movement of irregularly spaced gas pockets, usually through well-defined channels. As the gas pockets move upward, their former locations are filled with water. The quasi-steady state is a dynamic balance between the consolidation of the solid matrix (which decreases with time), the generation of gas (which very slowly decreases with time), and the upward movement of water and gas through the solid matrix. Except for local fluctuations, the densities are generally fairly uniform with depth. However, for large time, the densities in a thin layer near the surface are lower than the rest of the core. This is due to the transport of fine-grained particles from the interior to the surface; these fine-grained and cohesive particles then tend to form low-density aggregates near the surface.

From the bulk density as measured by the Density Profiler, ρ, and the density of the solid-water mixture as measured by the wet-dry procedure (no gas), ρ^*, the gas fraction can be determined, as demonstrated in Section 2.5. This gas fraction is shown in Figure 4.30(b) as a function of depth at different times from 2 days to 128 days. At 2 days, the gas fraction is fairly uniform with depth and averages about 1.5%. As time increases, the gas fraction increases and also becomes more nonuniform with depth and time. After 32 days, the gas fraction is reasonably constant with time when averaged over the depth and is approximately 11 to 14%. For the entire time period, the density of the solid-water matrix (no gas) continued to increase, although at a slower and slower rate, similar to the increase of density in the complete absence of gas.

The capping material was a 50/50 mixture by volume of topsoil and sand and had a mean particle size of 125 μm and an organic content of 1.3%. For a 20-cm core of the capping material alone (no base), the bulk densities as a function of depth at different times from 1 to 128 days are shown in Figure 4.31(a). This material was coarser than the Grasse River sediments and had a much higher

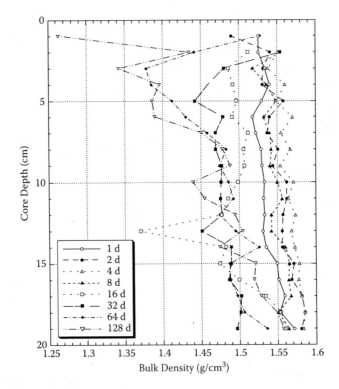

FIGURE 4.30(a)   Reconstructed Grasse River sediments: bulk density as a function of depth and time.

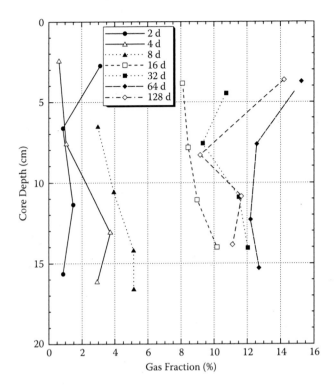

FIGURE 4.30(b)   Reconstructed Grasse River sediments: gas fraction as a function of depth and time.

density, generally between 1.9 and 2.0 g/cm³. Densities did not change appreciably with time for the first 4 days and then began to decrease thereafter. The rate of decrease first increased but then decreased with time; a steady state seemed to be approached for large time. Densities were fairly uniform with depth, except for larger time when a relatively low-density layer formed near the surface, as in the previous case and for the same reason.

Gas fractions as a function of depth at different times are shown in Figure 4.31(b). They are relatively low compared with those in the Grasse River sediments and approach a steady state of 3 to 5% after 64 to 128 days. During the experiment, the density of the solid-water matrix (no gas) did not change appreciably with time, as would be expected for a coarse-grained sediment with no gas present.

For the experiments with the capping material over a base of Grasse River sediments, the sediments were prepared as follows. The base was first formed by pouring thoroughly mixed Grasse River sediments into a coring tube and letting them consolidate. After 3 days, the sediments were reasonably firm, approximately 15 cm thick, and there was little suspended sediment. The capping material then was mixed thoroughly and poured slowly into the coring tube so as to minimize mixing of the capping material and base. After a few days of consolidation, the

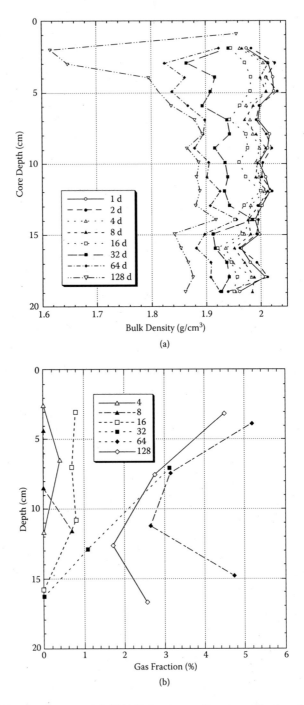

FIGURE 4.31 Capping material: (a) bulk density as a function of depth and time; and (b) gas fraction as a function of depth and time.

thickness of the cap was approximately 15 cm. Both layers then consolidated with time, with the local density and thickness of each layer changing with time due to water migration, gas generation, and gas movement.

Multiple replicate sediment cores were prepared and then sacrificed at times of 1, 2, 4, 8, 16, 32, and 128 days for consolidation and erosion tests. For the core that was allowed to consolidate for 128 days, the bulk densities as a function of depth and at different times are shown in Figure 4.32(a). As in the experiments with each sediment alone, the capping material had a higher density (between 1.8 and 2.0 g/cm^3) than did the base sediment (between 1.35 and 1.6 g/cm^3). The densities in both layers generally decreased monotonically with time.

The gas fraction was determined and is shown in Figure 4.32(b) as a function of depth at times of 4, 8, 16, 32, 64, and 128 days. As in the previous experiments, the gas fraction is much higher in the base sediment than in the capping material. The main reasons are (1) the higher and possibly different quality organic content (and hence higher gas generation) of the base, and (2) the coarser particle size of the cap; this coarseness allows gas to percolate more easily through the cap than through the base. A minor reason is that the base sediments, due to the preparation procedure, had been consolidating and generating gas for 3 days before the cap was added. By 128 days, the gas fraction in both layers seems to be in a reasonably steady state.

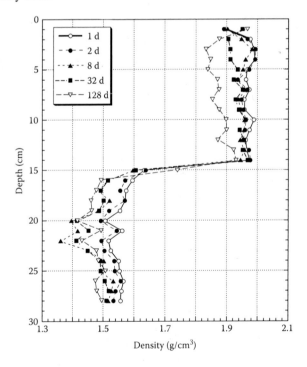

FIGURE 4.32(a)   Capping material over reconstructed Grasse River base: bulk density as a function of depth and time.

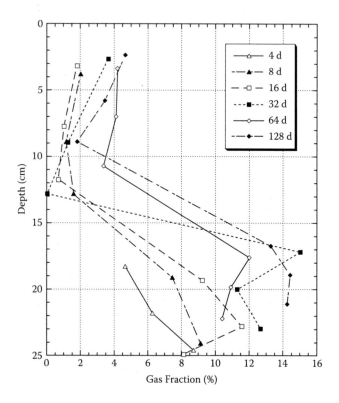

FIGURE 4.32(b)   Capping material over reconstructed Grasse River base: gas fraction as a function of depth and time.

To determine the gas fraction, the density ρ^* was measured for this core at different depths; this quantity at 128 days is shown in Figure 4.32(c), along with ρ as a function of depth at 2 and 128 days. In the measurements of ρ^*, a 1-cm layer of sediment was sampled. For the base sediments, ρ^* increases from approximately 1.55 g/cm^3 at 2 days (where it is approximately equal to ρ at 2 days because little gas is present at 2 days) to almost 1.7 g/cm^3 at 128 days. This increase in ρ^* is due to the vertical movement of water out of the solid-water matrix into the capping material above. The relatively large difference in ρ and ρ^* at 128 days is due to the large amounts of gas generated and contained in the base sediment.

In the capping material, ρ^* at 128 days is approximately the same as (or even a little less than) ρ^* at 2 days (equal to ρ at 2 days). Because the cap is relatively coarse (50% fine to medium sand), the consolidation of this material by itself is relatively small. The density ρ^* probably decreased somewhat due to the greater transport of water from the base sediment into the capping material as compared with that from the capping material into the overlying water.

From Figure 4.32 as well as from Figures 4.30 and 4.31, it can be seen that, during this time period of 128 days and for all three cases, the decrease in density due to gas generation more than offsets the increase in density due to solid-water

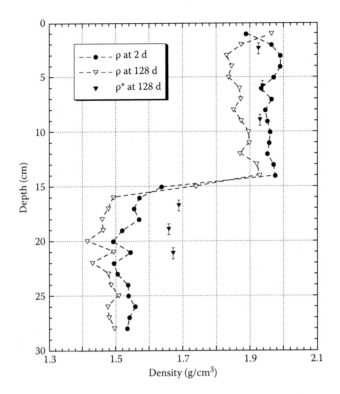

FIGURE 4.32(c) Capping material over reconstructed Grasse River base: densities as a function of depth at 2 and 128 days.

consolidation in the absence of gas. Comparison of Figures 4.30, 4.31, and 4.32 shows that the densities and gas fractions for the cap and base alone are quite similar to those for the cap over the base. Nevertheless, despite these similarities, there were differences in erosion rates between the cap alone and the cap over the base; this was attributed to the transport of small amounts of fine-grained particles from the base into the cap.

Contaminated sediments are typically fine-grained and contain significant amounts of organic matter; for these sediments, the presence of gas is probably the rule rather than the exception. Because of this, gas generation and the subsequent major effects of this gas on bulk densities and erosion rates must be considered and cannot be ignored.

4.6.2 Basic Theory of Consolidation

The basic equations presented here are conservation equations and are essentially the same as those described by Gibson et al. (1967), Diplas and Papanicolaou (1997), and Toorman (1996, 1999). The analysis is initially restricted to the first set of experiments described above, that is, an initially well-mixed sediment consisting of solid particles and water (but no gas) that is not in a steady state but that is consolidating due to the higher density of the solid particles relative to the pore waters, that is, self-weight consolidation.

Because consolidation is relatively slow, the acceleration of the pore fluid and solids can be ignored. With this assumption, the forces on the sediment (solids and water) within a control volume can be written as

$$\frac{\partial p_s}{\partial z} + \frac{\partial p_w}{\partial z} + g(\rho_w x_w + \rho_s x_s) = 0 \qquad (4.72)$$

where p_s is the effective vertical pressure on the solids, p_w is the fluid pressure, g is the acceleration due to gravity, ρ_s is the density of the solid matrix, ρ_w is the density of water, x_w is the volume fraction of the water (porosity), x_s is the volume fraction of the solids, and z is the vertical coordinate measured from the base of the sediment column, z = 0, to the surface, z = h. The third term in the above equation is the weight of the water and solids within the control volume. In this initial formulation, it is assumed that no gas is present so that

$$x_s + x_w = 1 \qquad (4.73)$$

Forces on the fluid within the control volume are given by

$$\frac{\partial p_w}{\partial z} + \rho_w g = -(u - u_s)\frac{\mu x_w}{K} \qquad (4.74)$$

where u is the water velocity, u_s is the solids velocity, μ is the viscosity of water, and $K(x_s)$ is the permeability of the sediment to water and is a function of x_s (or x_w through Equation 4.73). The last term in the above equation is due to the frictional force between the moving water and solids. Effective forces on the solid are obtained by subtracting Equation 4.74 from Equation 4.72 and are

$$\frac{\partial p_s}{\partial z} + g\Delta\rho x_s = +(u - u_s)\frac{\mu x_w}{K} \qquad (4.75)$$

where $\Delta\rho = \rho_s - \rho_w$. It is convenient to subtract the hydrostatic component of the fluid pressure

$$\frac{\partial p_0}{\partial z} = -\rho_w g \qquad (4.76)$$

from the fluid pressure. Equation 4.74 then becomes

$$\frac{\partial p}{\partial z} = -(u - u_s)\frac{\mu x_w}{K} \qquad (4.77)$$

where $p = p_w - p_0$.

Conservation equations for water and solids are

$$\frac{\partial x_w}{\partial t} + \frac{\partial(x_w u)}{\partial z} = 0 \qquad (4.78)$$

$$\frac{\partial x_s}{\partial t} + \frac{\partial(x_s u_s)}{\partial z} = 0 \qquad (4.79)$$

where t is time. By adding these equations, integrating the resulting equation over z, and using Equation 4.73, it follows that

$$x_s u_s + x_w u = \text{constant}$$

$$= 0 \qquad (4.80)$$

where the constant is zero if there is no flow of solids or water through the base of the sediment column, the normal situation when no ground-water flow is present.

Relations for $x_s(p_s)$ and $K(x_s)$, often called constitutive equations, are necessary to complete the set of equations. These are often prescribed as

$$x_s = x_{s0}(1 + a\, p_s^b) \qquad (4.81)$$

$$K = K_1\left(1 - \frac{x_s}{x_{s\infty}}\right)^{\delta} \qquad (4.82)$$

where a, b, K_1, $x_{s\infty}$, and δ are constants that depend on the properties of the solid particles. The quantity x_{s0} is the solids volume fraction at the surface of the sediments and is a function of time; it is not an independent parameter but is solved for as part of the calculation. Although Equations 4.81 and 4.82, with suitable choices for the constants, can give solutions that agree reasonably well with experimental results, other forms of these relations also are possible. This is discussed further below.

By means of Equations 4.73 and 4.80, the above equations can be rearranged to give

$$u = \frac{Kx_s}{\mu x_w}\left[\frac{\partial p_s}{\partial z} + g\Delta\rho x_s\right] \qquad (4.83)$$

$$u_s = -\frac{x_w}{x_s}u \qquad (4.84)$$

$$\frac{\partial x_s}{\partial t} = -\frac{\partial(x_s u_s)}{\partial z} \qquad (4.85)$$

$$x_w = 1 - x_s \tag{4.86}$$

$$p_s = \frac{1}{a}\left(\frac{x_s}{x_{s0}} - 1\right)^{1/b} \tag{4.87}$$

Boundary conditions for these equations are $p_s(h) = 0$ and $u(0) = 0$. The initial distribution of $x_s(z)$ also must be specified, either from laboratory measurements or from estimates of the properties of the *in situ* sediments. Once this is done, the above equations can be solved in order and then marched forward in time. If desired, Equation 4.77 then can be used to calculate the fluid pressure; however, this is not necessary for the determination of x_s or the other variables.

As written, Equation 4.85 is valid for a fixed coordinate system. It is sometimes convenient to use a coordinate system that moves with the solid matrix. This is done as follows. At the surface of the sediment, water volume is lost at the rate $(x_w u)_{z=h}$. Because solids are conserved, the thickness (volume per unit area) of the column then decreases at a rate

$$\frac{dh}{dt} = -x_w u$$

$$= x_s u_s \tag{4.88}$$

where the last equality follows with the use of Equation 4.80. Within the column, a typical volume element in the numerical calculations is defined such that $z_{i-1/2} < z < z_{i+1/2}$; all variables are calculated at the location z_i, where $z_i = (z_{i+1/2} + z_{i-1/2})/2$. The movement of the interfaces of the element at $z_{i+1/2}$ and $z_{i-1/2}$ can be calculated in the same manner as the movement of the surface, that is, by Equation 4.88. Once this is done, conservation of solids then requires that

$$x_{si}^{n+1} = x_{si}^n \frac{\left(z_{i+1/2}^n - z_{i-1/2}^n\right)}{\left(z_{i+1/2}^{n+1} - z_{i-1/2}^{n+1}\right)} \tag{4.89}$$

where the superscript "n" denotes the n-th time step. This equation then replaces Equation 4.85.

Numerical computations were made in the manner described above for the case illustrated in Figure 4.27, the Detroit River sediments. Good agreement between the numerical and experimental results was obtained for all times for each core and for different core lengths. In these calculations, only one set of parameters was used. This implies that, for this sediment, the model is valid for different depths of consolidating sediments and for all times and is therefore quite general. However, it should be emphasized that Equations 4.81 and 4.82 are semi-empirical relations and the specific values for the constants used in these equations were determined by calibration, that is, by adjusting the constants until the numerical and experimental results agreed. The numerical model is then predictive in that it can model the consolidation process for that sediment for a variety of conditions similar to those in the experiments. Other forms of the constitutive relations are also possible, were tried, and gave valid results, again indicating the empirical nature of the constitutive relations.

Parameters will change as the sediment type changes. To determine these parameters as functions of the bulk properties of sediments, a systematic study of consolidation for different sediments is needed.

4.6.3 CONSOLIDATION THEORY INCLUDING GAS

When gas is present, Equation 4.73 is replaced by

$$x_w + x_s + x_g = 1 \tag{4.90}$$

where x_g is the volume fraction of the gas. Because the weight of the gas within a control volume is negligible and if it is assumed that the force of the escaping gas on the solids and pore water is negligible, then Equations 4.72, 4.74, 4.75, and 4.77 remain the same. The mass balance equations for x_w and x_s (Equations 4.78 and 4.79) also remain the same. A mass balance equation for the gas is

$$\frac{\partial x_g}{\partial t} + \frac{\partial (x_g u_g)}{\partial z} = S_g \tag{4.91}$$

where u_g is the gas velocity and S_g is a source term due to the generation of gas by the decay of organic matter. S_g will depend on the amount and type of organic matter as well as the temperature. The gas velocity will depend on at least the permeability of the sediment to gas, on x_w, and possibly on x_g. When S_g and u_g are known, the above equation can be used to determine x_g.

By adding Equations 4.78, 4.79, and 4.91 and integrating the resulting equation over z, one obtains

$$x_w u + x_s u_s + x_g u_g = \int_0^z S_g dz \equiv I(z) \tag{4.92}$$

where it has been assumed that there is no flow of solids, water, or gas through the bottom of the water column.

It is assumed that Equations 4.75 and 4.81 are still valid when gas is present. With this assumption and the use of Equation 4.91, the governing equations, Equations 4.83 through 4.87, are now replaced by

$$u = \frac{Kx_s}{\mu x_w(1-x_g)}\left[\frac{\partial \rho}{\partial z} + g\Delta\rho x_s\right] - \left(\frac{x_g}{1-x_g}\right)u_s + \frac{I}{1-x_g} \tag{4.93}$$

$$u_s = -\frac{x_w}{x_s}u - \frac{x_g}{x_s}u_g + \frac{I}{x_s} \qquad (4.94)$$

$$\frac{\partial x_s}{\partial t} = -\frac{\partial}{\partial z}(x_s u_s) \qquad (4.95)$$

$$x_w = 1 - x_s - x_g \qquad (4.96)$$

$$p_s = \frac{1}{a}\left(\frac{x_s}{x_{s0}} - 1\right)^{1/b} \qquad (4.97)$$

Equation 4.88 becomes

$$\frac{dh}{dt} = x_s u_s$$

$$= -x_w u - x_g u_g + I(h) \qquad (4.98)$$

With the use of this equation, Equation 4.89 remains the same.

From an examination of the experimental results in Figures 4.30 and 4.31, it is reasonable to assume as a first approximation that within each layer (1) the rate of generation of gas, S_g, is a constant, independent of space and time; (2) x_g is independent of depth and is only a function of time; and (3) the gas velocity increases linearly with z, that is, $u_g = cz$. Equation 4.91 then reduces to

$$\frac{dx_g}{dt} + cx_g = S_g \qquad (4.99)$$

The solution to this equation is

$$x_g = \frac{S_g}{c}(1 - e^{-ct}) \qquad (4.100)$$

As $t \to 0$, $x_g = S_g t$ and therefore S_g can be determined from the experimental results for small time. As $t \to \infty$, $x_g(\infty) = S_g/c$ and therefore c can be determined from the steady-state results for x_g.

With these approximations, x_g is determined as a function of time and the numerical computation for u, u_s, x_w, and x_s proceeds from Equations 4.93 to 4.98 in the same manner as for the case when no gas is present. In this way and with the proper choices for the constants a and b, reasonable agreement can be found between the experimental and numerical calculations. When stratified layers are present, as in Figure 4.32, the numerical computations for each layer proceed in the same manner as those for a single layer. Matching conditions at the interface between the layers are that p_s and u are continuous across the interface.

APPENDIX A

Consider Equation 4.43 for each of the collision mechanisms of Brownian motion, fluid shear, and differential settling. As a first approximation, assume that A* and D* are constants. For differential settling, $\beta_d = \alpha_d \, ad^{2+m} = \alpha^* d^{2+m}$. With this substitution, Equation 4.43 reduces to

$$\frac{d(d)}{dt} = A^* \alpha^* Cd^2 \left[1 - \frac{D^*}{A^*} C\alpha^* d \right] \qquad (A.1)$$

and is independent of m. Normalize d and t as follows.

$$\eta = \frac{D^*}{A^*} C\alpha^* d \qquad (A.2)$$

$$\tau = \frac{A^{*2}}{D^*} t \qquad (A.3)$$

Equation A.1 then reduces to the characteristic equation

$$\frac{d\eta}{d\tau} = \eta^\alpha (1 - \eta) \qquad (A.4)$$

where $\alpha = 2$.

For fluid shear, β_f is $4G \, d^3/3$. With the substitutions,

$$d = z^\nu \qquad (A.5)$$

$$\nu = \frac{1}{2 - m} \qquad (A.6)$$

Equation 4.43 reduces to

$$\frac{vdz}{dt} = \frac{4}{3} A^* CGz^2 \left[1 - \frac{4}{3} \frac{D^*}{A^*} CGz \right] \qquad (A.7)$$

With the appropriate normalization, this equation also reduces to Equation A.4 with $\alpha = 2$.

For Brownian motion, β_b is a constant and not a function of d. With the substitutions,

$$d = z^\nu \qquad (A.8)$$

$$v = -\frac{1}{m+1} \tag{A.9}$$

and with the appropriate normalization, Equation 4.43 with $\beta = \beta_b$ also reduces to Equation A.4 with $\alpha = 2$.

When A^* and D^* are simple polynomials, Equation 4.43 can be reduced in a similar manner to Equation A.4 but with the value of α dependent on a_2 and b_2 and generally not equal to 2.

For Equation A.4, exact solutions are available when α is an integer. For $\alpha = 1, 2,$ and 3, the solutions are

$$\alpha = 1: \quad \tau = -\ell n\left(\frac{1-\eta}{\eta}\right) \tag{A.10}$$

$$\alpha = 2: \quad \tau = 2 - \frac{1}{\eta} - \ell n\left(\frac{1-\eta}{\eta}\right) \tag{A.11}$$

$$\alpha = 3: \quad \tau = 4 - \frac{1}{\eta} - \frac{1}{2\eta^2} - \ell n\left(\frac{1-\eta}{\eta}\right) \tag{A.12}$$

where the integration constant has been chosen so that $\eta(0) = \frac{1}{2}$. These solutions are shown in Figure 4.21.

APPENDIX B

As a first approximation to describe the erosion/deposition process for fine-grained, cohesive sediments, consider the following. Assume that the sediments consist of particles of uniform size and properties. The bottom sediments consist primarily of particles previously deposited and compacted for a relatively long period of time and have an erosion rate E_1 that is a function of the shear stress, particle size, and time after deposition. This bed is partially overlaid by newly deposited material. This newly deposited material is less compact than the underlying sediments and has an erosion rate E_2 (τ, d) that is greater than E_1. The newly deposited sediment covers a fraction X $(X < 1)$ of the bed to a depth $h^* = \delta/\rho$, where δ is the mass per unit area and ρ is the mass per unit volume of the deposited sediments.

With these assumptions, the erosion rate E can be written as

$$E = (1 - X)\,E_1 + X\,E_2 \tag{B.1}$$

The sediment-water flux is given by Equation 4.63 as before. By combining Equations 4.63 and B.1, one obtains

$$h\frac{dC}{dt} = (1-X)E_1 + XE_2 - pw_sC \qquad (B.2)$$

Similarly, it can be deduced that the net rate of change of newly deposited sediments can be described by

$$\delta\frac{dX}{dt} = -XE_2 + pw_sC \qquad (B.3)$$

It should be noted that a steady state can be achieved only if either $E_1 = 0$ (no erosion of old sediment) or $X = 1$ (the surface is entirely covered with newly deposited sediments).

A general solution to Equations B.2 and B.3 can be found but is cumbersome. This solution simplifies in two cases: (1) for $E_1 = E_2$ (equivalent to $X = 1$), and (2) for $E_1 = 0$. Case (1) is the same as that described previously, that is, the non-cohesive sediment. Case (2) corresponds to the previously deposited sediment being well compacted; no erosion of this sediment is therefore possible. In the following analysis of this case, it is assumed that $X < 1$. For $X = 1$, the results are the same as those above. For $0 < X < 1$ and for an initial concentration C_i, the solution of Equations B.2 and B.3 reduces to

$$C = (C_i - C_\infty)\exp(-t/t_2) + C_\infty \qquad (B.4)$$

where

$$C_\infty = \frac{C_i}{1 + \dfrac{pw_s\delta}{E_2h}} \qquad (B.5)$$

and

$$t_2 = \frac{\delta h}{E_2h + pw_s\delta} = \frac{1}{\dfrac{E_2}{\delta} + \dfrac{pw_s}{h}} \qquad (B.6)$$

Note that C_∞ is proportional to C_i, a result determined experimentally (Etter et al., 1968) for kaolinite clay (disaggregated particle size on the order of 5 μm) in distilled water. This result is not true for coarse-grained sediments or when the surface is covered with recently deposited fine-grained sediments ($X = 1$). In both of these cases, $C_\infty = E/pw_s$, from Equation 4.64.

5 Hydrodynamic Modeling

When considering the currents in surface waters as diverse as rivers, lakes, estuaries, and nearshore areas of the oceans, it is evident that large variations exist in the length and time scales describing these currents. Significant length scales vary from the vertical dimensions of the microstructure of stratified flows (as little as a few centimeters) to the size of the basin (up to several hundred kilometers), and time scales vary from a few seconds to many years. Although in principle the equations of fluid dynamics can describe the motions that include all these length and time scales, practical difficulties prohibit the use of the full equations of motion for problems involving any but the smallest length and time scales. Because of this, considerable effort and ingenuity have been expended to approximate these equations to obtain simpler equations and methods of solution.

The result is that many different numerical models of currents in surface waters currently exist. The primary difference between these models is usually the different length and time scales that the investigator believes is significant for the specific problem. For example, if the details of the flow in the vertical are not thought of as significant, one can use a two-dimensional, vertically integrated model; this may be either steady state or time dependent, depending on whether the time variation is considered significant. To investigate flows where vertical stratification due to temperature and/or salinity gradients is significant but horizontal flows in a transverse direction are not, a two-dimensional, horizontally integrated model is relatively simple and may be useful. More complex three-dimensional, time-dependent models may be necessary when flow fields vary significantly in all three directions and with time. In whatever numerical model chosen, for reasons of accuracy and stability, the grid sizes in both space and time must be smaller than the smallest space and time scales that are thought to be significant.

Fluid mechanics is a fascinating and diverse science; texts on the subject abound and should be consulted for descriptions of the fundamental processes and its many and diverse applications. Numerous texts also exist that emphasize civil engineering applications such as river flooding and control, the design of control structures, and the modeling of plumes from power plants or dredging operations. The present chapter is rather brief and does not discuss these subjects; its purpose is simply to give an overview of hydrodynamic modeling as applied to the transport of sediments and contaminants in surface waters. Most examples are rather elementary and are meant to illustrate interesting and significant features of flows that affect sediment and contaminant transport. A few more complex examples are given to illustrate the present state of the art in hydrodynamic modeling. In the following chapters, additional applications of hydrodynamic models are described as a complement to the analyses of specific sediment and contaminant transport problems. However, for a thorough understanding of the rich and fascinating field of fluid mechanics in surface waters, additional articles, texts, and conference proceedings should be consulted (e.g., Sorensen, 1978; Mei, 1983; Martin and McCutcheon, 1999; Spaulding, 2006).

In the present chapter, the basic three-dimensional, time-dependent conservation equations and boundary conditions that govern fluid transport are presented first. This is followed by brief discussions of eddy coefficients, the bottom shear stress due to currents and wave action, the surface stress due to winds, sigma coordinates, and the stability of the numerical difference equations. By integration of the three-dimensional equations over the water depth, simpler and more computationally efficient, vertically integrated (or vertically averaged) models result and are the topic of Section 5.2. These reduced models are comparatively easy to analyze and require little computer time, but they do not give details of the vertical variation of the flow. For some problems of sediment and contaminant transport, this detail is not necessary and a vertically integrated model is adequate and gives comparable results to those from a three-dimensional model. Two-dimensional, horizontally averaged models also have been developed and are useful for a qualitative understanding of the flow and for preliminary transport studies where vertical variations of the flow are significant but where flow in one horizontal direction can be neglected (e.g., a thermally stratified lake or a salinity stratified estuary). The basic equations and an example of this type of model are presented in Section 5.3. Three-dimensional, time-dependent models and applications of these models are described in Section 5.4. In the modeling of sediment transport, an important parameter for erosion is the bottom shear stress; this stress is due not only to currents but also to wave action. A simple model of wave action and an application to Lake Erie are described in Section 5.5.

5.1 GENERAL CONSIDERATIONS IN THE MODELING OF CURRENTS

5.1.1 BASIC EQUATIONS AND BOUNDARY CONDITIONS

The basic equations used in the modeling of currents are the usual hydrodynamic equations for conservation of mass, momentum, and energy, plus an equation of state. In waters with variable salinity, a conservation equation for salinity is also needed. In sufficiently general form for almost all modeling of surface waters, these equations are the following.

- Mass conservation:

$$\frac{\partial \rho}{\partial t} + \frac{\partial \rho u}{\partial x} + \frac{\partial \rho v}{\partial y} + \frac{\partial \rho w}{\partial z} = 0 \qquad (5.1)$$

- Conservation of momentum in the x-direction:

$$\frac{\partial u}{\partial t} + u\frac{\partial u}{\partial x} + v\frac{\partial u}{\partial y} + w\frac{\partial u}{\partial z} - fv = -\frac{1}{\rho}\frac{\partial p}{\partial x} + \frac{\partial}{\partial x}\left(A_H\frac{\partial u}{\partial x}\right)$$

$$+ \frac{\partial}{\partial y}\left(A_H\frac{\partial u}{\partial y}\right) + \frac{\partial}{\partial z}\left(A_v\frac{\partial u}{\partial z}\right) \qquad (5.2)$$

- Conservation of momentum in the y-direction:

$$\frac{\partial v}{\partial t} + u\frac{\partial v}{\partial x} + v\frac{\partial v}{\partial y} + w\frac{\partial v}{\partial z} + fu = -\frac{1}{\rho}\frac{\partial p}{\partial y} + \frac{\partial}{\partial x}\left(A_H\frac{\partial v}{\partial x}\right)$$

$$+ \frac{\partial}{\partial y}\left(A_H\frac{\partial v}{\partial y}\right) + \frac{\partial}{\partial z}\left(A_v\frac{\partial v}{\partial z}\right) \qquad (5.3)$$

- Conservation of momentum in the z-direction:

$$\frac{\partial w}{\partial t} + u\frac{\partial w}{\partial x} + v\frac{\partial w}{\partial y} + w\frac{\partial w}{\partial z} = -\frac{1}{\rho}\frac{\partial p}{\partial z} + g + \frac{\partial}{\partial x}\left(A_H\frac{\partial w}{\partial x}\right)$$

$$+ \frac{\partial}{\partial y}\left(A_H\frac{\partial w}{\partial y}\right) + \frac{\partial}{\partial z}\left(A_v\frac{\partial w}{\partial z}\right) \qquad (5.4)$$

- Energy conservation:

$$\frac{\partial T}{\partial t} + u\frac{\partial T}{\partial x} + v\frac{\partial T}{\partial y} + w\frac{\partial T}{\partial z} = \frac{\partial}{\partial x}\left(K_H\frac{\partial T}{\partial x}\right) + \frac{\partial}{\partial y}\left(K_H\frac{\partial T}{\partial y}\right) + \frac{\partial}{\partial z}\left(K_v\frac{\partial T}{\partial z}\right) + S_H \quad (5.5)$$

- Salinity conservation:

$$\frac{\partial S}{\partial t} + u\frac{\partial S}{\partial x} + v\frac{\partial S}{\partial y} + w\frac{\partial S}{\partial z} = \frac{\partial}{\partial x}\left(K_H\frac{\partial S}{\partial x}\right) + \frac{\partial}{\partial y}\left(K_H\frac{\partial S}{\partial y}\right) + \frac{\partial}{\partial z}\left(K_v\frac{\partial S}{\partial z}\right) \quad (5.6)$$

- An equation of state:

$$\rho = \rho(p,S,T) \qquad (5.7)$$

where u, v, and w are fluid velocities in the x-, y-, and z-directions, respectively (z is positive upward); t is time; f is the Coriolis parameter, which is assumed constant; p is the pressure; ρ is the density; A_H is the horizontal eddy viscosity and A_v is the vertical eddy viscosity; K_H is the horizontal eddy conductivity and K_v is the vertical eddy conductivity; g is the acceleration due to gravity; T is the temperature; S is salinity; and S_H is a heat source term. In this chapter, for simplicity and following convention, the symbol ρ denotes the density of water; in other chapters, ρ denotes the bulk density of the sediment, whereas ρ_w denotes the density of water.

Several approximations are implicit in these equations. These are (1) eddy coefficients are used to account for the turbulent diffusion of momentum, energy, and salinity; and (2) the kinetic energy of the fluid is small in comparison with the internal energy (which is proportional to the temperature) of the fluid so that energy transport (Equation 5.5) is dominant and can be described by the transport of internal energy (temperature) alone.

However, these equations are more general, and hence more complex and computer intensive, than usually necessary for most surface water dynamics. In most applications, they can be further simplified by making the following approximations: (1) vertical velocities are small in comparison with horizontal velocities so that a hydrostatic approximation is valid, and (2) variations in density are small and can be neglected except in the buoyancy term in the vertical momentum equation (the Boussinesque approximation). With these approximations, the above equations reduce to the following:

$$\frac{\partial u}{\partial x} + \frac{\partial v}{\partial y} + \frac{\partial w}{\partial z} = 0 \tag{5.8}$$

$$\frac{\partial u}{\partial t} + \frac{\partial u^2}{\partial x} + \frac{\partial uv}{\partial y} + \frac{\partial uw}{\partial z} - fv = -\frac{1}{\rho_r}\frac{\partial p}{\partial x} + \frac{\partial}{\partial x}\left(A_H \frac{\partial u}{\partial x}\right)$$
$$+ \frac{\partial}{\partial y}\left(A_H \frac{\partial u}{\partial y}\right) + \frac{\partial}{\partial z}\left(A_v \frac{\partial u}{\partial z}\right) \tag{5.9}$$

$$\frac{\partial v}{\partial t} + \frac{\partial uv}{\partial x} + \frac{\partial v^2}{\partial y} + \frac{\partial wv}{\partial z} + fu = -\frac{1}{\rho_r}\frac{\partial p}{\partial y} + \frac{\partial}{\partial x}\left(A_H \frac{\partial v}{\partial x}\right)$$
$$+ \frac{\partial}{\partial y}\left(A_H \frac{\partial v}{\partial y}\right) + \frac{\partial}{\partial z}\left(A_v \frac{\partial v}{\partial z}\right) \tag{5.10}$$

$$\frac{\partial p}{\partial z} = -\rho g \tag{5.11}$$

$$\frac{\partial T}{\partial t} + \frac{\partial uT}{\partial x} + \frac{\partial vT}{\partial y} + \frac{\partial wT}{\partial z} = \frac{\partial}{\partial x}\left(K_H \frac{\partial T}{\partial x}\right) + \frac{\partial}{\partial y}\left(K_H \frac{\partial T}{\partial y}\right) + \frac{\partial}{\partial z}\left(K_v \frac{\partial T}{\partial z}\right) + S_H \tag{5.12}$$

$$\frac{\partial S}{\partial t} + \frac{\partial uS}{\partial x} + \frac{\partial vS}{\partial y} + \frac{\partial wS}{\partial z} = \frac{\partial}{\partial x}\left(K_H \frac{\partial S}{\partial x}\right) + \frac{\partial}{\partial y}\left(K_H \frac{\partial S}{\partial y}\right) + \frac{\partial}{\partial z}\left(K_v \frac{\partial S}{\partial z}\right) \tag{5.13}$$

$$\rho = \rho(S,T) \qquad (5.14)$$

where ρ_r is a reference density.

The appropriate boundary conditions depend on the particular problem to be solved. At the free surface, $z = \eta(x,y,t)$, usual conditions include (1) the specification of a shear stress due to the wind,

$$\rho A_v \frac{\partial u}{\partial z} = \tau_x^w, \qquad \rho A_v \frac{\partial v}{\partial z} = \tau_y^w \qquad (5.15)$$

where τ_x^w and τ_y^w are the specified wind stresses in the x- and y-directions, respectively; (2) a kinematic condition on the free surface,

$$\frac{\partial \eta}{\partial t} + u \frac{\partial \eta}{\partial x} + v \frac{\partial \eta}{\partial y} - w = 0 \qquad (5.16)$$

(3) the pressure is continuous across the water-air interface and therefore the fluid pressure at the surface equals the local atmospheric pressure p_a,

$$p(x,y,\eta,t) = p_a \qquad (5.17)$$

and (4) a specification of the heat flux at the surface,

$$q = -\rho K_v \frac{\partial T}{\partial z} = H(T - T_a) \qquad (5.18)$$

where q is the energy flux, H is a surface heat transfer coefficient, and T_a is the air temperature. At the bottom, the conditions are (1) those of no fluid motion or a specification of shear stress in terms of either integrated mass flux or near-bottom velocity and (2) a specification of temperature or a specification of heat flux. Variations in these boundary conditions and additional boundary conditions are discussed in the following sections.

5.1.2 Eddy Coefficients

The numerical grid sizes for a problem usually are determined from considerations of the physical detail desired and computer limitations. Once the grid size is chosen, it is implicitly assumed that all physical processes smaller than this can either be neglected or approximately described by turbulent fluctuations. Turbulent fluctuations usually manifest themselves in an apparent increase in the viscous stresses of the basic flow. These additional stresses are known as Reynolds stresses. The total stress is the sum of the Reynolds stress and the usual molecular viscous stress. In turbulent flow, the latter is comparatively small and therefore can be neglected in most cases. Analogous to the coefficients of molecular viscosity, an eddy viscosity coefficient can be introduced (as has been done in the equations above) so that the shear stress is proportional to a velocity gradient. Similarly, an eddy diffusion coefficient can be introduced so that the heat and salinity fluxes are proportional to temperature and salinity gradients, respectively. In turbulent flow, these coefficients are not properties of the fluid as in laminar flow but depend on the flow itself, that is, on the processes generating the turbulence.

The determination of these turbulent eddy coefficients is a significant problem in hydrodynamic modeling. Two approaches to this problem are (1) the use of semi-empirical algebraic equations to relate the eddy coefficients to the local flow conditions and (2) the use of turbulence theory to relate the eddy coefficients to the turbulence kinetic energy and a turbulence length scale by means of transport equations for these quantities. Both approaches depend on laboratory and field measurements to quantify parameters that appear in the analyses. However, the use of turbulence theory, although more complex, is also more general and requires less adjustment of parameters for a specific problem.

In the first approach, semi-empirical algebraic relations are used that relate the eddy coefficients to processes (which generate turbulence) and to density changes (which reduce turbulence). Because the scale and intensity of the vertical and horizontal components of turbulence are generally quite different, it is convenient to consider these effects separately, as has been done in the equations presented above. The vertical eddy viscosity, A_v, and vertical eddy diffusivity, K_v, should in general vary throughout the system. Some of the more important generating processes of this vertical turbulence and causes for its variation include (1) the direct action of the wind stress and heat flux on the water surface, (2) the presence of vertical shear in currents due to horizontal pressure gradients, (3) the presence of internal waves, and (4) the effect of bottom irregularities and friction due to currents and waves.

If the density of the water increases with depth, stability effects will reduce the intensity of the turbulence. These effects depend on the Richardson number, defined by

$$Ri = -\frac{g\frac{\partial \rho}{\partial z}}{\rho\left(\frac{\partial \bar{u}}{\partial z}\right)^2} \qquad (5.19)$$

where $\bar{u}$ is the mean horizontal velocity. Various empirical equations have been developed that relate the eddy viscosity coefficient and the Richardson number. A typical relation is that developed by Munk and Anderson (1948):

$$A_v = A_{v0}(1 + 10\,Ri)^{-1/2} \qquad (5.20)$$

where A_{v0} is the value of A_v in a nonstratified flow. Values for A_{v0} are generally on the order of 1 to 50 cm²/s.

In a nonstratified flow, it is believed that the eddy diffusivity is approximately equal to the eddy viscosity. However, for a stratified flow, the mechanisms of turbulent transfer of momentum and heat are somewhat different, and this leads to different dependencies of these coefficients on the Richardson number. For example, a semi-empirical relation (Munk and Anderson, 1948) similar to Equation 5.20 suggests that

$$K_v = K_{v0} (1 + 3.33 \, Ri)^{-3/2} \qquad (5.21)$$

where K_{v0} is the value of the vertical eddy diffusivity in a nonstratified flow. Many additional relationships for A_v and K_v similar to the above two equations have been proposed and used.

Horizontal viscosity coefficients are generally much greater than the vertical coefficients. It is found from experiments that the values of the horizontal viscosity coefficient increase with the scale ℓ of the turbulent eddies. An empirical relation of this type is

$$A_H = a\varepsilon^{1/3}\ell^{4/3} \qquad (5.22)$$

where a is a constant and ε is the rate of energy dissipation (e.g., see Okubo, 1971; Csanady, 1973). Observations indicate values of 10^4 to 10^5 cm^2/s for A_H for the overall circulation in the Great Lakes (Hamblin, 1971), with smaller values indicated in the nearshore regions. In numerical models, A_H and K_H are often chosen as the minimum values required for numerical stability (see below).

Implicit in the use of algebraic equations for the eddy coefficients is the assumption that the production and dissipation of turbulent mixing are in local equilibrium (Bedford, 1985). However, in many cases, this is not an accurate assumption and the transport of turbulence must be considered. For this purpose, equations have been developed that describe the transport of turbulent kinetic energy, k, and viscous dissipation, ε. Alternately, because the viscous dissipation is proportional to $k^{3/4}/\ell$, where ℓ is a turbulence length scale, a transport equation for ℓ can be used instead of a transport equation for ε. In either case, the resulting model for turbulence is generally known as a k-ε (turbulence closure) model.

An example of a k-ε model that is widely used is the following (Mellor and Yamada, 1982; Galperin et al., 1988; Blumberg et al., 1992). In this model, the vertical mixing coefficients are given by

$$A_v = \hat{A}_v + \upsilon_M, \quad K_v = \hat{K}_v + \upsilon_H \qquad (5.23)$$

$$\hat{A}_v = q\ell S_M, \quad \hat{K}_v = q\ell S_H \qquad (5.24)$$

where $q^2/2$ is the turbulent kinetic energy, S_M and S_H are stability functions defined by solutions to algebraic equations given by Mellor and Yamada (1982) as modified by Galperin et al. (1988), and υ_M and υ_H are constants. The stability functions account for the reduced and enhanced vertical mixing in stable and unstable vertically density-stratified systems in a manner similar to Equations 5.20 and 5.21.

The variables q^2 and ℓ are determined from the following transport (or conservation) equations:

$$\frac{\partial q^2}{\partial t} + \frac{\partial(uq^2)}{\partial x} + \frac{\partial(vq^2)}{\partial y} + \frac{\partial(wq^2)}{\partial z} = \frac{\partial}{\partial z}\left[K_q \frac{\partial q^2}{\partial z}\right]$$

$$+2A_v\left[\left(\frac{\partial u}{\partial z}\right)^2 + \left(\frac{\partial v}{\partial z}\right)^2\right] + \frac{2g}{\rho_o}K_v\frac{\partial \rho}{\partial z} - 2\frac{q^3}{B_1\ell} + F_q \qquad (5.25)$$

$$\frac{\partial(q^2\ell)}{\partial t} + \frac{\partial(uq^2\ell)}{\partial x} + \frac{\partial(vq^2\ell)}{\partial y} + \frac{\partial(wq^2\ell)}{\partial z} = \frac{\partial}{\partial z}\left[K_q \frac{\partial(q^2\ell)}{\partial z}\right]$$

$$+E_1\ell\left\{A_v\left[\left(\frac{\partial u}{\partial z}\right)^2 + \left(\frac{\partial v}{\partial z}\right)^2\right] + \frac{g}{\rho_o}K_v\frac{\partial \rho}{\partial z}\right\} - \frac{q^3}{B_1\tilde{\omega}} + F_\ell \qquad (5.26)$$

where $K_q = 0.2q\ell$, the eddy diffusion coefficient for turbulent kinetic energy; F_q and F_ℓ represent horizontal diffusion of the turbulent kinetic energy and turbulence length scale and are parameterized in a manner analogous to Equation 5.22; $\tilde{\omega}$ is a wall proximity function defined as $\tilde{\omega} = 1 + E_2(\ell\,/\,\kappa L)^2$; $(L)^{-1} = (\eta - z)^{-1} + (h + z)^{-1}$; κ is the von Karman constant; h is the water depth; η is the free surface elevation; and E_1, E_2, and B_1 are empirical constants set in the closure model.

The above and similar k-ϵ models have been used extensively, and the coefficients appearing in them have been determined from laboratory experiments as well as from comparison of results of the numerical models with field measurements. Applications of these models are described in the following sections and chapters.

5.1.3 Bottom Shear Stress

5.1.3.1 Effects of Currents

A shear stress is produced at the sediment-water interface due to physical interactions between the flowing water and the bottom sediments. This stress depends in a nonlinear manner on the flow velocity and the bottom roughness. It has been investigated extensively and quantified approximately by means of laboratory experiments, field tests, and model calibrations. Results for the shear stress are usually reported as

$$\tau = \rho c_f q\mathbf{q} \qquad (5.27)$$

where c_f is a coefficient of friction and is dimensionless; q is a near-bed, or reference, velocity; and a bold symbol denotes a vector (i.e., the shear stress is a vector aligned with the flow velocity). The components of the stress can be written as

$$\tau_x = \rho c_f q u \tag{5.28}$$

$$\tau_y = \rho c_f q v \tag{5.29}$$

where $q = (u^2 + v^2)^{1/2}$.

The coefficient of friction depends on the grain size of the sediment bed; sediment bedforms such as mounds, ripples, and dunes; and any biota present. Typical values for c_f are between 0.002 and 0.005. In the absence of site-specific information, a value of 0.003 often is chosen for smooth, cohesive sediment beds. A more accurate value for c_f can be determined as an average over the entire sediment bed from calibration of modeled to measured currents when the latter are available.

From turbulent flow theory and assuming a logarithmic velocity profile, it can be shown that the coefficient of friction can be determined from

$$c_f = \frac{\kappa^2}{\left(\ln \dfrac{h}{2z_0} \right)^2} \tag{5.30}$$

where κ is von Karman's constant (0.41), z_0 is the effective bottom roughness, and h is a reference distance or depth. For three-dimensional models or two-dimensional, horizontally integrated models, h is the thickness of the lowest layer of the numerical grid; for vertically integrated models, h is the water depth. From this formula and by assuming a value for z_0, c_f can be calculated and will vary locally as a function of depth and z_0. However, the difficulty with this formula is that an effective value for z_0 is difficult to determine accurately. This is often done by model calibration.

For a more accurate determination of z_0 and c_f, field measurements of velocity profiles can be used (e.g., Cheng et al., 1999; Sea Engineering, 2004). As a specific example, consider the measurements and analyses made by the USGS and Sea Engineering, Inc., for the Fox River (Sea Engineering, 2004). For fully developed turbulent flow in a river, the near-bottom velocity profile is given by

$$u = \frac{u_*}{\kappa} \ell n \frac{z}{z_0} \tag{5.31}$$

where $u_* = (\tau/\rho)^{1/2}$. This is known as the universal logarithmic velocity profile, or law of the wall (e.g., Schlichting, 1955). It can be rewritten as

$$\ell n \, z = \frac{\kappa}{u_*} u + \ell n \, z_0 \qquad (5.32)$$

Once $u(z)$ is obtained from field measurements and $\ell n \, z$ is plotted as a function of u, then the z intercept of the line given by the above equation is z_0 and the slope of the line is κ/u_*.

To determine $u(z)$, more than 100 vertical profiles of current velocities were measured at various locations in the Fox during four separate flow events. From this, z_0 and u_* were determined. Because $\tau = \rho u_*^2$, the coefficient of friction then can be calculated from Equation 5.27 as

$$c_f = \frac{\tau}{\rho U_{avg}^2} \qquad (5.33)$$

where U_{avg} is the vertically averaged velocity. For each vertical profile, c_f as determined in this way is shown in Figure 5.1. Also shown is c_f as determined from the law of the wall (Equation 5.30) with z_0 equal to 0.2 cm, the average from all the measurements. There is a reasonable fit to the data except for shallow waters, where the law of the wall under-predicts the values of c_f compared with measured data. This discrepancy may be due to the larger irregularities (bottom roughness) in the local bathymetry near shore; this would increase z_0 and hence c_f. Because of this discrepancy, a semi-empirical equation given by

$$c_f = \frac{0.004}{1 - 1.9 \exp(-1.28h)} \qquad (5.34)$$

was chosen to approximate the data (Figure 5.1) and was later used in hydrodynamic modeling of the Fox River.

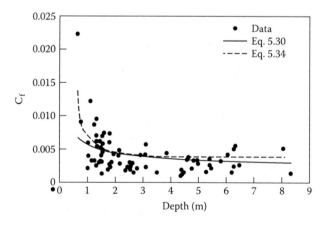

FIGURE 5.1 Bottom shear stress coefficient of friction, c_f, as a function of depth from measurements, from the law of the wall with $z_0 = 0.2$ cm, and from Equation 5.34. (*Source:* From Sea Engineering, 2004.)

In general, the total shear stress acting on the sediment bed is considered to consist of a shear stress due to skin friction and a shear stress due to form drag. The former is due to viscous forces acting tangential to the surface, whereas the latter is due to the normal stresses (mostly pressure) acting on the surface. Because form drag is caused by normal forces, it is not considered to contribute significantly to the shear stress causing sediment erosion. Equation 5.30 implicitly includes the effect of form drag when z_0 is an effective bottom roughness; this usually is determined as part of the hydrodynamic model calibration. Explicit equations for the form drag due to sand dunes in rivers have been developed (Einstein and Barbarossa, 1952; Engelund and Hansen, 1967; Wright and Parker, 2004); for sandy sediments, the bottom shear stress then can be explicitly separated into a frictional shear stress and a shear stress due to form drag and calculated in this way.

5.1.3.2 Effects of Waves and Currents

In general, bottom shear stresses in surface waters are generated by a combination of waves and currents. A highly nonlinear relationship exists between the two processes. Grant and Madsen (1979) have developed a detailed description of these processes. They assumed that the water column near the sediment bed can be separated into a wave boundary layer and a current boundary layer; the large scale of the current boundary layer makes the velocity gradient associated with it smaller than the velocity gradient for the wave boundary layer. Because of this, even if the magnitudes of the wave orbital velocity and the current velocity are the same, the shear stresses due to the waves will be greater than those due to the currents.

Soulsby et al. (1993) reviewed much of the experimental and theoretical work on wave and current interactions. The model chosen for presentation here (Christoffersen and Jonsson, 1985) was developed from the original Grant and Madsen (1979) work and was shown to reproduce experimental data well. In this model, the total bottom shear stress, τ, can be divided into the periodic wave component, τ_w, and the steady current component, τ_s:

$$\tau = \tau_w + \tau_s \qquad (5.35)$$

where the current and wave make an arbitrary angle, θ, with each other. The shear stress due to a current was represented as

$$\tau_s = \frac{1}{2} c_f \rho \, u^2 \qquad (5.36)$$

whereas the shear stress due to waves was written as

$$\tau_w = \frac{1}{2} c_w \rho \, u_{bm}^2 \qquad (5.37)$$

where c_w is the wave friction factor and u_{bm} is the bottom wave orbital velocity. By developing a model for the nonlinear interaction of waves and currents, Christoffersen and Jonsson (1985) were able to express the total bottom shear stress as

$$\tau = \frac{1}{2} c_w \rho u_{bm}^2 m \tag{5.38}$$

where

$$m = \sqrt{1 + \sigma^2 + 2\sigma |\cos(\theta)|} \tag{5.39}$$

and

$$\sigma = \frac{c_f u^2}{c_w u_{bm}^2} \tag{5.40}$$

The current friction factor is calculated from

$$\sqrt{\frac{2}{c_f}} = \frac{1}{\kappa} \ln \frac{30h}{ek_N} - \frac{1}{\kappa} \ln \frac{k_A}{k_N} \tag{5.41}$$

with the apparent roughness, k_A, defined as

$$\frac{k_A}{k_N} = 30 \frac{\delta_w}{k_N} \exp\left(-\frac{\kappa \delta_w}{\beta k_N} \sqrt{\frac{\sigma}{m}} \right) \tag{5.42}$$

and where δ_w, the wave boundary layer thickness, is calculated from

$$\frac{\delta_w}{k_N} = 0.45 \frac{\pi}{\sqrt{2}} \sqrt{\beta J} \tag{5.43}$$

The existence of the wave boundary layer causes the currents to experience a larger bed shear than they would in a condition with no waves. The roughness that is felt by the currents is therefore larger than the Nikuradse roughness, k_N. This is reflected through the apparent roughness, k_A.

The wave friction factor is given by

$$\frac{c_w}{m} = \frac{2\beta}{J} \tag{5.44}$$

where β is 0.747 and J is given by

$$J = \frac{u_{fm}}{k_N \omega_A} \tag{5.45}$$

where ω is the absolute wave frequency and u_{fm} is the wave friction velocity defined as

$$u_{fm} = \sqrt{\frac{mc_w}{2}} u_{bm} \qquad (5.46)$$

These methods have been applied successfully in modeling studies (e.g., Chroneer et al., 1996) and have been applied to numerous field studies. Drake and Cacchione (1992) provide an example of the application of these and similar techniques to calculating wave and current generated shear stresses in the field from velocity profiles.

The bottom boundary layer and shear stress are also influenced by suspended sediment stratification. A general model that includes wave-current interaction and stratification due to suspended sediment has been developed by Glenn and Grant (1987) and improved by Styles and Glenn (2000).

5.1.4 WIND STRESS

In the modeling of the wind-driven circulation in surface waters, it is necessary to know the horizontal shear stress imposed as a boundary condition at the surface. This stress, τ^w, is caused by the interaction of the turbulent air and water. The relation of this stress to the wind speed is difficult to determine from theoretical considerations, and its value is usually based on semi-empirical formulas and on observations. Similar to the shear stress at the sediment-water interface, a common relation assumed between these quantities is

$$\tau^w = \rho_a C_d W_a^{n-1} \mathbf{W}_a \qquad (5.47)$$

where C_d is a drag coefficient, ρ_a is the density of the air, W_a is the wind velocity at 10 m above the water surface, and n is an empirically determined exponent not necessarily an integer.

Wilson (1960) has analyzed data from many different sources and has given a best fit to the data. For W_a in units of centimeters per second (cm/s), ρ_a in units of grams per cubic centimeter (g/cm^3), and τ^w in units of dynes per square centimeter (dynes/cm^2), Wilson suggests a value of n = 2 and C_d = 0.00237 for strong winds and 0.00166 for light winds. From a comparison of field data and results of numerical models, Simons (1974) suggests values of n = 2 and C_d = 0.003.

An additional problem in determining the wind stress is that W_a in the above formula is the wind velocity at the specified location over the water. Unfortunately, wind velocities are generally measured on shore and not at the desired over-the-water location. For large bodies of water, over-the-water winds tend to be higher than the land values by as much as a factor of 1.5. For rivers and smaller lakes, over-the-water winds tend to be smaller than those measured at land locations far from the body of water. For these smaller bodies of water, a wind-shelter coefficient has been proposed (Cole and Buchak, 1995) that has a range between 0 and 1, depending on the shape and size of the water body.

5.1.5 SIGMA COORDINATES

In a Cartesian grid, the vertical distances between grid points depend only on z and not on x or y. In a variable-depth basin, the number of grid points between the sediment-water and air-water interfaces is therefore variable. Because of this, in a basin with large depth variations, it is difficult to obtain adequate resolution of the flow in the vertical in both the deep and shallow parts of the basin. This problem is minimized by the use of a sigma coordinate in the vertical. The dimensionless sigma coordinate is defined as

$$\sigma = \frac{h_0 + z}{h_0 + \eta} \qquad (5.48)$$

where z is the physical vertical coordinate measured in the upward direction from the undisturbed water surface, $h_0(x, y)$ is the depth of the undisturbed water, $\eta(x, y, t)$ is the displacement of the water surface above $z = 0$, and $0 < \sigma < 1$. The advantage of the sigma coordinate is that it divides the water column into the same number of layers in the vertical independent of the water depth, $h = h_0 + \eta$. A disadvantage is that the transformation can produce errors in computing the pressure gradient term; these may be significant when steep bottom slopes are present. Another disadvantage is that the small distances between grid points in the vertical that are often generated can decrease the numerical stability of the equations.

The source of the pressure gradient difficulty can be understood as follows. In σ coordinates, the x-component of the pressure gradient is written as

$$\left(\frac{\partial p}{\partial x}\right)_z = \left(\frac{\partial p}{\partial x}\right)_\sigma - \frac{\sigma}{h}\frac{\partial h}{\partial x}\left(\frac{\partial p}{\partial \sigma}\right)_x \qquad (5.49)$$

The potential error in calculating this gradient is due to the numerical truncation errors that occur from the terms on the right-hand side when these are approximated by finite differences. The first term on the right-hand side is the pressure gradient along lines of constant σ and depends on changes in bottom topography. The second term is the pressure gradient in the vertical across lines of constant σ. When steep bottom slopes are present, both of these terms may be large, comparable in magnitude, but opposite in sign. Errors then arise due to small truncation errors in each of these terms, that is, small differences between large numbers.

This type of error has been extensively investigated (Haney, 1991; Mellor et al., 1994, 1998). The general conclusion from these studies is that it can be minimized by the use of fine grids in both the horizontal and vertical directions. As an example, an investigation of this type of error for the modeling of currents using the hydrodynamic model EFDC in the Lower Duwamish Waterway (Arega and Hayter, 2007) is summarized in Section 5.4.

5.1.6 Numerical Stability

In the numerical modeling of currents, the basic differential equations are approximated by difference equations and then solved by various numerical procedures. These difference equations involve grid sizes in space (e.g., Δx) and time (Δt). For numerical efficiency, one would like to use space and time steps as large as are consistent with the accuracy and physical detail desired. However, other restrictions on the allowable space and time steps are dictated by the stability of the calculation procedure. These restrictions, of course, depend on the particular numerical scheme used but, when present, can usually be related to the physical space and time scales of the problem.

For example, consider an explicit, forward-time, central-space scheme. Simple one-dimensional theory indicates that limits on the time step Δt are given approximately by the following:

1. $\Delta t < \Delta x/(gh)^{1/2}$, a restriction indicating that the numerical time step must be less than the time it takes a surface gravity wave (speed of $(gh)^{1/2}$) to travel the horizontal distance between two grid points Δx
2. $\Delta t < \Delta x/u$, the time step must be less than the time it takes a fluid particle to be convected horizontally a distance Δx
3. $\Delta t < (\Delta z)^2/2A_v$, the time step must be less than the time for turbulent diffusion between two grid points in the vertical
4. $\Delta t < (\Delta x)^2/2A_H$, the same argument as (3) but applied to horizontal diffusion
5. $\Delta t < 2\pi/f$, the time step must be less than the inertial period due to rotation
6. $\Delta t < \Delta x/u_i$, where u_i is the speed of an internal wave, an argument similar to (1) above but for internal waves

As an example, consider parameters appropriate for a large lake: u = 50 cm/s, h = 20 m, Δx = 2 km, Δz = 1 m, A_v = 10 cm^2/s, A_h = 10^5 cm^2/s, f = 10^{-4}/s, and g = 980 cm^2/s. Corresponding to the above listing, the limiting time steps are then approximately (1) 140 s, (2) 4000 s, (3) 500 s, (4) 2 × 10^5 s, (5) 6 × 10^4 s, and (6) 4 × 10^5 s. The limiting time step depends on the smallest allowed time, 140 s in this example. However, this restriction was implicitly based on a constant-depth basin, or average depth. For a variable depth basin where the largest and smallest depths are 20 and 1 m, respectively, and where a σ-coordinate is used with 20 grid points in the vertical (as above), the smallest Δz is 0.05 m, and the limiting time step described by (3) is now only 1.25 s instead of 500 s.

All the above rules are equivalent to stating that the numerical time step must be less than the time it takes for a disturbance to propagate between two grid points. Each of these restrictions can be eliminated by various numerical procedures. However, each numerical procedure has its own difficulties, and which procedure is most advantageous depends on the particular problem being studied.

5.2 TWO-DIMENSIONAL, VERTICALLY INTEGRATED, TIME-DEPENDENT MODELS

5.2.1 BASIC EQUATIONS AND APPROXIMATIONS

Two-dimensional, vertically integrated conservation equations are obtained by integrating the three-dimensional equations of motion (Equations 5.8 through 5.14) over the water depth from $z = -h_o(x,y)$ to $z = \eta(x,y,t)$, where h_o is the equilibrium water depth and η is the surface displacement from equilibrium (e.g., see Ziegler and Lick, 1986). From Equations 5.8 through 5.11, the resulting equations for mass and momentum conservation are

$$\frac{\partial \eta}{\partial t} + \frac{\partial U}{\partial x} + \frac{\partial V}{\partial y} = 0 \tag{5.50}$$

$$\frac{\partial U}{\partial t} + gh\frac{\partial \eta}{\partial x} - fV = \tau_x^w - \tau_x + A_H\left(\frac{\partial^2 U}{\partial x^2} + \frac{\partial^2 U}{\partial y^2}\right) - \frac{\partial(U^2/h)}{\partial x} - \frac{\partial(UV/h)}{\partial y} \tag{5.51}$$

$$\frac{\partial V}{\partial t} + gh\frac{\partial \eta}{\partial y} + fU = \tau_y^w - \tau_y + A_H\left(\frac{\partial^2 V}{\partial x^2} + \frac{\partial^2 V}{\partial y^2}\right) - \frac{\partial(UV/h)}{\partial x} - \frac{\partial(V^2/h)}{\partial y} \tag{5.52}$$

where g is the acceleration due to gravity, $h = h_o + \eta$ is the total water depth, and U and V are vertically integrated velocities defined by

$$U = \int_{-h_o}^{\eta} u \, dz \tag{5.53}$$

$$V = \int_{-h_o}^{\eta} v \, dz \tag{5.54}$$

where u and v are the velocities in the x- and y-directions, respectively, and z is the vertical coordinate. Equations 5.50 through 5.52 are also often written in terms of the vertically averaged velocities, defined as

$$U_{avg} = \frac{1}{h}\int_{-h_o}^{\eta} u \, dz = \frac{U}{h} \tag{5.55}$$

$$V_{avg} = \frac{1}{h}\int_{-h_o}^{\eta} v \, dz = \frac{V}{h} \tag{5.56}$$

When this is done, the resulting equations are referred to as vertically averaged equations. Whether written in terms of vertically integrated or vertically averaged velocities, both sets of conservation equations lead to essentially the same results. The bottom stress is represented by τ, whose components are given by

$$\tau_x = c_f q \frac{U}{h} \qquad (5.57)$$

$$\tau_y = c_f q \frac{V}{h} \qquad (5.58)$$

where q is the velocity magnitude and c_f is a bottom shear stress coefficient of friction, as in Equation 5.30.

Numerous vertically integrated (or vertically averaged), two-dimensional, time-dependent hydrodynamic models have been developed. Some are publicly available; some are not. Examples of publicly available models include TABS-2 (USACE/WES; Thomas and McAnally, 1985); HSCTM-2D (EPA CEAM; Hayter et al., 1999); FESWMS-2D (USGS; Froelich, 1989); and SEDZL and SEDZLJ (Ziegler and Lick, 1986; Jones and Lick, 2001a) as well as two-dimensional versions of primarily three-dimensional models (see Section 5.4). SEDZL and SEDZLJ were primarily developed as sediment transport models, and that aspect of the models is discussed in some detail in Chapter 6; both these models are essentially the same as far as hydrodynamics are concerned. Results of hydrodynamic calculations with SEDZL and SEDZLJ are presented here to illustrate the two-dimensional modeling of currents and for further use in Chapter 6 in the modeling of sediment transport. To illustrate the similarities and differences in the currents in rivers and lakes, calculated results are shown for the Lower Fox River and Lake Erie.

5.2.2 THE LOWER FOX RIVER

Because of the extensive hydrodynamic and sediment data available for the Lower Fox, this river was chosen as an application of SEDZLJ (Jones and Lick, 2000, 2001a). The specific area of the river considered was from the DePere Dam to Green Bay (Figure 1.2), a distance of approximately 11 km. A bathymetric map of this portion of the river is shown in Figure 5.2. The upstream region of the river is fairly wide (more than 0.5 km in some regions), with shallow pools in the nearshore regions as little as 1 m in depth. Previous dredging in the center of the channel created a channel 7 to 8 m deep; this has partially filled up with time but, in the upstream portion, was still up to 5 m deep in 2002.

The flow into this section of the Lower Fox is controlled primarily by the DePere Dam. In the calculations, it was assumed that approximately 10% of any given flow at this dam concurrently flows in through the East River; however, during some large events, this flow can be considerably higher. Flow contributions due to runoff and other smaller tributaries were considered negligible. Flow rates

FIGURE 5.2   Bathymetry of Lower Fox River from DePere Dam to Green Bay. Depth in meters. For clarity, distances in vertical direction are scaled by a factor of 3 compared to the horizontal.

during a typical year vary from 30 to 280 m^3/s. The highest flow rate in the past 80 years was about 650 m^3/s. High flow events, in general, are caused by opening the dams or by large storms.

In the numerical calculations, a 30- by 90-m rectangular grid was generated to discretize the river. The water depth at each node was determined from NOAA navigational charts. A no-slip boundary condition was imposed along the shoreline to approximate the very low flow areas in the shallow regions near the shores of the river.

Constant flow rate calculations were first made to determine the general characteristics of the hydrodynamics of the river under various flow conditions. Calculations also were made for variable flow (flood) conditions. Results for a constant flow rate over the DePere Dam of 280 m^3/s, a 99.7 percentile (or once-in-one-year) flow, are shown here.

The bottom shear stress coefficient of friction was calculated from Equation 5.30 with a z_0 of 100 μm and is shown in Figure 5.3. The minimum value is 0.002, whereas the maximum value is 0.0032 and is located in the shallow regions. Figure 5.4 shows the velocity vectors and shear stress contours. As the flow enters upstream at the DePere Dam, it makes a 90° turn into a wide, shallow region. The region just below the dam has velocities of 35 cm/s and shear stresses up to 0.5 N/m^2 that rapidly drop off as the flow turns and slows. The currents and shear stresses are higher in the channel than in the nearshore. Midway downstream, the river makes a 45° turn and narrows. Due to the narrow cross-section, the flow accelerates to 50 cm/s in some regions and produces bottom shear stresses up to 0.5 N/m^2. At the East River, more water is added to the flow and the center channel narrows; this results in a maximum velocity of 65 cm/s and a corresponding shear stress of 0.8 N/m^2. The flow slows as the river widens into Green Bay. These and similar calculations are used as the basis for the modeling of sediment transport in the Fox River, as discussed in Section 6.4.

At a later time, between May 2003 and July 2004, the USGS conducted four field surveys that included measurements of vertical velocity profiles at 30 locations in the Fox River from DePere Dam to Green Bay. This period included four flow periods: low flow (100 m^3/s), which was dominated by seiche motion from

FIGURE 5.3   Coefficient of friction, c_f, with $z_0 = 100$ μm. (*Source*: From Jones and Lick, 2000. With permission.)

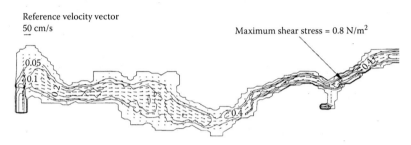

FIGURE 5.4   Velocity vectors and shear stress contours for a constant flow rate of 280 m^3/s. (*Source*: From Jones and Lick, 2000. With permission.)

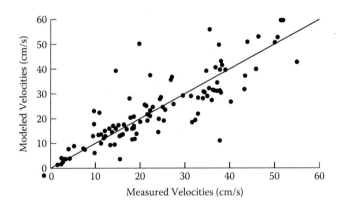

FIGURE 5.5   Measured and calculated vertically averaged velocities for four events between May 2003 and July 2004. (*Source*: From Sea Engineering, 2004.)

Green Bay; two moderate flows (200 to 300 m^3/s); and high flow (400 to 450 m^3/s). Calculated shear stresses ranged from 0.01 to 1.38 N/m^2.

Essentially the same model as described above (but with more accurate descriptions of the bottom shear stress and water depths) was then applied to these four flow periods (Sea Engineering, 2004). Good agreement (correlation coefficients of 0.80 or greater) between the measured and calculated vertically averaged velocities was obtained (Figure 5.5). In this figure, the furthest outliers generally occurred during periods of rapid flow variations or upstream flow at the mouth.

5.2.3 WIND-DRIVEN CURRENTS IN LAKE ERIE

Lake Erie (Figure 2.2) is relatively shallow compared to the other Great Lakes. Topographically, it is convenient to divide it into a Western Basin (average depth of 7 m, or 24 ft); a Central Basin (average depth of 19 m, or 60 ft); and an Eastern Basin (average depth of 26 m, or 80 ft). Because of the differences in the depths of the basins, the hydrodynamics, wave action, and sediment and contaminant transport are distinctly different for each basin. For this lake, a series of numerical calculations of the wind-driven currents, wave action, and the resulting sediment transport has been made (Lick et al., 1994a). The purpose was to illustrate the basic characteristics of these processes. Results for the wind-driven currents are shown here, for wave action in Section 5.5, and for sediment transport in Section 6.5.

Calculations of wind-driven currents were made for a variety of wind directions and magnitudes, and also for a major storm. For the calculation presented here, the period of the event was 13 days. It was assumed that the wind direction was constant from the southwest (the dominant wind direction) throughout the period. However, the wind speed was assumed to be variable and was constant at 5 mph for 2 days, constant at 22.5 mph (moderately high winds) for 1 day, and then constant at 5 mph for 10 days. Results of calculations at the end of 3 days (end of high wind period) are shown in Figure 5.6 (currents) and Figure 5.7 (bottom shear stresses).

In Figure 5.6, it can be seen that the nearshore flows are in the direction of the wind, whereas the offshore flows are opposed to the wind. This is typical of wind-driven currents in lakes. Vertically averaged velocities are approximately 5 to 10 cm/s. In Figure 5.7, constant bottom shear stresses (due to waves and currents) of 0.1, 1.1, 2.1, and 3.1 N/m² are plotted. A shear stress of 0.1 N/m² is approximately the stress at which sediments just begin to be resuspended. The stresses in most of the Central and Eastern basins but only a small part of the Western Basin

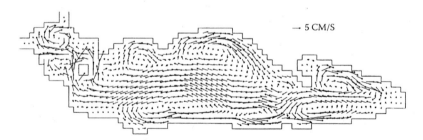

FIGURE 5.6   Lake Erie. Vertically averaged, wind-driven currents at end of storm. Winds are from the southwest at 22.5 mph for a period of 1 day. (*Source*: From Lick et al., 1994a. With permission.)

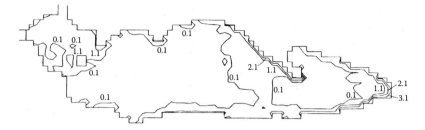

FIGURE 5.7 Lake Erie. Bottom shear stress (N/m^2) at end of storm. Winds are from the southwest at 22.5 mph for a period of 1 day. (*Source*: From Lick et al., 1994a. With permission.)

are below 0.1 N/m^2; little or no resuspension therefore occurs in these regions for this moderate wind. The highest shear stresses and therefore highest resuspensions occur in narrow zones along the northeastern shore of the Central Basin and the eastern shore of the Eastern Basin where the wave action is strongest. For all wind speeds, the shear stress is primarily due to wave action, whereas the effects of currents on the shear stress are generally only a small correction.

The fact that the highest resuspension generally occurs in the nearshore areas of Lake Erie as well as other lakes is due to the high wave action and resulting high shear stresses in these regions. This is in contrast to the Fox and other rivers where the highest shear stresses and highest resuspension generally occur in the central, deeper parts of the river; this difference is due to the higher currents that are present there and that are the main cause of the bottom shear stress in rivers.

5.3 TWO-DIMENSIONAL, HORIZONTALLY INTEGRATED, TIME-DEPENDENT MODELS

Two-dimensional, horizontally integrated (or horizontally averaged), time-dependent models often are used to predict longitudinal and vertical variations in velocities, temperatures, and/or salinities in stratified reservoirs and estuaries, that is, in systems where vertical variations are significant but where horizontal velocities transverse to the longitudinal direction can be neglected compared with longitudinal and vertical velocities. The conservation equations for this type of model are obtained by integrating the three-dimensional equations in the transverse direction. Numerous models of this type exist, for example, the Laterally Averaged Reservoir Model (Buchak and Edinger, 1984a, b); the CE-QUAL-W2 model (Cole and Buchak, 1995); and a model by Blumberg (1975, 1977). The latter two models have been used most extensively.

In small lakes and reservoirs, the Coriolis force and the associated transverse velocities can normally be neglected. However, in Lake Erie, because of its size, this is no longer correct. In this case, a modified version of a two-dimensional, time-dependent model that included the Coriolis force was used to investigate the temperatures and currents in Lake Erie (Heinrich et al., 1981). In this investigation, the emphasis was on the general characteristics of thermocline formation, maintenance, and decay, and therefore the time scales of interest were weeks and months rather than hours or even days. The purpose was to approximately reproduce and understand the main features of the observed spatial and temporal distributions of temperatures and currents in Lake Erie as an example of a large stratified lake. In particular, calculations were made to predict conditions in a vertical cross-section of the lake from Port Stanley in the north to Ashtabula in the south (a north–south line just west of the Ohio–Pennsylvania border, Figures 2.2 and 5.8). Some of the results of this investigation are described here.

5.3.1 Basic Equations and Approximations

In the analysis, it was assumed that properties of the flow varied only in the vertical direction, z, and one horizontal direction in the plane of the cross-section, x, but not in the other horizontal direction perpendicular to the cross-section, y. Because of this, derivatives of u, v, w, and T with respect to y could be neglected. Coriolis forces were included. These forces induce velocities in the y-direction that are approximately independent of y but are dependent on x and z. A net flux in the y-direction therefore results if a zero pressure gradient in the y-direction is assumed. Over sufficiently long time scales such that averaged seiche effects are negligible (and these are the time scales that are of interest here), this net flux in the y-direction must be zero. In the analysis, a pressure gradient in the y-direction independent of x and z was therefore assumed such that this net flux in the y-direction was zero. Because of this assumption, velocities in the transverse direction are present and are functions of x, z, and t. The model is a simplification to two dimensions of a three-dimensional, time-dependent model previously developed and used (Paul and Lick, 1974).

The basic equations used in the analysis are the usual hydrodynamic equations for conservation of mass, momentum, and energy, plus an equation of state. With the above assumptions, these are as follows:

$$\left(\frac{\partial u}{\partial x} + \frac{\partial w}{\partial z} \right) = 0 \tag{5.59}$$

$$\frac{\partial u}{\partial t} + \frac{\partial u^2}{\partial x} + \frac{\partial uw}{\partial z} - fv = -\frac{1}{\rho_r}\frac{\partial p}{\partial x} + \frac{\partial}{\partial x}\left(A_H \frac{\partial u}{\partial x} \right) + \frac{\partial}{\partial z}\left(A_V \frac{\partial u}{\partial z} \right) \tag{5.60}$$

$$\frac{\partial v}{\partial t} + \frac{\partial uv}{\partial x} + \frac{\partial vw}{\partial z} + fu = -\frac{1}{\rho_r}\frac{\partial p}{\partial y} + \frac{\partial}{\partial x}\left(A_H \frac{\partial v}{\partial x} \right) + \frac{\partial}{\partial z}\left(A_V \frac{\partial v}{\partial z} \right) \tag{5.61}$$

$$\frac{\partial p}{\partial z} = -\rho g \tag{5.62}$$

$$\frac{\partial T}{\partial t} + \frac{\partial uT}{\partial z} + \frac{\partial wT}{\partial z} = \frac{\partial}{\partial x}\left((K_H \frac{\partial T}{\partial x} \right) + \frac{\partial}{\partial z}\left(K_v \frac{\partial T}{\partial z} \right) \tag{5.63}$$

$$\rho = \rho(T) \tag{5.64}$$

$$\iint\limits_R v \, dx \, dz = 0 \tag{5.65}$$

where R denotes the area of the two-dimensional cross-section in the x-z plane.

In the analysis, the horizontal eddy viscosity and eddy diffusion coefficients were assumed constant. From previous one-dimensional analyses and in analogy to Equation 5.21, the vertical eddy diffusivity was approximated as

$$K_v = \frac{K_S(c_1 + c_2 e^{+z/D})}{\left(1 + \sigma_2 \dfrac{g\alpha D^2 e^{-2z/D}}{\tau^w} \dfrac{\partial T}{\partial z}\right)^{3/2}} \tag{5.66}$$

$$K_S = K_S(\tau^w, h) \tag{5.67}$$

where the terms dependent on z were meant to approximate the generation of shear by surface waves and near-surface currents. Similarly, the vertical eddy viscosity was approximated as

$$A_v = \frac{A_S(c_1 + c_2 e^{+z/D})}{\left(1 + \sigma_1 \dfrac{g\alpha D^2 e^{-2z/D}}{\tau^w} \dfrac{\partial T}{\partial z}\right)^{1/2}} \tag{5.68}$$

where α is the thermal expansion coefficient for water; c_1, c_2, and D are empirical constants and D is the same order of magnitude as the wave length of the surface waves; A_S is assumed equal to K_S; both are functions of the wind stress, τ^w, and water depth, h; and σ_1 and σ_2 are constants. When the temperatures were calculated, it sometimes was found that the colder, denser water was above warmer, less dense water. In this statically unstable situation, the unstable region was uniformly mixed.

At the water-air interface, z = 0, a heat flux and a stress due to the wind was specified

$$-\rho K_v \frac{\partial T}{\partial z} = q \tag{5.69}$$

$$\rho A_v \frac{\partial u}{\partial z} = \tau_x^w, \quad \rho A_v \frac{\partial v}{\partial z} = \tau_y^w \tag{5.70}$$

At the sediment-water interface, the conditions were those of no fluid motion and either zero heat flux or continuity of heat flux and temperature across the sediment-water interface.

5.3.2 TIME-DEPENDENT THERMAL STRATIFICATION IN LAKE ERIE

Calculations were first made in an effort to understand the general effects of various boundary conditions and parameters. A more realistic calculation then was made to model general and essential characteristics of stratified flow during the time of formation, maintenance, and decay of a thermocline. This latter calculation extended over a period of 180 days, from early spring to fall. The following was assumed: (1) the heat flux was constant at 0.003 cal/cm²/s for 50 days and then decreased linearly so that it became zero at 120 days and negative thereafter, an approximation to real conditions; and (2) the wind stress was constant at 0.075 N/m² and directed to the right (south) from 0 to 50 days, to the left (north) from 50 to 70 days, to the right again from 70 to 130 days, to the left from 130 to 150 days, and then to the right from 150 to 180 days, an approximation to show and include effects due to varying wind direction. Other parameters were as follows; $A_S = K_S = 15$ cm²/s, D = 900 cm, $A_H = 10^6$ cm²/s, $K_H = 3 \times 10^5$ cm²/s, $\sigma_1 = \sigma_2 = 0.001875$, $c_1 = 0.5$, $c_2 = 1.5$, and no heat transfer to the bottom. The initial temperature was 4°C throughout.

Some results of the computations are shown in Figures 5.8 to 5.11 for times of 50, 60, 120, and 150 days. The general features of the temperature and current distributions are similar to what is observed in Lake Erie during the spring, summer, and fall periods. Figure 5.8 shows the temperatures at 50 days. A thermocline has formed at a depth of 16 to 18 m and is deeper in the downwind direction. The hypolimnion temperatures are 10 to 11°C. The epilimnion temperature is 13°C in the middle of the lake and increases by 1 to 2°C in the nearshore areas. In a narrow (Ekman) layer near the surface, there is a flow in the direction of the wind. Below this, there is a broad return flow in the opposite direction.

The results for the temperature at 60 days (10 days after wind reversal) are shown in Figure 5.9. The thermocline is quite flat. The epilimnion temperatures

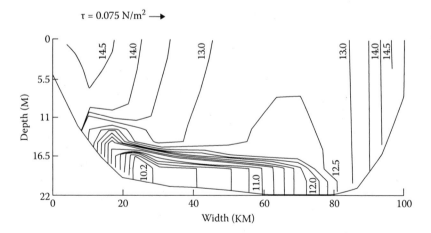

FIGURE 5.8 Temperature distribution after 50 days for a vertical cross-section of Lake Erie. Wind is from left to right. (*Source*: From Heinrich et al., 1981. With permission.)

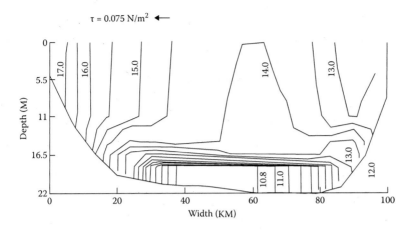

FIGURE 5.9   Temperature distribution at 60 days. Heat flux is decaying linearly and wind direction is from right to left from 50 to 60 days. (*Source*: From Heinrich et al., 1981. With permission.)

have increased by approximately 1.5°C, whereas the hypolimnion temperatures have increased by less than 0.5°C. Wind reversal causes a time-dependent shift in the orientation of the thermocline so that the depth of the thermocline tends to increase in the downwind direction; that is, the hypolimnion waters are shifted in the upwind direction. This motion of the hypolimnion is slow, and 10 to 20 days are needed before a new quasi-steady state is reached. Although the location of the hypolimnion is significantly affected by wind reversal, the temperatures in the hypolimnion change little.

The wind was again reversed at 70 days and remained constant until 120 days. At this time, the temperatures and velocities are shown in Figure 5.10. The epilimnion temperatures (Figure 5.10(a)) have increased to 19°C, whereas the hypolimnion temperatures have increased to only 12 to 14°C. Currents in the x-z plane (Figure 5.10(b)) are now more restricted to the epilimnion, whereas the coastal currents in the y-direction (Figure 5.10(c)) have increased in strength.

In the central basin, the thermocline is generally located quite close to the bottom. The hypolimnion volume is therefore relatively small compared to the epilimnion volume. Because of this, the temperatures in the hypolimnion are particularly sensitive to heat fluxes from the epilimnion to the hypolimnion. When stratification is strongest, the vertical flux of heat by diffusion is reduced, and the convective flux of heat from the epilimnion to the hypolimnion in the nearshore regions (where stratification is less) becomes significant.

After 120 days, the heat flux is negative and cooling takes place. Cooling occurs more rapidly near shore, eliminating the warm nearshore areas. At 150 days (Figure 5.11), a thermocline exists but is quite weak and getting weaker. By 180 days, nearshore areas are somewhat cooler than offshore areas; however, practically no vertical stratification exists and the flow is similar to that in a constant-temperature basin.

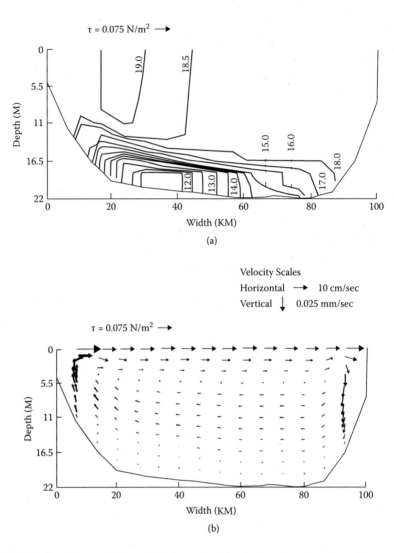

FIGURE 5.10 Flow variables at 120 days. Heat flux is decaying linearly and is zero at this time. Winds are from left to right. (a) Temperature distribution, (b) currents in the x-z plane. (*Source*: From Heinrich et al., 1981. With permission.)

This investigation was later extended by means of a three-dimensional, time-dependent model by Heinrich et al. (1983). Close agreement between the results of the two models was generally observed. The agreement was best for short times, on the order of or less than 60 days. For longer times, the stratification predicted by the three-dimensional model was not as strong as that predicted by the two-dimensional model. The reason was that convection of heat near shore from the epilimnion to the hypolimnion was greater for the three-dimensional model; this tended to be most significant when stratification was strongest (i.e., during the summer).

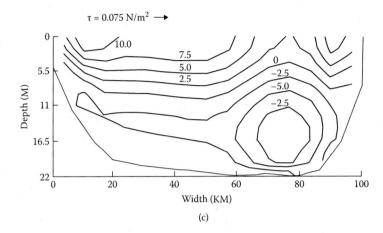

τ = 0.075 N/m² →

FIGURE 5.10 (CONTINUED)   (c) Flow variables at 120 days. Heat flux is decaying linearly and is zero at this time. Winds are from left to right. v velocities. (*Source*: From Heinrich et al., 1981. With permission.)

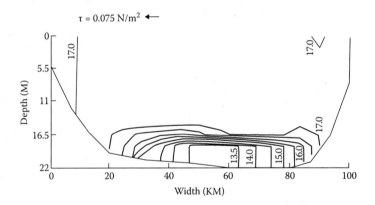

τ = 0.075 N/m² ←

FIGURE 5.11   Temperature distribution at 150 days. Wind direction is from right to left from 130 to 150 days. (*Source*: From Heinrich et al., 1981. With permission.)

5.4 THREE-DIMENSIONAL, TIME-DEPENDENT MODELS

Three-dimensional, time-dependent models are becoming widely used (e.g., see the proceedings on Estuarine and Coastal Modeling edited by Spaulding (2006)). In almost all these models, the basic equations are those described in Section 5.1. Differences between the models are because of different numerical differencing procedures, the treatment of mode splitting, and different approximations to turbulence. Some of the most widely used three-dimensional models are ECOM (Estuarine and Coastal Model; Blumberg and Mellor, 1987); CH3D (Johnson et al., 1993); and EFDC (Environmental Fluid Dynamics Code; Hamrick, 1992a, b).

ECOM is a modification and extension of the Princeton Ocean Model (POM). It incorporates the Mellor and Yamada turbulent closure scheme, uses Cartesian or curvilinear coordinates in the horizontal, and uses a σ-coordinate transformation in the vertical. To reduce computational time, the model uses a mode-splitting technique such that the faster-moving, external barotropic modes are calculated with a relatively small time step, whereas the slower-moving, internal baroclinic modes are calculated with a relatively large time step. Applications of the model include Chesapeake Bay (Blumberg and Goodrich, 1990); the coastal ocean (Blumberg et al., 1993); and New York Harbor (Blumberg et al., 1999).

EFDC is similar to ECOM in its major assumptions, with some differences in its numerical differencing procedures. It has been applied to a diversity of problems in surface water hydrodynamics (Hamrick, 1992a); sediment transport (Tetra Tech, 2000); and contaminant transport (Tetra Tech, 1999). Specific applications are described, for example, by Hamrick (1992b), Hamrick (1994), Hamrick and Wu (1997), Wu et al. (1997), and Arega and Hayter (2007). This latter application is concerned with the modeling of currents in the Lower Duwamish Waterway (LDW) and is briefly described below.

In all the problems mentioned so far, the vertical velocities were generally small compared with the horizontal velocities. When this is true, the hydrostatic approximation is valid and the equations and solution procedure are considerably simplified. However, there are problems of interest where vertical velocities are comparable with horizontal velocities, and the full three-dimensional equations of motion must therefore be used. Examples of this are the flows and scours around partially submerged hydraulic structures such as bridge piers, dikes, abutments, and pipelines. A problem of this type is discussed after the LDW example.

5.4.1 LOWER DUWAMISH WATERWAY

As an interesting and challenging application of EFDC, consider the hydrodynamic modeling of the Lower Duwamish Waterway (Arega and Hayter, 2007). The LDW is an estuary that flows north from the Green River through the industrial area of Seattle, Washington, and into Elliot Bay (Figure 5.12(a)). Elliot Bay is located on the eastern shore of Puget Sound. The LDW is a heavily used shipping port, is contaminated due to shipping and industrial operations, but is also a significant habitat for salmon and other wildlife. It is a Superfund site. Most of the LDW is dredged for shipping, with dredging extending upstream from Elliot Bay to the Turning Basin. The water depths in the dredged sections vary from 15 m at the mouth of the LDW to 4 m at 9.5 km upstream. However, depths in the modeled area range from about 194 m to 1.5 m (Figure 5.12(b)). Average and spring tidal ranges are 2.3 and 3.4 m, respectively. The LDW is often highly stratified, with a well-defined salt wedge in the navigation channel when fresh water flows in the Green River are greater than 28 m^3/s; for lower flows, the vertical stratification is more gradual or negligible.

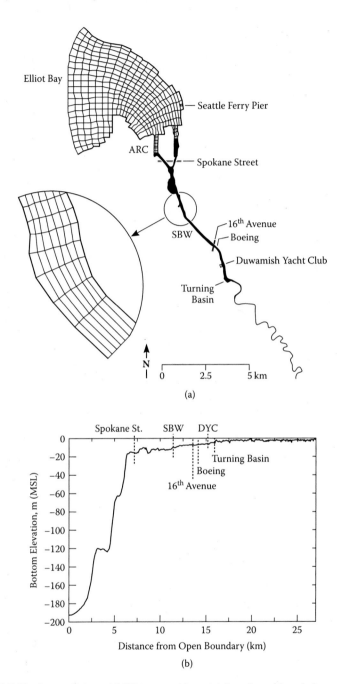

FIGURE 5.12 Lower Duwamish Waterway: (a) model domain, grid, and observation stations; and (b) depth variation along thalweg through the west inlet. (*Source:* From Arega and Hayter, 2007.)

A major purpose of the study was to test the capability of the EFDC model in predicting currents and salinity in a highly stratified environment within a bathymetry that had a large range of water depths and steep bottom slopes. A second purpose was to investigate and model the currents and salinity as a basis for a water quality model.

5.4.1.1 Numerical Error due to Use of Sigma Coordinates

Many, probably most, three-dimensional hydrodynamic models for surface waters use sigma coordinates in the vertical. However, as stated previously, a well-known difficulty with sigma coordinates is the truncation error in the numerical representation of the pressure gradient; this is especially noticeable when steep bottom slopes are present (Haney, 1991; Mellor et al., 1994, 1998). This truncation error is minimized by the use of fine grids in both the vertical and horizontal directions.

To investigate this problem, calculations were made for the LDW with no external forcing (tides, freshwater inflow, or wind). Under these conditions, no currents should be present except for those due to pressure gradient truncation errors caused by the use of sigma coordinates.

As an example of the calculated results, induced velocities near the surface and near the bottom at the location in Elliot Bay where the steepest bottom slope (and largest induced currents) occurs are shown in Figure 5.13 for numerical simulations with 5, 10, and 20 vertical layers. For 5 and 10 layers, the induced velocities

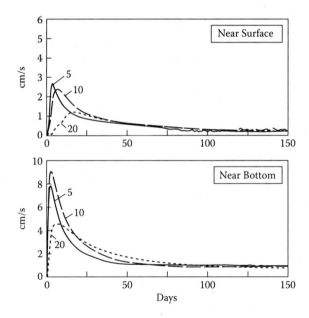

FIGURE 5.13   Currents near surface and near bottom in bay due to pressure gradient errors as a function of time. Calculations with 5, 10, and 20 vertical layers are shown. (*Source*: From Arega and Hayter, 2007.)

reach their peak in approximately 5 days and then decay slowly with time; for 20 layers, the induced velocities are about half those for 5 and 10 layers and reach their peak in about 20 days. The induced currents were initially generated and were largest in the area where the bathymetric slope was greatest; this disturbance then propagated into Elliot Bay, where it dissipated. The maximum induced velocities were about 8 cm/s for 5 and 10 layers but less than 5 cm/s for 20 layers. These velocities should be compared with maximum measured tidal velocities of about 80 cm/s in the LDW. After the initial increase with time, the induced velocities then decrease and become relatively small. The error due to the use of the sigma coordinate is therefore tolerable, even for the LDW, where quite steep gradients were present.

5.4.1.2 Model of Currents and Salinities

In the general modeling of currents and salinities, the 20-layer simulation was used because the induced errors due to the sigma coordinates were less than for 5 or 10 layers, but also because a fine grid was necessary to resolve the highly stratified flow; 1995 grid cells in the horizontal were used (Figure 5.12(a)). The simulations were made for Julian days 238 to 283 in 1996. Outer boundary conditions in Elliot Bay for water surface elevation were obtained by an inverse procedure; that is, six harmonic components were chosen to represent tidal forcing, and their amplitudes were chosen such that the differences between observed and predicted amplitudes at two interior locations (the Seattle Ferry Pier and the Spokane Avenue Bridge) were minimized. A constant salinity of 31 ppt (parts per thousand) was assumed at the outer boundary in Elliot Bay, whereas zero salinity was assumed for the river inflow.

For validation of the modeling, current measurements were made at the SBW and Boeing stations for near-surface, middle, and near-bottom depths. Comparisons of calculated and measured currents near surface and near bottom at the Boeing station are shown in Figure 5.14. Despite difficulties and uncertainties in averaging measurements and calculations in the vertical, good agreement between calculated and measured currents was obtained at this station as well as at other stations. This also was verified by statistical analyses of all calculated and measured currents.

Data for salinity were collected at three levels (1 m below the surface, 5 m above the bottom, and 1 m above the bottom) at the Spokane and 16th Avenue stations. Figure 5.15 shows comparisons of the measured and calculated salinities near surface and near bottom at the 16th Avenue station. Some over-prediction during low tide can be observed, but there is generally good agreement between measured and calculated results.

Calculated results for the salinity in a vertical section along the thalweg in Elliot Bay are shown in Figure 5.16. The time is approximately an hour before flood tide. Fresh water is flowing from the right into the bay near the surface and gradually mixing with the salt water flowing in near the bottom from Puget

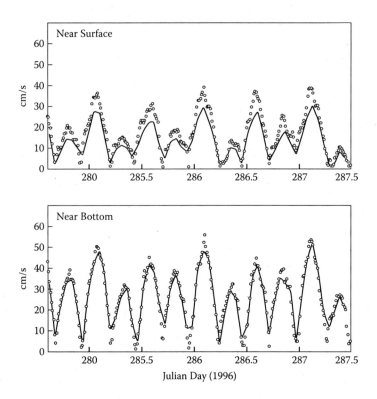

FIGURE 5.14   Comparison of computed and measured currents near surface and near bottom at the Boeing station. Practical salinity units = psu = ppt. (*Source*: From Arega and Hayter, 2007.)

Sound. An investigation of the effects of various parameters on flow, salinity, and sediment transport in a stratified estuary is presented in Section 6.6.

5.4.2 FLOW AROUND PARTIALLY SUBMERGED CYLINDRICAL BRIDGE PIERS

Because of its importance in transportation, the flow and scour around partially submerged cylindrical bridge piers in flowing streams has been extensively investigated (see reviews by Breusers and Raudkivi, 1991; Whitehouse, 1998; and Melville, 1999). Some features of this flow are illustrated here. It will be seen that vertical velocities are comparable to horizontal velocities so that the full three-dimensional equations (Equations 5.1 through 5.5) must be used.

A schematic of the flow around a cylindrical bridge pier is shown in Figure 5.17. The flow generally has a time-dependent surface at the water-air interface and a time-dependent surface at the sediment-water interface due to sediment scour. The flow approaching the pier forms a bow wave at the free surface on the upstream side; further along the sides and back of the pier, the free surface is drawn down as the flow accelerates around the pier. The flow that approaches the middle of the pier is forced to decelerate as it approaches the cylinder. The

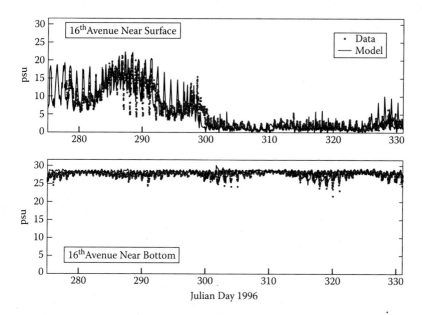

FIGURE 5.15   Comparison of computed and measured salinities near surface and near bottom at the 16th Avenue Bridge station. Practical salinity units = psu = ppt. (*Source*: From Arega and Hayter, 2007.)

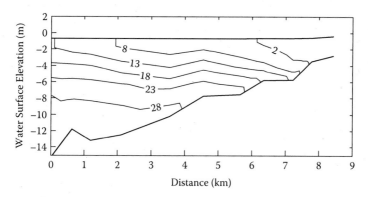

FIGURE 5.16   Calculated salinities in vertical section along thalweg of Elliot Bay. (*Source:* From Arega and Hayter, 2007.)

associated dynamic component of the pressure is highest in the upper part of the column and decreases downward toward the bed. This downward pressure gradient causes a downward flow along the upstream face of the cylinder. The interaction between this downward flow and the horizontal boundary layer near the sediment-water interface results in the formation of a horseshoe vortex at the base of the pier. This horseshoe vortex, which has a horizontal axis of rotation, erodes and transports particles from the sediment bed. This causes movement of

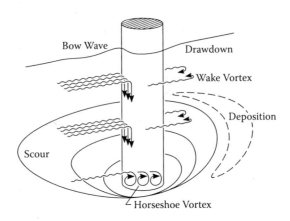

FIGURE 5.17   Flow around cylindrical pier with scour. (*Source:* Modified from Richardson and Panchang, 1998.)

the sediment-water interface until a steady state is reached. The horseshoe vortex extends downstream past the sides of the pier for a short distance before it loses its identity and becomes a part of the general flow turbulence. Along the middle of the pier, the flow separates at the sides of the pier and forms wake vortices (vortices with vertical axes of rotation) that are shed periodically downstream at lower stream velocities but are shed aperiodically at higher stream velocities.

The general features of this flow are well known due to experimental and theoretical studies in the fields of aerodynamics and hydrodynamics (Daraghi, 1989; Graf and Yulistiyanto, 1998; Deng and Piquet, 1992; Richardson and Panchang, 1998; Ali and Karim, 2002). As a preliminary to calculations of the scour around a cylindrical pier, Sudarsan et al. (2003) have made calculations of the flow, including a sediment-water interface that is moving due to sediment scour; an example from this study is given here.

In the hydrodynamic calculations, the computational model CFD 2000 by Adaptive Research was used (CFD, 2000). The model is fully three-dimensional and time dependent and uses a k-ε model to approximate turbulence. In the example given here, the diameter of the cylinder was assumed to be 80 cm, the depth of the water was 140 cm, and the free stream flow velocity was 31 cm/s. This velocity corresponds to a bed shear stress somewhat less than 0.3 N/m^2, the assumed critical shear stress for the sediments; that is, there was no erosion of sediments far upstream of the pier. The radius of the outer extent of the computational domain was set at 800 cm; from numerical experiments, this was sufficient to minimize boundary effects on the flow and scour.

For no scour, Figure 5.18 shows the shear stress at the planar sediment-water interface at steady state. Shear stresses are less than 0.3 N/m^2 upstream, decrease gradually as the flow slows along the centerline until the pier is reached, and then increase rapidly as the flow accelerates around the pier. Maximum shear stresses are about 1.1 N/m^2. Shear stresses above 0.3 N/m^2 indicate areas where erosion will occur initially. The rate of erosion depends on the shear stress, sediment

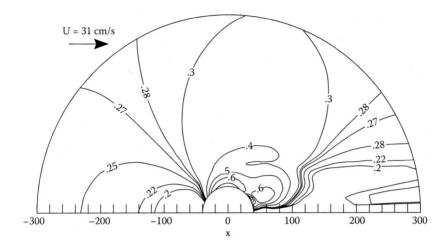

FIGURE 5.18   Bed shear stress (N/m²) with no scour.

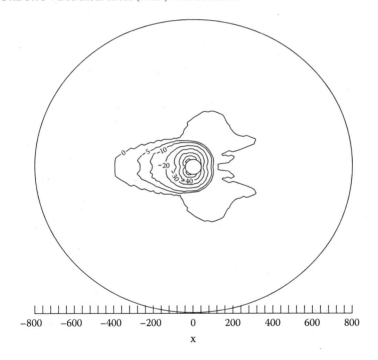

FIGURE 5.19   Depth contours of sediment-water surface. Angle of repose is 36°.

properties, local bed slope, and angle of repose (Section 3.6). Erosion will continue until an equilibrium shape of the scour hole is reached; at this time, shear
stresses throughout the domain will be less than 0.3 N/m². A preliminary result
(not including deposition) of the depth of scour at steady state is shown in Figure 5.19. For this calculation, the angle of repose was assumed to be 36°, in the

range of what is normally observed for sands. The maximum depth of scour is more than 60 cm, comparable to what is observed. The depth of scour is asymmetric, with the scour hole extending upstream further than downstream or in the transverse direction. This asymmetry depends on the angle of repose, with the asymmetry increasing as the angle of repose decreases. Deposition will occur behind the scour hole in the downstream direction.

5.5 WAVE ACTION

5.5.1 WAVE GENERATION

A simple procedure for predicting wave action is the SMB (Sverdrup, Munk, and Bretschneider) method; this procedure and an example application to Lake Erie are summarized here. The procedure includes wave generation due to wind stress but does not consider wave diffraction or refraction, processes that can be significant in shallow waters with complex topography and/or bathymetry. For these more complex cases, numerical models such as STWAVE (Smith et al., 1999) can be used.

For the SMB model, Bretschneider (1958) obtained graphical relations for the nondimensional parameters of mean depth, wind fetch, wave height, and period. Ijima and Tang (1966) have transformed these graphical relations to the following more convenient numerical forms (U.S. Army CERC, 1973):

$$\frac{gH_s}{U^2} = 0.283 \tanh\left[0.53\left(\frac{gD}{U^2}\right)^{0.75}\right] \tanh\left[\frac{0.0125\left(\frac{gF}{U^2}\right)^{0.42}}{\tanh\left[0.53\left(\frac{gD}{U^2}\right)^{0.75}\right]}\right] \quad (5.71)$$

$$\frac{gT_s}{2\pi U} = 1.2 \tanh\left[0.833\left(\frac{gD}{U^2}\right)^{0.375}\right] \tanh\left[\frac{0.077\left(\frac{gF}{U^2}\right)^{0.25}}{\tanh\left[0.833\left(\frac{gD}{U^2}\right)^{0.375}\right]}\right] \quad (5.72)$$

where H_s is the significant wave height, T_s is the significant wave period, U is the wind speed, F is the fetch length, D is the mean depth along the fetch, and g is the acceleration due to gravity. The significant height is the average height of the one-third highest waves in an observed wave train and is approximately equal to the average height observed by trained personnel. The significant wave period is the average period corresponding to the highest one third of the waves.

For a progressive wave of small amplitude, a, propagating in the x-direction in water of depth, d, linear wave theory shows the following. The surface displacement

η is given by $\eta(x,t) = a \cos(kx - \sigma t)$ where $k = 2T/L$ is the wave number, $\sigma = 2\pi/T$ is the wave frequency, L is the wave length, and T is the wave period. The velocity potential is given by

$$\phi(x,y,t) = \frac{ag}{\sigma} \frac{\cosh k(y+d)}{\cosh kd} \cos(kx - \sigma t) \qquad (5.73)$$

The dispersion relation that relates the phase velocity c to the wave number is

$$c^2 = (\frac{g}{k}) \tanh kd \qquad (5.74)$$

It follows that the wave length is given by

$$L = \frac{gT_s^2}{2\pi} \tanh\left(\frac{2\pi d}{L}\right) \qquad (5.75)$$

The maximum orbital velocity at the bottom is given by

$$u_{bm} = \frac{\pi H_s}{T_s} \frac{1}{\sinh\left(\dfrac{2\pi d}{L}\right)} \qquad (5.76)$$

In the presence of roughness elements on the bottom surface, the flow characteristics of the boundary layer are altered because of waves and vortices. However, for a smooth bottom as assumed below for Lake Erie calculations (Thomas et al., 1976), the skin friction, c_w, is given by

$$c_w = 1/R \qquad (5.77)$$

where R is a Reynolds number and is given by

$$R = \frac{u_{bm}\delta^*}{\nu} \qquad (5.78)$$

where ν is the kinematic viscosity and $\delta^* = (\nu/\sigma)^{1/2}$ is the wave displacement thickness. For a rough bottom, c_w is more complex and depends on the effective bottom roughness (Section 5.1).

5.5.2 LAKE ERIE

The bottom topography of Lake Erie is shown in Figure 2.2(a). In numerical calculations of wave action (Kang et al., 1982), a square horizontal grid with $\Delta x = \Delta y = 1.6$ km was used. At each grid point, the significant wave height and wave period were determined through Equations 5.71 and 5.72 for given values of wind speed, fetch length, and mean water depth. The fetch length was defined as the distance in the direction of the wind between the shoreline and the grid point. The mean depth was obtained by averaging the water depth along the fetch.

A steady wind with a speed of 40 km/hr (25 mph) was assumed. This is a moderately strong wind for Lake Erie. Calculations were made for wind directions from the southwest (240°) and north (15°), as these correspond approximately to the two dominant wind directions. Some of these results are summarized here. Numerical calculations were continued to a water depth of 1.8 m or to the limiting depth of breaking waves, d = 1.28 H_s.

5.5.2.1 A Southwest Wind

The southwest wind is the prevailing wind direction in Lake Erie. Numerical results for this wind are shown in Figures 5.20(a) and (b). In the Western Basin, the significant wave heights (Figure 5.20(a)) range up to 1.5 m, and the wave periods

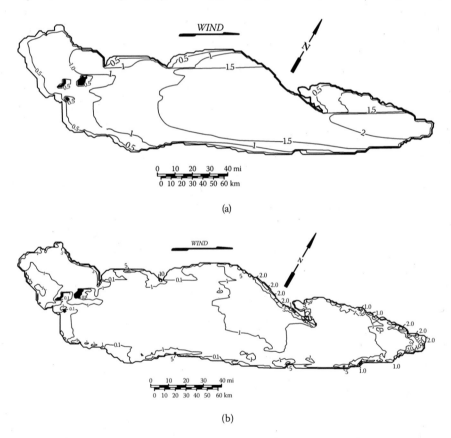

(a)

(b)

FIGURE 5.20   Lake Erie. Southwest wind: (a) significant wave height (m), and (b) bottom shear stress (N/m²). (*Source*: From Kang et al., 1982, *J. Great Lakes Res.* With permission.)

range up to 4.5 s. The bottom orbital velocities were generally in the range of 1 to 50 cm/s. However, along the north shore, particularly at the tip of Pelee Point, the velocities are higher, occasionally more than 100 cm/s. Figure 5.20(b) shows the bottom shear stresses. Even in the middle of the Western Basin, the shear stresses are relatively strong, with $\tau > 0.1$ N/m^2.

In the Central and Eastern basins, the fetches for a southwest wind are the largest due to the orientation of Lake Erie. Because of this, the surface waves generated by this wind are the largest. Wave heights range up to 2.2 m, and wave periods range up to 6 s. They increase eastward or northeastward as the fetch length increases. The larger waves can especially be seen in the Eastern Basin, where the fetches are greater than in the other basins. The bottom orbital velocities increase eastward in the relatively flat Central Basin; in the Eastern Basin, they first decrease eastward as the depth increases and then increase with decreasing depth. The bottom shear stresses (Figure 5.20(b)) are very strong in the nearshore regions of the north shore between Erieau and Long Point and the coastal regions of the Eastern Basin and quite weak in the deeper parts of the basins.

5.5.2.2 A North Wind

For a north wind, both wave height and wave period increase southward with increasing wind fetch. For the Western Basin, strong shear-stress zones ($\tau > 0.5$ N/m^2) lie in the nearshore regions along the southern shore. Waves in the Central and Eastern basins are generally larger than those in the Western Basin, as these basins are deeper and the wind fetches are much greater. The bottom shear stresses are relatively weak in most of the Central and Eastern basins. Strong orbital velocities and bottom shear stresses are found in the narrow, nearshore regions along the south shore and in the shallow basin behind Long Point.

5.5.2.3 Relation of Wave Action to Sediment Texture

Lake Erie sediments are fine-grained (silt and clay-sized) materials derived from erosion of shoreline bluffs and from river inputs. Based on extensive textural analyses of Lake Erie sediments (samples at 275 different locations), Thomas et al. (1976) determined a mean grain-size distribution as shown in Figure 2.2(b). By comparison of this figure with the shear stress distribution in Figure 5.20(b), the similarity is quite apparent and indicates the close relation between the bottom shear stress that causes sediment erosion and the particle size of the bottom sediments; that is, in areas of high shear stress, the sediments tend to be coarse, whereas in areas of low shear stress, the sediments are finer.

6 Modeling Sediment Transport

In previous chapters, many of the basic and most significant sediment transport processes were discussed. In the present chapter, these ideas are applied to the modeling of sediment transport. In Section 6.1, a brief but general overview of sediment transport models is given. In Section 6.2, transport as suspended load and/or bedload is discussed with the purpose of describing a unified approach for modeling erosion. Simple applications of sediment transport models are then described in Section 6.3. More complex applications of sediment transport models to rivers, lakes, and estuaries are presented in Sections 6.4 through 6.6; the purpose is to illustrate some of the significant and interesting characteristics of sediment transport in different types of surface water systems as well as to illustrate the capabilities and limitations of different models.

6.1 OVERVIEW OF MODELS

Numerous models of sediment transport exist. They differ in (1) the number of space and time dimensions used to describe the transport and (2) how they describe and quantify various processes and quantities that are thought to be significant in affecting transport. Some of the processes and quantities that may be significant include (1) erosion rates, (2) particle/floc size distributions (i.e., the number of sediment size classes), (3) settling speeds, (4) deposition rates, (5) flocculation of particles, (6) bed consolidation, (7) erosion into suspended load and/or bedload, and (8) bed armoring. In practice, most sediment transport models do not include accurate descriptions of all of these processes. The final choice of space and time dimensions, what processes to include in a model, and how to approximate the processes that are included is a compromise between the significance of each process; an understanding of and ability to quantify each process; the desired accuracy of the solution; the data available for process description, for specification of boundary and initial conditions, and for verification; and the amount of computation required.

6.1.1 DIMENSIONS

In the modeling of sediment transport, it is necessary to describe the transport of sediments in the overlying water as well as the dynamics (erosion, deposition, consolidation) of the sediment bed. In reality and for generality, these descriptions should be three-dimensional in space as well as time dependent. However, if this is done, the resulting models are quite complex and computer intensive and sometimes may be unnecessary. Simpler models can be obtained by reducing the number of space dimensions and sometimes by assuming a steady state.

For the problems considered here, two or three space dimensions as well as time dependence are generally necessary. Because of this, steady-state and one-dimensional models will not be considered. For the transport of sediments in the overlying water, it will be shown later in this chapter that, in many cases, two-dimensional, vertically integrated transport models give results that are almost identical to three-dimensional models and are therefore often sufficient to accurately describe sediment transport. For the most complex problems, three-dimensional, time-dependent transport models are necessary.

In many models, erosion and deposition are described by simple parameters that are constant in space and time. A model of sediment bed dynamics is then not necessary. However, as emphasized in Chapter 3, erosion rates are highly variable in the horizontal direction, in the vertical direction (depth in the sediment), and with time (due to changes in sediment properties caused by erosion, deposition, and consolidation). Because of this and for quantitative predictions, a three-dimensional, time-dependent model of sediment bed properties and dynamics is usually necessary.

6.1.2 QUANTITIES THAT SIGNIFICANTLY AFFECT SEDIMENT TRANSPORT

6.1.2.1 Erosion Rates

Because erosion is a fundamental process that dominates sediment transport and because of its high variability in space and time, it is essential to understand quantitatively and be able to predict this quantity throughout a system as a function of the applied shear stress and sediment properties. In general, for sediments throughout a system, erosion rates cannot be determined from theory and must therefore be determined from laboratory and field measurements. This was discussed extensively in Chapter 3.

In models, various approximations to describe erosion rates have been used. At its simplest, the erosion rate is approximated as a resuspension velocity, v_r, that is constant in space and time. This parameter then is estimated by adjusting v_r until results of the overall transport model agree with field observations. In this approximation, v_r is strictly an empirical parameter, does not reflect the physics of sediment erosion, and has no predictive ability.

A widely used and more justifiable approximation for erosion rates is to assume that

$$E = a(\tau - \tau_c) \tag{6.1}$$

where a and τ_c are constants. This is a linear approximation to Equations 3.22 and 3.23. The parameters a and τ_c are usually empirical parameters chosen by parameterization based on comparisons of calculated and measured suspended sediment concentrations. For small erosion rates or for small changes in erosion rates, the above equation may be a justifiable approximation because, over a small range, any set of data can be approximated as a straight line. However, the choices for a and τ_c are crucial, and these parameters should be obtained from laboratory and field data — not from model calibration.

In the limit of fine-grained sediments, the amount of erosion for a particular shear stress is limited so that the concept of an erosion potential, ε, is valid (Section 3.1). This amount of erosion occurs over a limited time, T, typically on the order of an hour, so that an approximate erosion rate can be determined from ε/T; after this time, $E = 0$ until the shear stress increases. Several sediment transport models (e.g., SEDZL) have used this concept. However, for real sediments, sediment properties often change rapidly with depth and time; the SEDZL model does not include this variability (except for bulk density). Because of this, it is only quantitatively valid for fine-grained sediments that have uniform properties throughout; however, it will give qualitatively correct results for other types of sediments.

When sediment properties change rapidly and in a nonuniform manner in time and space (which is most of the time), the most accurate procedure for determining erosion rates is by using Sedflume for existing *in situ* sediments and a combination of laboratory tests with Sedflume and consolidation and bed armoring theories to predict the erosion rates of recently deposited sediments as they consolidate with time. Equation 3.23 can then be used to approximate the erosion rates as a function of shear stress. The use of Sedflume data and space- and time-variable sediment properties are incorporated into the SEDZLJ transport model.

Although Sedflume determines erosion rates as a function of the applied shear stress, an additional difficulty in the modeling of sediment transport is the accurate determination of the bottom shear stress. As a first approximation, this stress is the same as the shear stress used in the modeling of the hydrodynamics (Section 5.1). However, as stated there, the hydrodynamic shear stress is due to frictional drag and form drag. Only the former is thought to contribute to the shear stress causing sediment resuspension. This distinction between friction and form drag is significant when sand dunes are present, and the two stresses then can be determined independently. For fine-grained, cohesive sediments, dunes and ripples tend not to be present, form drag is thought to be negligible, and the total drag is essentially the same as frictional drag.

6.1.2.2 Particle/Floc Size Distributions

In Chapter 2, it was emphasized that large variations in particle sizes typically exist in real sediments throughout a surface water system, often by two to three orders of magnitude. However, as an approximation in many sediment transport models, only one size class is assumed. This is quite often necessary when only meager data for model input and verification are available or when knowledge and/or data are insufficient to accurately characterize the transport processes.

This assumption also may be reasonable when changes in environmental conditions in space and time are relatively small.

However, this assumption is not valid when there are large variations in environmental conditions and/or when flocculation is significant, for example, (1) during large storms or floods, (2) when there are large spatial and temporal changes in flow velocities, or (3) when calculations over long time periods or large spatial distances are required. For these cases as well as others, several size classes are necessary for the accurate determination of suspended sediment concentrations and especially the net and gross amounts of sediment eroded as a function of space and time. Three size classes are often necessary and sufficient.

6.1.2.3 Settling Speeds

Modelers often state that settling speeds used in their models were obtained from laboratory and/or field data. However, as noted in Chapter 2, the values for settling speeds for sediments in a system generally range over several orders of magnitude. The appropriate value to use for an effective settling speed is therefore difficult to determine or even define. To illustrate this, the value for the settling speed determines where and to what extent suspended sediments deposit and accumulate on the bottom. For settling speeds that differ by an order of magnitude, the location where they deposit also will differ by an order of magnitude, for example, from a few kilometers downstream in a river to tens of kilometers downstream. Because a wide range of settling speeds is possible depending on the particle/floc properties and the flow regime, a wide range of settling speeds is also necessary in a model for a valid approximation to the vertical flux, transport, and deposition of sediments throughout a system.

In most models, the actual value that is used for the settling speed is determined by parameterization, that is, by adjusting its value until the calculated and observed values of suspended sediment concentration agree. As noted previously in Section 1.2, non-unique solutions can result by use of this procedure. As another example of this difficulty, consider the specification of settling speeds as illustrated in several texts on water quality modeling (e.g., Thomann and Mueller, 1987; Chapra, 1997). In these texts, the almost universal choice for a settling speed is 2.5 m/day; this seems to be based on earlier articles by Thomann and Di Toro (1983) and O'Connor (1988). From Stokes law, this settling speed corresponds to a particle size of about 5 µm. By comparison, median particle sizes for sediments in the Detroit River, Fox River, and Santa Barbara Slough are 12, 20, and 35 µm, respectively (Section 2.1), whereas cores from the Kalamazoo River show median sizes as a function of depth that range from 15 to 340 µm (Section 3.2). For the latter five values of particle size, the corresponding settling speeds (from Stokes law) are 11, 31, 95, 18, and 9000 m/day, respectively.

The settling speed of 2.5 m/day was not determined from laboratory or field measurements but was estimated based on previous modeling exercises. The corresponding particle diameter of 5 µm seems quite low compared to those for real sediments. It is also somewhat surprising that one settling speed seems to work for a variety of problems. The fact is that a settling speed of 2.5 m/day is not unique or necessary; that is, a wide range of settling speeds can be used and will give the observed suspended sediment concentration, as long as the erosion rate (or equivalent) is modified appropriately, just as is indicated by Equation 1.2. However, if this is done, as stated in Section 1.2 and summarized by Equation 1.2, multiple solutions are then possible and a unique solution cannot be determined from calibration of the model using the suspended solids concentration alone. The amounts and depths of erosion/deposition will vary, depending on the choice of settling speed. The depth of erosion/deposition is an important quantity that a water quality model should be able to predict accurately; it should not depend on a somewhat arbitrary choice of settling speed.

When three or more size classes are assumed, the average settling speed for each size class can be used. When three size classes and their average settling speeds are determined from laboratory and/or field data, the uncertainty of parameterization is substantially decreased. Whenever possible, this should be done.

6.1.2.4 Deposition Rates

Deposition rates and the parameters on which they depend are discussed in Section 4.5. Because of limited understanding of this quantity, the rates that are used in modeling are usually parameterized using Equation 4.61 or a similar equation. A better approach is suggested in Section 4.5.

6.1.2.5 Flocculation of Particles

In the above sections, the effects of flocculation have not been explicitly stated. However, as described in Chapter 4, flocculation can modify floc sizes and settling speeds by orders of magnitude. Because of this, flocculation must be considered in the accurate modeling of particle/floc size distributions, settling speeds, and deposition rates when fine-grained sediments are present. The quantitative understanding of the flocculation of sedimentary particles is relatively new, and the quantitative determination of many of the parameters necessary for the modeling of flocculation has been done for only a few types of sediments. Because of this, most sediment transport models do not include flocculation. A few exceptions will be noted in the following. Now that a simple model of flocculation is available (see Section 4.4), variations in floc sizes, settling speeds, and deposition rates due to flocculation now can be efficiently included in overall transport models.

6.1.2.6 Consolidation

When coarse-grained particles are deposited, little consolidation occurs and the bulk density of the sediments is almost independent of space and time. In this case, erosion rates are dependent only on particle size. However, when fine-grained particles are deposited, considerable consolidation of the sediments can occur, the bulk density usually (but not always) increases with depth and time, and the erosion rate (which is a sensitive function of the bulk density of the sediments, Section 3.3) usually decreases with sediment depth and time. As indicated in Section 4.6, the presence and generation of gas in the sediments have significant effects on consolidation and erosion rates but usually are not measured or even considered. Modeling of consolidation can be done, but the accuracy of this modeling depends on laboratory experiments of consolidation (Section 4.6).

6.1.2.7 Erosion into Suspended Load and/or Bedload

Most water quality transport models assume there is suspended load and ignore bedload. If sediments are primarily coarse, noncohesive sediments, some sediment transport models consider bedload only. If sediments include both coarse-grained and fine-grained particles, both suspended load and bedload may be significant and need to be considered. This often is done by treating suspended load and bedload as independent quantities. However, upon deposition, both the suspended load and bedload can modify the bulk properties and hence the erosion rates of the surficial sediments. This, in turn, affects the suspended load and bedload; that is, suspended load and bedload are interactive quantities and should be treated as such. This is discussed in the next section.

6.1.2.8 Bed Armoring

Bed armoring can significantly affect erosion rates, often by one to two orders of magnitude. A model of this process is described in the following section, whereas applications of this model to illustrate some of the characteristics of bed armoring are given in Sections 6.3 and 6.4.

6.2 TRANSPORT AS SUSPENDED LOAD AND BEDLOAD

6.2.1 SUSPENDED LOAD

The three-dimensional, time-dependent conservation of mass equation for the transport of suspended sediments in a turbulent flow is

$$\frac{\partial C}{\partial t} + \frac{\partial (uC)}{\partial x} + \frac{\partial (vC)}{\partial y} + \frac{\partial [(w - w_s)C]}{\partial z}$$

$$= \frac{\partial}{\partial x}\left(D_H \frac{\partial C}{\partial x}\right) + \frac{\partial}{\partial y}\left(D_H \frac{\partial C}{\partial y}\right) + \frac{\partial}{\partial z}\left(D_v \frac{\partial C}{\partial z}\right) + S \qquad (6.2)$$

where C is the mass concentration of sediments, t is time, x and y are horizontal coordinates, z is the vertical coordinate (positive upwards), u and v are the sediment (fluid) velocities in the x- and y-directions, w is the fluid velocity in the z-direction, w_s is the settling speed of the sediment relative to the fluid and is generally positive, D_H is the horizontal eddy diffusivity, D_v is the vertical eddy diffusivity, and S is a source term.

In many cases, a two-dimensional, vertically integrated, time-dependent conservation equation is sufficient. For this case, vertical integration of the above equation gives

$$\frac{\partial(hC)}{\partial t} + \frac{\partial(UC)}{\partial x} + \frac{\partial(VC)}{\partial y} = D_H \left[\frac{\partial}{\partial x}\left(h\frac{\partial C}{\partial x}\right) + \frac{\partial}{\partial y}\left(h\frac{\partial C}{\partial y}\right)\right] + Q \qquad (6.3)$$

where C is now the suspended sediment concentration averaged over depth, h is the local water depth, U and V are vertically integrated velocities defined by Equations 5.53 and 5.54, D_H is assumed constant, and Q is the net flux of sediments into suspended load from the sediment bed — that is, Q is calculated as erosion flux into suspended load, E_s, minus the deposition flux from suspended load, D_s, or

$$Q = E_s - D_s \qquad (6.4)$$

Two-dimensional calculations based on these latter two equations are valid when the sediments are well mixed in the vertical. When this is not the case, an additional calculation to approximate the vertical distribution of C is sometimes made so as to more accurately determine the suspended concentration near the bed and hence to more accurately determine the deposition rate. This is done by assuming a quasi-steady, one-dimensional balance between settling and vertical diffusion. A first approximation to the vertical distribution of sediments can then be shown to be

$$C(z) = C_o \exp\left(-\frac{w_s z}{D_v}\right) \qquad (6.5)$$

where C_o is the near-bed concentration at z = 0. This equation can be integrated over the water depth to give a relation between the average suspended sediment concentration and C_o. The concentration C_o then can be determined at any location using this relation and C(x,y,t) from Equation 6.3.

When different-size classes are considered, the above equations apply to each size class; the terms S and Q then must include transformations from one size class to another, for example, due to flocculation or precipitation/dissolution.

6.2.2 BEDLOAD

For the description of bedload transport, many procedures and approximate semiempirical equations are available (Meyer-Peter and Muller, 1948; Bagnold, 1956; Engelund and Hansen, 1967; Van Rijn, 1993; Wu et al., 2000). The procedure described by Van Rijn (1993) is used here.

The mass balance equation for particles moving in bedload (similar to Equation 6.3) can be written as

$$\frac{\partial h_b C_b}{\partial t} + \frac{\partial h_b q_{bx}}{\partial x} + \frac{\partial h_b q_{by}}{\partial y} = Q_b \qquad (6.6)$$

where C_b is the sediment concentration in bedload, q_{bx} and q_{by} are the horizontal bedload fluxes in the x- and y-directions, h_b is the thickness of the bedload layer, and Q_b is the net vertical flux of sediments between the sediment bed and bedload. The horizontal bedload flux is calculated as

$$q_b = u_b C_b \qquad (6.7)$$

where u_b is the bedload velocity in the direction of interest. The bedload velocity and thickness can be calculated from the empirical relation

$$u_b = 1.5T^{0.6}\left[\left((\rho_s - 1)gd\right)\right]^{0.5} \qquad (6.8)$$

$$h_b = 3dd_*^{0.6}T^{0.9} \qquad (6.9)$$

where d_* is the nondimensional particle diameter defined as $d_* = d[(\rho_s-1)g/v^2]^{1/3}$, d is the particle diameter, and ρ_s is the density of the sedimentary particle. The transport parameter, T, is defined as

$$T = \frac{\tau - \tau_c}{\tau_c} \qquad (6.10)$$

In Equation 6.7, q_b is the flux (mass/area/time) of sediment in bedload. To obtain the mass/time of sediment being transported in bedload, q_b must be multiplied by the area of the bedload layer — that is, by h_b × width of the layer.

The net flux of sediments between the bottom sediments and bedload, Q_b, is calculated as the erosion of sediments into bedload, E_b, minus the deposition of sediments from bedload, D_b, and is

$$Q_b = E_b - D_b \qquad (6.11)$$

where D_b is given by

$$D_b = p\, w_s\, C_b \qquad (6.12)$$

and p is the probability of deposition.

In steady-state equilibrium, the concentration of sediments in bedload, C_e, is due to a dynamic equilibrium between erosion and deposition, that is,

$$E_b = p\, w_s\, C_e \qquad (6.13)$$

From this, p can be written as

$$p = \frac{E_b}{w_s C_e} \qquad (6.14)$$

The equilibrium concentration has been investigated by several authors; the formulation by Van Rijn (1993) will be used here and is

$$C_e = 0.117 \frac{\rho_s T}{d_*} \qquad (6.15)$$

Once E_b, w_s, and C_e are known as functions of particle diameter and shear stress, p can be calculated from Equation 6.14. It then is assumed that this probability is also valid for the nonsteady case so that the deposition rate can be calculated in this case, also. This procedure guarantees that the time-dependent solution will always tend toward the correct steady-state solution as time increases.

The equilibrium concentration, C_e, is based on experiments with uniform sediments. In general, the sediment bed contains, and must be represented by, more than one size class. In this case, the erosion rate for a particular size class is given by $f_k E_b$, where f_k is the fraction by mass of the size class k in the surficial sediments. It follows that the probability of deposition for size class k is then given by

$$p_k = \frac{f_k E_b}{w_{sk} f_k C_{ek}} = \frac{E_b}{w_{sk} C_{ek}} \qquad (6.16)$$

As in Equation 6.14, it is implicitly assumed in this equation that there is a dynamic equilibrium between erosion and deposition for each size class k.

6.2.3 EROSION INTO SUSPENDED LOAD AND/OR BEDLOAD

As bottom sediments are eroded, a fraction of these sediments is suspended into the overlying water and transported as suspended load; the remainder of the eroded sediments moves by rolling and/or saltation in a thin layer near the bed — that is, in bedload. The fraction of the eroded sediments going into each of the transport modes depends on particle size and shear stress.

For fine-grained particles (which are generally cohesive), erosion occurs both as individual particles and in the form of small aggregates or chunks of particles. The individual particles generally move as suspended load. The aggregates tend to move downstream near the bed but generally seem to disintegrate into small particles in the high-stress boundary layer near the bed as they move downstream. These disaggregated particles then move as suspended load. This disaggregation-after-erosion process is not quantitatively understood. For this reason, it is assumed here that fine-grained sediments less than about 200 μm are completely transported as suspended load.

Coarser, noncohesive particles (defined here as those particles with diameters greater than about 200 μm) can be transported as both suspended load and bedload, with the fraction in each dependent on particle diameter and shear stress. For particles of particular size, the shear stress at which suspended load (or sediment suspension) is initiated is defined as τ_{cs}. This shear stress can be calculated from (Van Rijn, 1993)

$$\tau_{cs} = \begin{cases} \dfrac{1}{\rho_w}\left(\dfrac{4w_s}{d_*}\right)^2 & \text{for } d \leq 400\,\mu m \\[4mm] \dfrac{1}{\rho_w}(0.4w_s)^2 & \text{for } d > 400\,\mu m \end{cases} \qquad (6.17)$$

The variation of τ_{cs} as a function of d is shown in Figure 6.1 and can be compared there with $\tau_c(d)$.

For $\tau > \tau_{cs}$, sediments are transported as both bedload and suspended load, with the fraction of suspended load to total load transport increasing from 0 to 1 as τ increases. Guy et al. (1966) have quantitatively demonstrated this by means of detailed flume measurements of suspended load and bedload transport for sediments ranging in median diameter, d_{50}, from 190 to 930 μm. They found that the fraction of suspended load transport to total load transport, q_s/q_t, increases as the ratio of shear velocity (defined as $u_* = \sqrt{\tau/\rho_w}$) to settling velocity increases. This

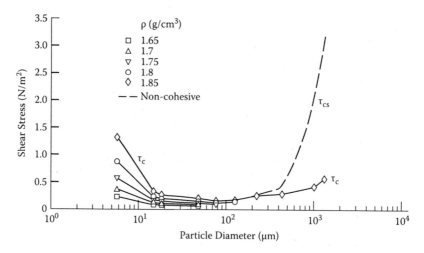

FIGURE 6.1 Critical shear stresses for erosion and suspension of quartz particles.

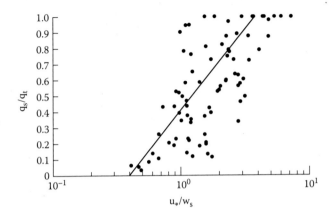

FIGURE 6.2   Suspended flux as a fraction of total flux. Straight line is approximation by Equation 6.18. Data from Guy et al. (1966). (*Source:* From Jones and Lick, 2001a.)

fraction is shown as a function of u_*/w_s in Figure 6.2. Their data can be approximated by

$$\frac{q_s}{q_t} = \begin{cases} 0 & \text{for } \tau \le \tau_{cs} \\[2mm] \dfrac{\ln(u_*/w_s) - \ln(\sqrt{\tau_{cs}/\rho_w}/w_s)}{\ln(4) - \ln(\sqrt{\tau_{cs}/\rho_w}/w_s)} & \text{for } \tau > \tau_{cs} \text{ and } \dfrac{u_*}{w_s} < 4 \\[2mm] 1 & \text{for } \dfrac{u_*}{w_s} > 4 \end{cases} \qquad (6.18)$$

This approximation is shown as a straight line in the figure.

By multiplying the total erosion flux of a particular size class by q_s/q_t, the erosion flux of that size class into suspended load, E_s, can be calculated. The erosion flux into bedload, E_b, can be calculated by multiplying the total erosion flux of the size class by $(1 - q_s/q_t)$. Erosion fluxes for any size class k can be calculated as

$$\left. \begin{aligned} E_{s,k} &= \frac{q_s}{q_t}(f_k E) \\[4mm] E_{b,k} &= \left(1 - \frac{q_s}{q_t}\right)(f_k E) \end{aligned} \right\} \quad \text{for } \tau \ge \tau_{c,k}$$

$$\left. \begin{aligned} E_{s,k} &= 0 \\[3mm] E_{b,k} &= 0 \end{aligned} \right\} \quad \text{for } \tau < \tau_{c,k} \qquad (6.19)$$

6.2.4 BED ARMORING

A decrease in sediment erosion rates with time can occur due to (1) the consolidation of cohesive sediments with depth and time, (2) the deposition of coarser sediments on the sediment bed during a flow event, and (3) the erosion of finer sediments from the surficial sediment, leaving coarser sediments behind, again during a flow event. As far as the first process is concerned, the existing in situ changes in erosion rates with depth can be determined by Sedflume measurements from sediment cores. The time-dependent consolidation of sediments and decrease of erosion rates of sediments after deposition can be determined approximately from laboratory consolidation studies along with theoretical analyses (Section 4.6).

Here the concern is with bed armoring due to processes (2) and (3). To describe and model these processes, it is necessary to assume that a thin mixing layer, or active layer, is formed at the surface of the bed. The existence and properties of this layer have been discussed by several researchers (Borah et al., 1983; Van Niekerk et al., 1992; Parker et al., 2000). The presence of this active layer permits the interaction of depositing and eroding sediments to occur in a discrete layer without the deposited sediments modifying the properties of the undisturbed sediments below. Van Niekerk et al. (1992) have suggested that the thickness, T_a, can be approximated by

$$T_a = 2d_{50} \frac{\tau}{\tau_c} \tag{6.20}$$

where d_{50} is the median particle diameter. This formulation takes into account the deeper penetration of turbulence into the bed with increasing shear stress. In calculations, d_{50} is often approximated by the average diameter of the sediments in the surficial layer.

In the modeling of sediment bed dynamics, the thickness of the active layer is specified by the above equation and remains constant until conditions change during the erosion/deposition process. When sediments are eroded from this layer, an equal amount of sediment must be transferred into this layer from the layer below to keep the thickness of the active layer constant; this transfer generally modifies the properties of the active layer. When sediments are deposited into the active layer, an equal amount of sediment is transferred from the active layer into the layer below; a small change in properties of this lower layer may then occur.

6.3 SIMPLE APPLICATIONS

Bed armoring and different particle size distributions can have large (orders of magnitude) effects on sediment transport. Three examples are presented here to illustrate this and also to demonstrate the modeling of sediment transport when these effects are significant. The examples are concerned with transport, particle size redistribution, and coarsening in (1) a straight channel, (2) an expansion region, and (3) a curved channel. An example that illustrates the effects of aggregation and disaggregation on the vertical transport and distribution of flocs is also given.

6.3.1 Transport and Coarsening in a Straight Channel

Little and Mayer (1972) made an elegant study of the transport and coarsening of sediments in a straight channel. In their experiments, a flume 12.2 m long and 0.6 m wide was used and was filled with a distribution of sand and gravel sediments. The mean size of the sediment particles was about 1000 μm, but there was a wide distribution of sizes around this mean (Figure 6.3). Clear water was run over the sediment bed at a flow rate of 0.016 m³/s. The eroded sediment was collected at the outlet of the flume, and the sediment transport rate was determined from this. Due to bed armoring, the transport rate decreased with time. When the rate had decreased to about 1% of its value at the beginning, the experiment was ended. This occurred in 75.5 hr.

The experiment was approximated by means of SEDZLJ. In the modeling, erosion rates from Sedflume data, multiple sediment size classes, a unified treatment of suspended load and bedload, and bed armoring were included (Jones and Lick, 2000, 2001a). The hydrodynamics and sediment transport were approximated as two-dimensional and time dependent; 13 elements with a downstream dimension of 100 cm and a cross-stream dimension of 60 cm were used to discretize the domain. The sediment bed was assumed to be three-dimensional and time dependent and consisted of nine size classes that were selected to accurately represent the sediment bed in the experiment (Figure 6.3). Data from the Roberts et al. (1998) Sedflume studies on quartz were used to define the erosion

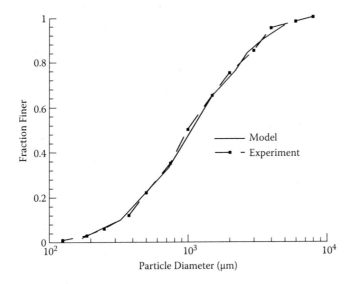

FIGURE 6.3   Particle size distributions for experiment and model at the beginning of the experiment. (*Source:* From Jones and Lick, 2001a.)

rates and critical shear stresses for these sediments (Table 6.1). The coefficient of hydrodynamic friction, c_f, was set such that the measured shear stress of 1.0 N/m² was reproduced in the model. The active layer was held at a constant thickness of 0.5 cm, which is consistent with Equation 6.20.

The experimental and calculated transport rates (kg/m/s) at the outlet of the flume are plotted in Figure 6.4 as a function of time and decrease by about two orders of magnitude during the course of the experiment. The model shows good agreement with the experimental data for the entire time. The average particle size in the active layer of the model and erosion/deposition rates at the end of the flume are plotted in Figure 6.5 as a function of time. In the first few hours,

TABLE 6.1
Sediment Size Class Properties

Particle Size	Initial bed percentage by mass	w_s(cm/s)	τ_c(N/m²)	τ_{cs}(N/m²)
125	2	0.9	0.15	0.15
222	8	2.25	0.24	0.26
432	23	5.2	0.33	0.45
1020	32	11.30	0.425	2.12
2000	11	18.01	0.93	5.36
2400	8	20.18	0.97	6.73
3000	6	23.07	1.2	8.79
4000	6	27.25	1.6	12.26
6000	4	34.13	2.48	19.2

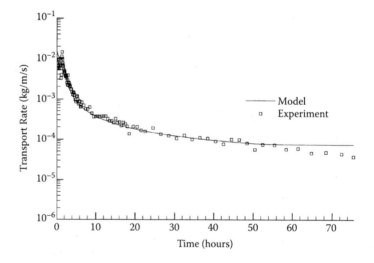

FIGURE 6.4   Measured and calculated transport rates for flow in a straight channel as a function of time. (*Source:* From Jones and Lick, 2001a.)

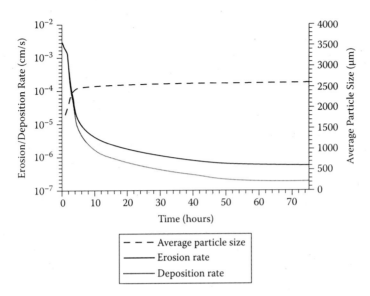

FIGURE 6.5   Average particle size in the active layer, erosion rate, and deposition rate at the end of the flume as a function of time. (*Source:* From Jones and Lick, 2001a.)

there is a rapid increase in the average particle size from 1600 to 2500 μm; this is followed by a much slower rate of increase to a little above 2500 μm by the end of the experiment. Associated with this increase in particle size is a more than three-orders-of-magnitude decrease in the erosion rate. The reason for this decrease is that the finer particle sizes are eroded from the sediment bed, whereas the coarser particles are left behind, thereby increasing the average particle size of the bed and decreasing the erosion rate. As a result, the suspended and bed-load concentrations decrease rapidly with time and are responsible for the rapid decrease in the deposition rate. Figure 6.6 demonstrates that initially the transport is almost equally bedload and suspended load, but as the bed coarsens, the transport becomes almost exclusively due to bedload. In general, the calculated results show good overall agreement with the data and trends observed in the Little and Mayer experiments.

6.3.2 Transport in an Expansion Region

To more fully understand the effects of bed coarsening and different particle size distributions, the transport in an expansion region was also modeled and analyzed by means of SEDZLJ (Jones and Lick, 2001a, b). The expansion region (Figure 6.7(a)) begins with a 2.75-meter wide channel that extends 10 m downstream. At this point, a 28.8° expansion begins and extends 5 m further downstream, where the channel then has a constant width of 8.25 m. The depth of water is 2 m throughout. An inlet flow rate of 2.5 m³/s and a zero sediment concentration were specified at the entrance to the channel, whereas an open boundary condition of

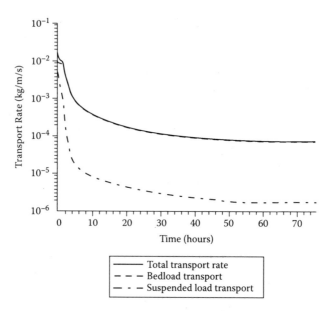

FIGURE 6.6   Calculated suspended load and bedload transport rates as a function of time. (*Source:* From Jones and Lick, 2001a.)

no reflected waves was used at the end of the channel. A no-slip condition was specified at the sidewalls.

Calculations were made (1) for a sediment bed consisting of particles with a uniform size of 726 μm and (2) for a bed initially consisting of a 50/50 mixture of two particle size classes (432 μm and 1020 μm) with an average particle size of 726 μm. In each case, calculations were made with and without bedload transport present. When no bedload transport was present, it was assumed that all of the eroded sediment went into suspended load. In all examples, the quartz erosion data by Roberts et al. (1998) were used; the sediment bulk densities were assumed to be 1.8 g/cm³.

For the first case of sediments with a uniform size of 726 μm, it was assumed that w_s = 8.6 cm/s, τ_c = 0.36 N/m², and τ_{cs} = 1.22 N/m². For a constant flow rate of 2.5 m³/s, a water depth of 2 m, and a surface sediment roughness of 726 μm, the coefficient of friction is 0.0034. For this c_f, Figure 6.7(a) shows the steady-state velocity vectors and shear stress contours. The maximum velocity is 69 cm/s, whereas the maximum shear stress is 1.6 N/m² and drops below 0.2 N/m² downstream.

In the calculations, after an initial transient of about 20 min, a quasi-steady state was approached where sediments were still eroding and depositing but doing so at a reasonably constant rate. Because all particles have the same size, no bed armoring occurred. For the case with bedload, Figure 6.7(b) shows the suspended sediment concentration profile at this time. The maximum suspended concentration is 70 mg/L; this rapidly decreases as the expansion begins, the flow velocity decreases, and the suspended sediments go into bedload and then deposit on the bed further downstream. Figure 6.7(c) shows the bedload concentration at this same time. The maximum bedload concentration is approximately 58,000 mg/L, within 1% of the value predicted by Van Rijn's (1993) empirical equation. It should be noted that the thickness of the bedload is only 0.3 cm (approximately 4 particle diameters). The bedload concentration and transport rapidly decrease in the downstream direction as the shear stress drops below τ_c (about 0.36 N/m^2). The net change in bed thickness after 20 min is shown in Figure 6.7(d). In the center of the upstream channel, there is about 10 cm of net erosion. These eroded sediments then deposit as the expansion begins and the shear stress decreases below the critical shear stress for suspension (1.22 N/m^2); further downstream, below the critical shear stress for erosion (0.36 N/m^2); a maximum deposition of 15 cm occurs. This pattern of large and rapid variations in erosion/deposition is usual where rapid changes in flow velocities occur, especially for bedload.

For this same case without bedload, the erosion rates are the same as above because the particle size is constant. However, now all the material that was originally eroding into bedload is assumed to go into suspended load. Calculations show that, as a result, the maximum suspended sediment concentration is now much higher than before and increases to more than 1300 mg/L, a factor of almost 20 greater than with bedload. Because this case does not include bedload sediments that stay near and deposit on the bottom more readily, sediments are transported further before depositing. Just before the beginning of the expansion, the local net erosion increases to more than 100 cm, whereas just downstream of the expansion, the local deposition increases to more than 100 cm.

The second case (with and without bedload) was assumed to have a sediment bed initially consisting of 50% 432-µm and 50% 1020-µm particles and therefore

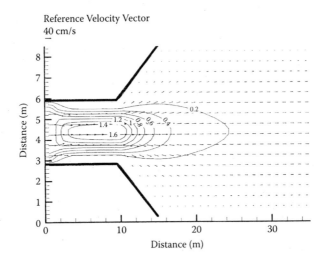

FIGURE 6.7 Transport in an expansion region. Particle size of 726 µm. Calculations include bedload. (a) Velocity vectors and shear stress contours (N/m^2). (*Source:* From Jones and Lick, 2001a.)

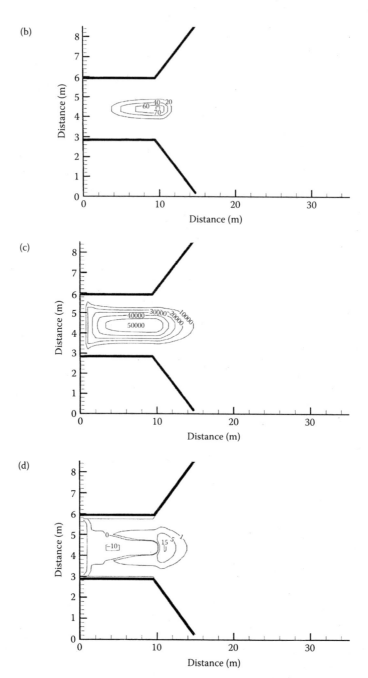

FIGURE 6.7 (CONTINUED) Transport in an expansion region. Particle size of 726 μm. Calculations include bedload. (b) Suspended sediment concentration (mg/L); (c) bedload sediment concentration (mg/L); and (d) net change in sediment bed thickness (cm). (*Source:* From Jones and Lick, 2001a.)

with an average particle size of 726 μm, the same as above. However, as erosion and deposition occur, the fraction of particles in each size class and the average particle size will change in space and time. For particles with a diameter of 432 μm, it was assumed that $w_s = 5.2$ cm/s, $\tau_c = 0.33$ N/m², and $\tau_{cs} = 0.45$ N/m². For the 1020 μm particles, $w_s = 11.3$ cm/s, $\tau_c = 0.425$ N/m², and $\tau_{cs} = 2.12$ N/m².

For the calculation with bedload transport, the suspended sediment concentration after 20 min is shown in Figure 6.8(a). The maximum concentration of 30 mg/L is not only smaller than the previous case with bedload (70 mg/L) as shown in Figure 6.7(a) but is also further upstream. The bedload concentration (Figure 6.8(b)) has a maximum concentration (34,000 mg/L) that is lower than in the previous case (58,000 mg/L), but the shape of the contours is similar. The reason for the similarity in shape is the strong dependence of the bedload concentration and transport on local shear stress.

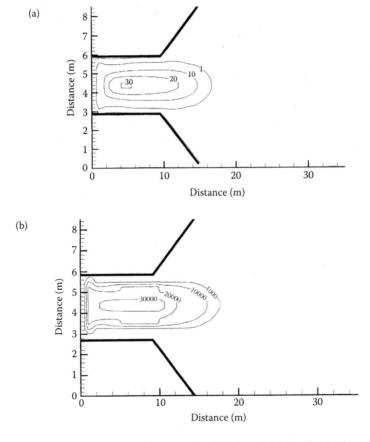

FIGURE 6.8 Transport in an expansion region. Initial particle size distribution of 50% 432 μm and 50% 1020 μm. Calculations include bedload. (a) Suspended sediment concentrations (mg/L); (b) bedload sediment concentrations (mg/L). (*Source:* From Jones and Lick, 2001a.)

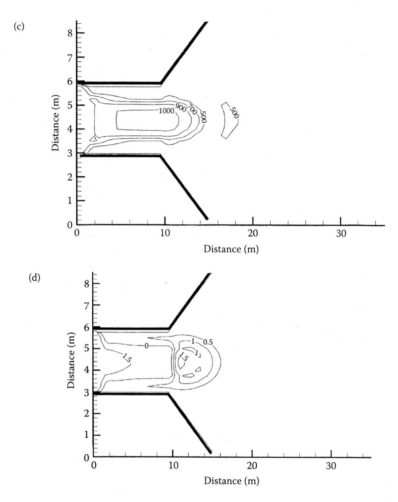

FIGURE 6.8 (CONTINUED)   Transport in an expansion region. Initial particle size distribution of 50% 432 μm and 50% 1020 μm. Calculations include bedload. (c) Average particle size (μm) in the active layer; and (d) net change in sediment bed thickness (cm). (*Source:* From Jones and Lick, 2001a.)

The reason for the differences between the two cases becomes apparent when the average particle size of the active layer for the second case is examined (Figure 6.8(c)). At the inlet, the average particle size remains at its original value of 726 μm. The reason is that there is erosion of all particle sizes and, with clear water inflow, there is no deposition of different particle sizes from upstream that could change the composition of the sediment bed. The eroded 1020-μm particles are deposited a short distance downstream; the eroded 432-μm particles tend to stay suspended longer and are transported further downstream. As a result, the sediment bed rapidly coarsens to more than 1000 μm and the erosion rate decreases in the inlet and beginning of the expansion region. In the latter part of the expansion region, the finer 432-μm particles that were eroded upstream can now deposit; the average particle size therefore decreases below the initial average particle size of 726 μm to about 450 μm. Figure 6.8(d) shows the net change in bed thickness during this time. In the channel, the coarsening of the bed occurs rapidly, allowing little net erosion there. The magnitudes of the maximum erosion and deposition are now only 1.5 cm at the upstream and expansion regions, respectively.

In the present case, the erosion and deposition rates decrease with time due to bed coarsening so that little net transport occurs as time increases. In a more realistic situation, there would be a flux of sediments from upstream that would modify the results shown here. However, the qualitative behavior of the sediment bed would be essentially the same.

When this case is run without bedload, the same trends as in the previous case are observed. The maximum suspended load concentration is increased by greater than an order of magnitude. Coarsening still takes place, but to a smaller degree. This means higher overall erosion rates, which in turn increase the suspended load concentration.

These examples illustrate the major changes in suspended and bedload sediment concentrations, erosion rates, and sediment transport due to changes in particle size distributions and the inclusion of bedload and bed armoring. All are significant and need to be included in sediment transport modeling.

6.3.3 Transport in a Curved Channel

Another example that quantitatively illustrates interesting and significant features of sediment transport is the transport and coarsening in a curved channel. In experiments by Yen and Lee (1995), 20 cm of noncohesive, nonuniform-size sand were placed in a 180° curved channel with 11.5-m entrance and exit lengths; these sediments were then eroded, transported, and deposited by a time-varying flow. The inner radius of the curved part of the channel was 4 m, and the channel width was 1 m. The water depth was 5.44 cm, and the base flow was 0.02 m³/s. For each experiment (five in all), the flow increased linearly from the base to a maximum (which was different for each run) and then decreased linearly back to the base flow.

As with flow in an annular flume (see Chapter 3), the primary flow in the curved part of the channel is in the direction of the centerline of the channel; however, there are small secondary currents due to centrifugal forces. These are radially outward along the upper surface, downward along the outer bank, inward along the bottom, and upward along the inner bank. The net result is a helical motion for fluid elements and suspended particles as they traverse around the bend. As with the annular flume, the shear stresses increase in the radial direction and are greater near the outer wall than at the inner wall. Because of these secondary currents and stress variations, the sediment transport is considerably modified in the bend of the channel as compared with the straight parts of the channel.

Results of five different experimental runs were reported. The sediments were noncohesive and mostly fine to coarse sands. Their particle size distribution

TABLE 6.2
Particle Size Distribution

d (mm)	0.25	0.42	0.84	1.19	2.00	3.36	4.76	8.52
Percent in size class	6.6	10.6	25.4	15.1	20.1	13.0	4.9	4.5

is shown in Table 6.2. For the flow rates in the experiments, only the 0.25-mm particles (which were a small part of the total) had the potential to travel as suspended load. The remainder could only travel as bedload. For the first run (the experiment that is summarized here), the experimental runtime was 180 min, the peak flow rate was 0.075 m^3/s, and the time during the experiment that the 0.25-mm particles could be resuspended was 157 min.

For these experiments, the hydrodynamics and sediment transport were treated as three-dimensional and time dependent and were modeled by means of EFDC, with modifications to the sediment bed dynamics as in SEDZLJ (James et al., 2005). Bed armoring was not included in the modeling. Because only a small fraction of the sediment could be resuspended, the transport was insensitive to the assumptions for resuspension. However, the transport was sensitive to the description of bedload. Because of this, five different bedload formulations (Meyer-Peter and Muller, 1948; Bagnold, 1956; Engelund and Hansen, 1967; Van Rijn, 1993; and Wu et al., 2000) were used and compared for Run 1. The formula by Wu et al. (2000) gave the best comparisons between the model and experiments and was used thereafter in all the calculations. Eight size classes were used in the results shown here.

In the experiments, measurements were made of surface elevation and surficial particle size distribution at 165 locations along different cross-sections of the channel. These quantities for Run 1 are shown in Figures 6.9(a) and 6.10(a). Figure 6.9 is a plot of $\Delta z/h_0$ in the channel, where Δz is the change in elevation and h_0 is the origi-

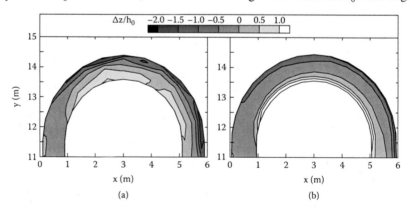

FIGURE 6.9 Transport in a curved channel. Change in sediment bed thickness, $\Delta z/h_0$: (a) measured and (b) calculated. (*Source:* From James et al., 2005.)

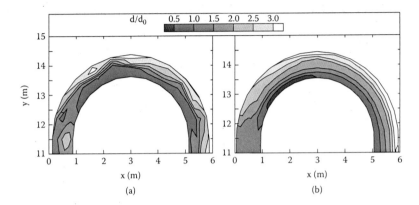

FIGURE 6.10 Transport in a curved channel. Particle size distribution of surficial layer, d/d_0: (a) measured and (b) calculated. (*Source:* From James et al., 2005.)

nal undisturbed depth of the water (5.44 cm). Figure 6.10 is a plot of d/d_0, where d is the local surficial average particle size and d_0 is the overall average particle size. Consistent with the hydrodynamics, net erosion and larger particle sizes are evident toward the outer edge, whereas net deposition and finer particle sizes are shown toward the inner edge.

Calculated results for these same quantities are shown in Figures 6.9(b) and 6.10(b). The calculated results compared with the measured show somewhat less erosion near the outer wall and more deposition near the inner wall as well as greater size gradation; however, the calculated results are qualitatively correct and reasonably accurate.

6.3.4 THE VERTICAL TRANSPORT AND DISTRIBUTION OF FLOCS

As flocs are transported vertically by settling and turbulent diffusion, their sizes and densities are modified by aggregation and disaggregation. In the upper part of the water column, fluid turbulence and sediment concentrations are relatively low; this leads to an increase in floc sizes and higher settling speeds. In the lower part of the water column, especially near the sediment-water interface, fluid turbulence and sediment concentrations tend to be high; this leads to smaller floc sizes and lower settling speeds.

As a first approximation to illustrate and quantify these effects, calculations were made for a one-dimensional, time-dependent description of this transport (Lick et al., 1992). In this case, the appropriate transport equation is the simplification of the conservation of mass equation, Equation 6.2, to one direction. For flocs of size class i with concentration C_i, this equation becomes

$$\frac{\partial C_i}{\partial t} - \frac{\partial}{\partial z}(w_{si} C_i) = \frac{\partial}{\partial z}\left(D_v \frac{\partial C_i}{\partial z}\right) + S_i \qquad (6.21)$$

where z is distance measured vertically upward from the sediment-water interface; w_{si} is the settling speed of the i-th component and is a function of floc size and density; D_v is the eddy diffusivity; and S_i is the source term due to flocculation and is given by $m_i \, dn_i/dt$, where dn_i/dt is determined as described in Section 4.4. There is an equation of this type for each floc size class. All equations are coupled through the source term, S_i, and all equations must therefore be solved simultaneously.

The steady-state distribution of floc sizes and concentrations is illustrated here. Consider first the case of a steady state with no flocculation. In this case, each component in the above equation can be treated separately. A solution to this equation is then:

$$C = C_0 \exp\left(-\frac{w_s z}{D_v}\right) + \frac{F}{w_s} \tag{6.22}$$

where C_0 is a reference concentration and F is an integration constant that corresponds to a constant flux of sediment in the negative z-direction. It can be seen that the concentration decays exponentially above the bottom with a decay distance of $z^* = D_v/w_s$. For the case of zero flux, $C = C_0 \exp(-w_s z/D_v)$.

When flocculation occurs, simple analytic solutions are no longer possible. In this case, Equation 6.21 was solved numerically for each size class in a time-dependent manner until a steady state was obtained. For the example illustrated here, parameters chosen were a water depth of 10 m, an initial sediment concentration of 5 mg/L independent of depth, and zero flux of sediment at the sediment-water and air-water interfaces. Turbulent shear stresses were assumed to be relatively low in most of the water column (approximately 0.04 N/m²) and to increase rapidly in the bottom meter to a maximum of 0.16 N/m² near the sediment-water interface. Ten size classes of flocs were assumed. A variable grid was used in the numerical calculations for increased accuracy near the sediment-water interface. Results for the steady-state concentration and median floc size are shown as a function of depth in Figure 6.11. It can be seen that the concentration increases from less than 1 mg/L at the top of the water column to about 20 mg/L at the sediment-water interface. The median diameter of the flocs is approximately 1000 μm at the top of the column but decreases to about 64 μm at the sediment-water interface due to increased shear and also increased sediment concentration as the sediment-water interface is approached.

In this calculation, zero flux conditions at the sediment-water and air-water interfaces were assumed. A more realistic bottom boundary condition would specify erosion and deposition fluxes at the sediment-water interface, fluxes that would depend on the local turbulent shear stress and suspended sediment concentration and would be different for each size class. In general, this would require the coupling of the problem described here to a more general three-dimensional fluid and sediment transport calculation.

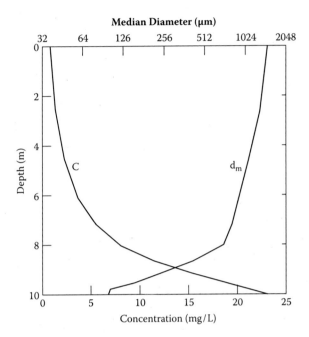

FIGURE 6.11   Steady-state sediment concentration and median floc diameter as a function of depth. (*Source:* From Lick et al., 1992. With permission.)

6.4 RIVERS

Sediment transport in rivers varies widely within and between rivers, depending on bathymetry, flow rates, sediment properties, and sediment inflow. The transports in the Lower Fox and Saginaw rivers are discussed here to illustrate various interesting features of sediment transport in rivers and the modeling of this transport.

6.4.1 SEDIMENT TRANSPORT IN THE LOWER FOX RIVER

An introduction to the PCB contamination problem in the Lower Fox River was given in Section 1.1, and the hydrodynamics were discussed in Section 5.2. The emphasis here is on sediment erosion, deposition, and transport. The specific area that was modeled was from the DePere Dam to Green Bay (bathymetry shown in Figure 5.2). Much of the contaminated sediments is buried here, especially in the upstream area that was previously dredged and is now filling in.

Flow rates during a typical year vary from 30 to 280 m^3/s. In 1989, extensive measurements of flow rates and suspended sediment concentrations were made. The highest flow rate during that year (about 425 m^3/s) was well above normal and was a once-in-5-years flow event. Several sediment transport events during that year were modeled (Jones and Lick 2000, 2001a); the period with the highest flow (from May 22 to June 20) is discussed here.

The model used for the hydrodynamic and sediment transport calculations was SEDZLJ. Erosion rates were obtained from Sedflume data. Three size classes of particles were assumed. The fraction in each size class varied spatially and with time and, because of bed armoring, this modified the erosion rates. Because of its significant effects on erosion and transport, the changes in particle size distribution are emphasized here.

6.4.1.1 Model Parameters

A 30-m by 90-m rectangular grid was generated to discretize the river. Suspended sediment concentrations were measured and averaged daily at the DePere Dam and at Green Bay. It was assumed that the incoming suspended sediment concentrations at the East River were equal to those at the DePere Dam. Seiche motion in Green Bay was neglected.

In an investigation by McNeil et al. (1996), erosion rates as a function of shear stress and depth in the sediment were determined for 30 sediment cores from the entire Lower Fox River by means of Sedflume measurements. Of these 30 cores, three cores (numbers 8, 11, and 14) were chosen to approximate the sediments in the part of the Fox below DePere Dam; their erosion rates as a function of depth and shear stress are shown in Figures 6.12(a), (b), and (c). Core number 8 (Figure 6.12(a)) was selected to describe the properties of the fine-grained nearshore regions with particle sizes on the order of 20 μm; core number 11 (Figure 6.12(b)) was selected to describe the properties of intermediate regions of the river with silt-sized particles on the order of 50 to 100 μm; and core number 14

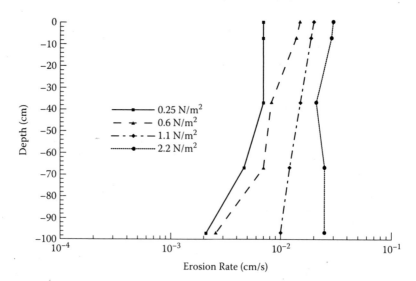

FIGURE 6.12   Erosion rate as a function of depth with shear stress (N/m^2) as a parameter from cores in the Lower Fox River: (a) core number 8. (*Source:* From Jones and Lick, 2001a.)

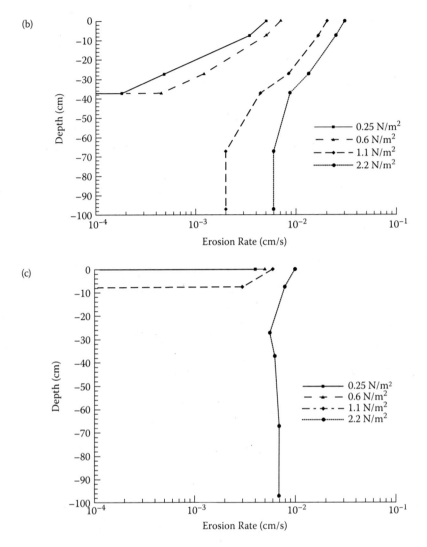

FIGURE 6.12 (CONTINUED) Erosion rate as a function of depth with shear stress (N/m²) as a parameter from cores in the Lower Fox River: (b) core number 11 and (c) core number 14. (*Source:* From Jones and Lick, 2001a.)

(Figure 6.12(c)) was selected to describe the properties of the center channel of the river with sand particles larger than 100 μm. Figure 6.13 shows the assumed initial horizontal distribution of the average particle size of the top 10 cm of each core. Three size classes (Table 6.3) were used to describe the initial distribution of particle sizes in the three cores.

In any investigation of erosion rates by means of Sedflume or other measurement device, the number of cores is typically quite limited. Because of this, there

FIGURE 6.13 Initial particle size distribution (μm) in the Lower Fox River. (*Source:* From Jones and Lick, 2001a.)

TABLE 6.3
Sediment Parameters for the Fox River

Particle Size	w_s(cm/s)	τ_c(N/m²)	τ_{cs}(N/m²)
5	0.0014	0.06	0.06
50	0.1388	0.13	0.13
300	2.99	0.28	0.32

is the question of how best to use Sedflume data so as to adequately and analytically approximate erosion rates throughout a river for use in a transport model. A general procedure for doing this is described in a subsection below and was used in the present modeling.

In the Fox River, it is typical for a thin layer of easily erodible sediments to be present on the top of the sediment bed. Based on visual observations, this layer was estimated to be on the order of 0.5 cm thick and consisted of fine material with an initial composition of 50% 5-μm and 50% 50-μm size particles (i.e., clay and silt particles). By means of model calibration, it was determined during modeling that the optimum value for the thickness of this layer was about 0.3 cm. Size distributions with depth for the other layers were determined from measured size distributions.

Erosion rates for recently deposited sediments were determined from Sedflume measurements on reconstructed Fox River sediments (Jepsen et al., 1997) and are shown in Figure 6.14 as a function of particle size. The erosion rate for any newly deposited sediment, with an average particle size of 5 to 300 μm, was interpolated from these values. The shear stresses, τ_c and τ_{cs}, were determined from Figure 6.1. All sediment size classes were assumed to be disaggregated particle sizes with no flocculation occurring.

6.4.1.2 A Time-Varying Flow

For the period from May 22 to June 20 of 1989, the flow rate and the suspended sediment concentrations at the DePere Dam and at Green Bay are shown as functions of time in Figure 6.15. On May 22, the flow rate is about 50 m³/s; it quickly increases to

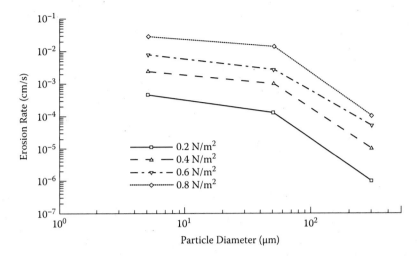

FIGURE 6.14 Erosion rate as a function of particle diameter with shear stress as a parameter. (*Source:* From Jones and Lick, 2001a.)

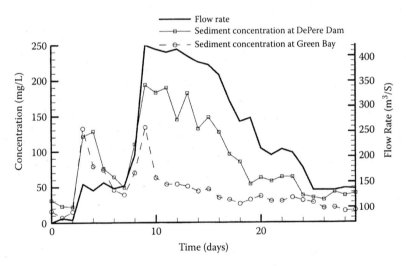

FIGURE 6.15 Flow rate and suspended sediment concentrations for the June 1989 flow. (*Source:* From Jones and Lick, 2001a.)

about 150 m³/s and stays constant for 6 days before increasing to 425 m³/s. The flow rate then decreases more or less monotonically for the rest of the period.

The suspended sediment concentration at the DePere Dam is initially about 30 mg/L and then increases to a peak of 130 mg/L as the flow rate increases to 150 m³/s; the concentration then decreases over the next 6 days, despite the fact that the flow is reasonably constant. When the flow rate then increases to 425 m³/s, the suspended sediment concentration at the dam increases, peaks at 190 mg/L, and then gradually decreases to about 50 mg/L as the flow rate decreases. The concentration at the mouth of the river at Green Bay behaves in a similar manner to that at the DePere Dam, although the first peak is a little larger and the second peak and concentrations thereafter are much lower. It is quite clear that the suspended sediment concentration depends on the flow rate but is not a unique function of the flow rate. This is primarily due to differential settling and bed armoring and is discussed in more detail below and in the following subsection.

In the calculation of the flow, the flow rate was initially assumed to be zero throughout the river and was then increased slowly to 70 m³/s, corresponding to the first day of the calculation period. After this, the flow rates were specified as measured. The computed sediment concentrations at the USGS sampling station at the mouth are compared with the USGS field measurements in Figure 6.16. Good agreement between the two was obtained.

For the first 3 days, because of the low flow, calculations show that little erosion occurs in the river and the sediments appearing at the mouth are primarily due to inflow at the dam and the East River. As the flow increases to 150 m³/s, erosion begins to occur in the narrow upstream and the downstream parts of the river. The maximum velocity in the narrow upstream part of the river is 20 cm/s, producing a shear stress of 0.1 N/m² (enough to erode the 5-μm particles). Downstream, the maximum velocity is 30 cm/s, producing a maximum shear stress of 0.24 N/m² (enough to erode the 5- and 50-μm particles). After day 8, the flow increases rapidly for the next 2 days to 425 m³/s. The model concentration peaks at 141 mg/L, a difference of less than 10% from the measured concentration of 132 mg/L.

On day 12 (when the flow is relatively constant at more than 400 m³/s), the maximum velocities and shear stresses are about 40 cm/s and 0.5 N/m², respectively, in the upstream part of the river (Figure 6.17(a)). In the downstream part, the velocities and shear stresses are higher, and a maximum velocity of 80 cm/s

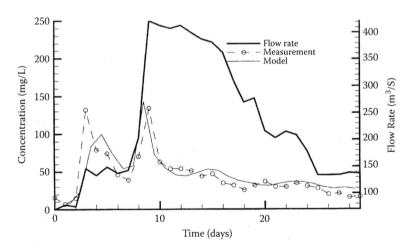

FIGURE 6.16   Flow rate and measured and modeled suspended sediment concentrations for the June 1989 flow. (*Source:* From Jones and Lick, 2001a.)

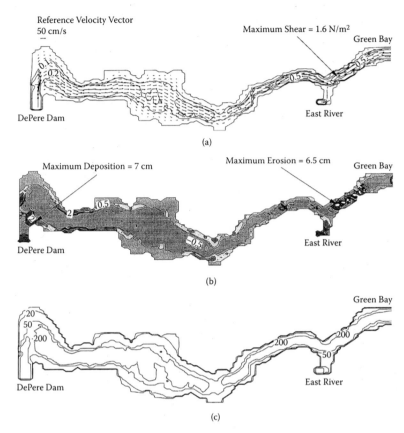

FIGURE 6.17   Day 12 of the June 1989 flow: (a) velocity vectors and shear stress contours (N/m^2); (b) net change in sediment bed thickness (cm); and (c) average particle size (μm) of the active layer. (*Source:* From Jones and Lick, 2001a.)

and a maximum shear stress of 1.6 N/m^2 occur. Figure 6.17(b) is a plot of the net change in bed thickness on day 12 and shows that there generally is erosion in the channel but deposition in the nearshore. The maximum erosion of 6.5 cm occurs just below the East River and the maximum deposition of 7 cm occurs just below the dam, where the flow slows and turns. The average particle size of the active layer after 12 days is shown in Figure 6.17(c). By comparison with the initial size distribution (Figure 6.13), much of the center channel has either coarsened or been eroded down to the fine sand particle sizes above 200 μm. Due to deposition of the finer particles, the particle sizes in the nearshore regions remain below 20 μm.

After the second peak in the sediment concentration, the concentrations at the dam and at the mouth slowly decrease. During this period, there is only minimal additional erosion in the river (due to bed armoring) and deposition almost everywhere. Most of the suspended sediment concentration at the mouth is due to flow of the sediments from the Dam and East River. By day 30, a maximum erosion of 7.5 cm has occurred downstream (only 1 cm more than at day 12), whereas a

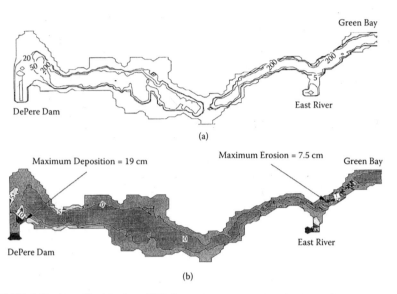

FIGURE 6.18   Day 30 of the June 1989 flow: (a) average particle size (μm) of the active layer; and (b) net change in sediment bed thickness (cm). (*Source:* From Jones and Lick, 2001a.)

maximum deposition of 19 cm has occurred just below the dam (12 cm more than at day 12, Figure 6.18(a)). The distribution of the average particle size of the active layer in the river (Figure 6.18(b)) shows that the particle sizes have decreased in some of the wider portions of the river where finer sediments have begun to deposit during the lower flow rates.

6.4.2 UPSTREAM BOUNDARY CONDITION FOR SEDIMENT CONCENTRATION

In the modeling of sediment transport in a river, important boundary conditions for the model are the flow rate and suspended sediment concentration as functions of time at the upstream boundary of the segment of the river being considered. Typically, the flow rate is measured and is therefore known as a function of time, but, more often than not, the upstream sediment concentration as a function of time is not measured and is not known. To remedy this, historical data on sediment concentration as a function of flow rate are often used. An example of this type of data is shown in Figure 6.19. The concentration seems to increase as the flow rate increases, but there is a large variation in C at any particular flow rate. The idea is then to determine a formula that approximates sediment concentration, C, as a function of flow rate, Q, from these data. This is essentially the same problem as determining the sediment discharge from a river as a function of its flow rate; this functional relation C(Q) is commonly called a rating curve.

As a first approximation, it is generally assumed that

$$C = aQ^n \tag{6.23}$$

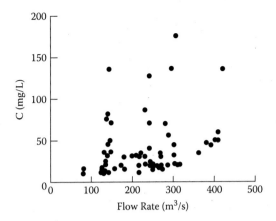

FIGURE 6.19   Suspended sediment concentrations as a function of flow rate for the Fox River.

However, because of the large amount of "scatter" usually present (as is illustrated for the example in Figure 6.19), no simple and accurate dependence of sediment concentration on flow rate is usually obvious. In fact, this is true for most rivers, and the attempt to find such a relation has generally been unsuccessful. Although conventional wisdom states that sediment concentration must be a unique function of flow rate, this is generally not true, as noted above for the Fox.

As a basis for a second and more accurate approximation, it has often been observed that there is a hysteresis effect in the rating curve; that is, for the same flow rate, the sediment concentration is higher when the flow rate is increasing (rising limb) than when the flow rate is decreasing (falling limb). An indication of this can be seen for the Fox in Figure 6.15. However, although this is often approximately true, accurately quantifying this relation usually has not been successful.

For a better understanding of the problem, consider the modeling of sediment concentration in the Fox River. By means of this modeling, the sediment concentration in the Fox has been shown to be deterministic and predictable, unlike the inference from Figure 6.19 that C depends on Q but seems to have a large random component. But Figure 6.19 is data from the Fox! It was obtained from Figure 6.15 (sediment concentration at Green Bay) and from similar modeling and data for two other, smaller events in 1989 and 1990. To interpret these data more easily, the data can be divided into separate events, as shown in Figure 6.20. In particular, consider the data for the June 1989 event that was modeled and described above. For that event, C(t) is quite deterministic, can be calculated by means of the sediment transport model, is dependent on the flow rate, but is not a unique function of Q. The other two events have been modeled in the same way and also show a deterministic, but not unique, relation between C and Q.

From data and calculations such as these, several general observations and conclusions can be made. For example, when two closely spaced flow events with

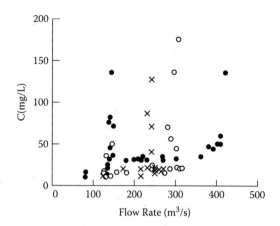

FIGURE 6.20   Suspended sediment concentrations as a function of flow rate for the Fox River for different flow events: ● June 1989; × March 23 to April 10, 1989; ○ March 12 to March 31, 1990.

similar flow rates occur, the first event has a significantly higher maximum sediment concentration than the second (i.e., a non-unique C(Q)). The reason is that, before the first event, a large amount of easily erodible sediment often exists from previous depositions; this is then swept away during the first event and is therefore not present for erosion when the second event occurs. By the same reasoning, (1) the sediment concentration will be higher when the flow rate is increasing than when the flow rate is decreasing (the hysteresis effect); and (2) for two closely spaced events with different flow rates, the first will have a larger proportional dependence of C on flow rate than the second.

As described in detail for the Fox River, the suspended sediment concentration at the end of one section of a river (for example, at Green Bay) is dependent on the erosion and deposition rates within that section (and these change with time as erosion and deposition change) but also is dependent on conditions upstream of that section (for example, upstream of DePere Dam). When these dependencies are considered, it is not surprising that C is not a unique function of Q and that the first two approximations described at the beginning of this subsection are quite limited in their accuracy.

Because C is not a unique function of Q, C cannot be uniquely or accurately determined as a function of Q from plots such as Figure 6.19. However, C(t) is deterministic and therefore can be determined by means of a sediment transport model once the upstream boundary condition is known. To determine C(t) at the upstream boundary when it is not measured, the following approximate procedure is suggested and is based on the above ideas.

As an example, for the segment from DePere Dam to the mouth of the Fox River at Green Bay, the sediment transport model should first be run for the river segment immediately above DePere Dam. For this latter segment, conditions at the dam are relatively insensitive to conditions at the upstream boundary of the upstream segment. Because only C(t) at the dam is needed for this calculation, the upstream segment can be modeled relatively crudely. From this, C(t) at the dam can then be determined and used as the upstream condition for the segment below the dam. This type of analysis also will help determine the variation in particle size as a function of time at the upstream boundary, another quantity that is important in the modeling of sediment transport but which is generally not measured.

6.4.3 USE OF SEDFLUME DATA IN MODELING EROSION RATES

As shown in the model calculations above as well as by many field measurements and laboratory tests, erosion rates change by orders of magnitude in both the horizontal and vertical directions throughout a system as well as with time as conditions vary and bed armoring occurs. Of course, in any investigation of erosion rates by means of Sedflume, the number of Sedflume cores is typically quite limited. As a result, there is the question of how best to use Sedflume data so as to adequately and analytically approximate erosion rates throughout a river for use in a sediment transport model.

To use Sedflume results accurately and efficiently, the following procedure is suggested and is essentially what was done on the Fox River (Jones and Lick, 2001a). To begin, prescribe the general horizontal distribution of grain sizes (three or more size classes) throughout the river based on preliminary surveys or estimates of grain size (including Sedflume results). This distribution is related to the variation in the hydraulic regime. With this distribution as a preliminary estimate, run a few sediment transport calculations (e.g., at small, medium, and high flow rates) to see how particle sizes are redistributed throughout the river during these flows. On the basis of the resulting distributions, choose three or more areas and approximate $E(\tau, z)$ in each area as a function of depth (including stratification when present). In each area, $E(\tau, z)$ should be obtained as an approximation from the cores in that area. In this way, $E(\tau, z)$ will vary horizontally, vertically, and with time as the calculation proceeds.

In each area, it is convenient for numerical purposes to determine an analytic approximation for $E(\tau)$. For this purpose, Equation 3.21 or Equation 3.23 is recommended. Because of the vertical stratification normally encountered, the parameters τ_c, τ_{cn}, and n will be functions of depth, that is, approximately constant within a layer but varying from one layer to the next. These parameters can be determined from Sedflume data within a layer (but not by averaging data from several layers). As a first approximation, n = 2. For more accuracy, n can be determined from Sedflume results, but n should generally be 2 or greater. The parameter τ_c depends on the bulk properties of the sediment, especially particle size and bulk density.

As an example, consider erosion rates for a core from Woods Pond in the Housatonic River (Gailani et al., 2006), as shown in Figure 6.21. Equation 3.23 will be used to approximate the data at each depth. For x = 10 cm, a reasonable approximation to the data is given by $\tau_{cn} = 0.1$ N/m^2, $\tau_c = 0.4$ N/m^2, and n = 2.4. For x = 30 cm, reasonable parameters are given by $\tau_{cn} = 0.1$ N/m^2, $\tau_c = 0.8$ N/m^2, and n = 2.6.

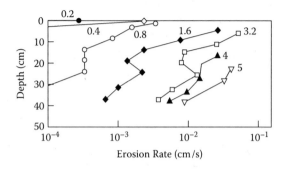

FIGURE 6.21   Erosion rate as a function of depth with shear stress (N/m²) as a parameter; core WP06. (*Source:* From Gailani et al., 2006.)

TABLE 6.4
Analytic Approximation to Sedflume Erosion Rate Data for Core WP06

Depth in sediment (cm)	τ (N/m²)	Measured E (cm/s)	Calculated E (cm/s)
10	0.8	7.6×10^{-4}	7×10^{-4}
	1.6	6×10^{-3}	4.8×10^{-3}
	3.2	2×10^{-2}	2.7×10^{-2}
30	1.6	1×10^{-3}	7×10^{-4}
	3.2	8×10^{-3}	4.8×10^{-3}
	4.0	1×10^{-2}	8.7×10^{-3}
	5.0	2×10^{-2}	1.6×10^{-2}

For these parameters, the calculated and measured values for the erosion rates are given in Table 6.4. The calculated and measured values agree with each other to within a factor of two. This can be done at each depth and, in this manner, $\tau_c(x)$ and $n(x)$ can be determined for each core. In this particular case, τ_c varies from 0.4 to 0.8 and n varies from 2.4 to 2.6 as ρ varies from 1.15 to 1.25 g/cm³ and d varies from 20 to 40 μm.

In the normal procedure for measuring erosion rates by means of Sedflume, erosion measurements are made in cycles, with a measurement of erosion rate at one depth and one shear stress followed by a measurement at the next lower depth and a higher shear stress; this is done sequentially from the lowest value of the shear stress to the highest. The cycle is then repeated. Because bulk densities of sediments usually increase with depth (whereas erosion rates decrease with depth), this procedure does not measure the erosion rates for sediments at the same density but at densities that are generally increasing. If this type of approximation is used to determine the coefficients in Equation 3.21 or Equation 3.23, this will bias the data toward increased densities (and hence lower erosion rates) as the shear stress increases; this inevitably leads to n being smaller than it should

(e.g., an n less than 2). It is therefore important to approximate the data by interpolation of $E(\tau)$ at one depth (i.e., at one density), as described above.

6.4.4 EFFECTS OF GRID SIZE

For the period from May 22, 1989 to June 20, 1989, the flow rate at DePere Dam
and the suspended sediment concentrations at DePere Dam and Green Bay are
shown in Figure 6.15. Because the sediment concentration at DePere Dam is
always greater than that at Green Bay, this indicates that this segment of the river
is net depositional during this period. From the flow rates and the differences
in sediment concentrations between the dam and the bay, an estimate of the net
deposition of sediments during this period can be made; the result is approximately 2×10^{10} g. Because the area of this segment of the river is approximately
5×10^6 m^2 = 5×10^{10} cm^2 (Tracy and Keane, 2000), this indicates an average
deposition per unit area of 0.4 g/cm^2.

This number can be converted to depth of deposition if the average sediment
bulk density is known or assumed. For a bulk density of 1.6 g/cm^2 (a reasonable
estimate), the solids fraction is about 0.4 and the water fraction is 0.6. A deposition of 1 g/cm^2 of solid then corresponds to a depth of deposition of about 1 cm.
This estimate of a net deposition of 0.4 g/cm^2 or 0.4 cm for this 30-day, high-
flow period compares well with estimates of 0.6 to 2.5 cm/yr from radioisotope
profiles, 2 to 25 cm/yr from bathymetry data, and 1 to 4.3 cm/yr from dredging
records (Limno-Tech, 1999).

The net changes in sediment bed thickness over the 30-day period as predicted by the model are shown in Figure 6.18(a). A maximum erosion of 7.5 cm
occurs in the narrow section of the Fox just below the East River junction, where
the river is narrow and moving rapidly. A maximum deposition of 19 cm occurs
just below DePere Dam in a relatively shallow and slow-moving part of the river.
Except for these few, rather isolated areas, net erosion is generally negligible in
the deeper, central channel of the river, whereas net deposition is generally 0 to
5 cm in the shallow, nearshore areas of the river.

As calculated above, the average change in sediment bed thickness over the
entire segment of the river is approximately 0.4 g/cm^2. Another way of interpreting this is to realize that a one-segment transport model (after calibration with
observations) would "predict" a deposition of approximately 0.4 g/cm^2. This is
to be compared with the results for the fine-grained model, that is, local erosions
up to 7.5 g/cm^2; local depositions up to 19 g/cm^2; erosion generally in the deeper,
central channel of the river; deposition only in the shallow, nearshore areas; and
a gradual change in sedimentation pattern from higher deposition upstream to
lower deposition and higher erosion in the downstream direction. This general
pattern of erosion/deposition as predicted by the fine-grained model is obviously
quite different from the results from the one segment model.

If a model with two segments in the longitudinal direction (but averaged across
the river) were used for this part of the river, the calibrated model would predict
an average deposition somewhat greater than 0.4 g/cm^2 in the upper segment and

an average deposition less than 0.4 g/cm², or possibly a slight erosion (less than 0.5 g/cm²), in the lower segment.

If a model with multiple segments along the river but averaged across the river were used, the model would, at best, essentially average the results of the fine-grid model over each cross-sectional segment of the river. Just below DePere Dam, the average deposition would then be about 5 g/cm². However, for the rest of the river above the midpoint, there would generally be net deposition of 1 g/cm² or less; this would decrease in the downstream direction. Below the midpoint, most cells would show some deposition, with a few cells possibly showing a small amount of erosion, probably a centimeter or less.

Because of calibration and adjustment of parameters for each of the models, the general dependence on time of the suspended sediment concentration at Green Bay for each model could be made to more or less agree with the measured concentration. However, as demonstrated, the details of the erosion/deposition patterns would be quite different. By comparison with the fine-grid model, the other models predict primarily small amounts of deposition with little or no erosion. For larger flow events, it is anticipated that areas of erosion would be larger, depths of erosion would be greater, and the differences between the models would be even greater than those shown here.

6.4.5 SEDIMENT TRANSPORT IN THE SAGINAW RIVER

The Saginaw River is a major drainage river for much of central and eastern Michigan; it flows north and empties into Saginaw Bay. It is not controlled by dams and locks; however, a channel is dredged through the lower part of the river to allow freighter traffic. Because of the disposal of contaminants from industries in the past, the bottom sediments of the Saginaw are currently contaminated.

As part of the U.S. Environmental Protection Agency's Assessment and Remediation of Contaminated Sediments (ARCS) project, an investigation was made of the extent, causes, and possible actions to remediate this contamination. One of the main concerns was with PCBs (polychlorinated biphenyls) contained in the bottom sediments. To determine the erosion and transport of these PCBs, sediment transport and the resulting bathymetric changes in the part of the lower Saginaw that extends from just below Middle Ground Island to Saginaw Bay (Figure 6.22), a distance of about 10 km, were investigated (Cardenas et al., 1995). The emphasis here is on the effects of big events on bathymetric changes and on long-term predictions of sediment transport; PCB transports in the river and Saginaw Bay are discussed in Chapter 8.

As can be seen from Figure 6.22, there are shallow areas near shore with depths of 1 m or less, except where turning basins or docks are located; here the river is dredged nearly side to side. The channel is relatively deep and steep-sided. The average width of the river is approximately 250 m, with a maximum width of 600 m. Flow rates from 1940 to 1990 have been estimated by the USGS and are shown in Figure 1.5. The maximum flow rate occurred in 1948, with a magnitude of 1930 m³/s. A flow of almost the same magnitude occurred in 1986. The

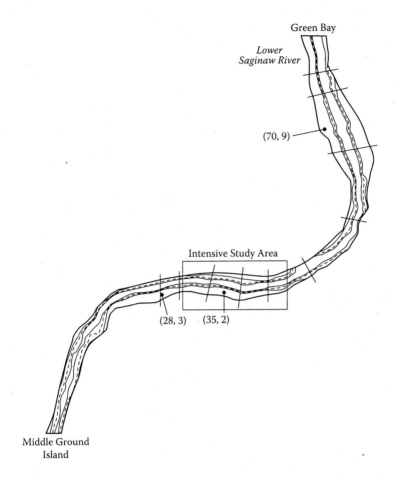

FIGURE 6.22   Bathymetry, transects, core locations, and intensive study area for the Saginaw River. Depth contours are shown for 3, 5, and 7 m. (*Source:* From Cardenas et al., 1995. With permission.)

99.97-percentile (once in 10 years) flow rate was 1527 m^3/s, the 99.7-percentile (once in 1 year) flow rate was 982 m^3/s, the 50-percentile flow rate was 57 m^3/s, and the minimum flow rate was 7 m^3/s. Over this period, the flow rate varies by more than two orders of magnitude. The river is slow flowing most of the time but has very high flow rates for very short periods of time. This is characteristic of a natural, uncontrolled river and is in contrast to a controlled river such as the Fox River, for which the flow rates vary only over one order of magnitude.

In the numerical calculations, SEDZL was used. The numerical grid consisted of 900 curvilinear elements, 90 along the river and 10 across the river. The input data for the model consisted of upstream flow rates, upstream sediment concentrations, and water surface elevations at the mouth of the river. Hydrodynamic and sediment transport calculations were made for the period from 28 August 1991 to 13 May 1992. The flow rate for this period is shown in Figure 6.23. At the

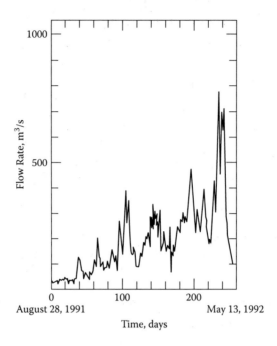

FIGURE 6.23   Flow rate on the Saginaw River from August 28, 1991, to May 13, 1992. (*Source:* From Cardenas et al., 1995. With permission.)

beginning of the period, the flows are relatively low. A series of flow events then occurs, with each event having a somewhat greater magnitude and duration. The largest flows are those on December 10 (day 105), with a peak flow rate of about 400 m^3/s; on March 11 (day 196), with a peak flow rate of about 480 m^3/s; and on April 19 (day 235), the peak of the spring runoff, with a peak flow rate of about 800 m^3/s. This is somewhat below the maximum flow rate of 982 m^3/s, expected once per year.

From surveys of suspended sediment concentrations done in 1991, the inflow sediment concentration was approximated as a function of flow rate as follows. For flows below 150 m^3/s, the concentration was assumed to be 30 mg/L. For flows greater than 150 m^3/s, the concentration was assumed to be a linear function of the flow rate, that is, C = 0.2 Q, where C is the sediment concentration in milligrams per liter (mg/L) and Q is the flow rate in cubic meters per second (m^3/s). For the purpose of verifying the model calculations of changes in sediment bed thickness, measurements of bottom elevation at each of nine transects in the Saginaw River (see Figure 6.22 for the locations of the transects) were taken on August 28, 1991, and May 13, 1992.

From field observations, it is known that the sizes of sediment particles in the Saginaw vary from less than 1 μm to more than 1 mm but are predominately in the silt and sand size range (10 to 100 μm). To accurately model sediment transport, three size classes of sediment were used: fine, medium, and coarse. Fine-grained sediments (generally less than 10 μm) have low settling speeds (less than 10^{-2} cm/s), are transported through the river without appreciable deposition on the bottom, and hence do not contribute significantly to changes in the bathymetry. In addition, these fine sediments make up a significant fraction of the suspended solids only during low flow periods. The medium-size sediments (on the order of 10 to 100 μm) are present during all flows and generally are the major fraction of the suspended solids. It was assumed that these sediments were cohesive and therefore flocculated, with their settling speed having a quasi-steady-state dependence on the state of flocculation (Sections 4.3 and 4.4). It also was assumed that their critical shear stress for deposition was 0.01 N/m^2. Coarse-grained sediments (generally greater than 100 μm) are a small fraction of the total load at low and medium flow rates. However, during large storms and high runoff events, large amounts of coarse-grained material can be washed from the surrounding drainage basin and may become a significant fraction of the total suspended solids load in the river. It was assumed that this coarse-grained material was noncohesive, did not flocculate, had an average settling speed of 1 mm/s, and had a critical shear stress for deposition of 0.02 N/m^2.

To take into consideration the dependence of the size fractions of the incoming sediments on the flow rate, the following approximations were made. For the Saginaw, the fine-grained sediments make up a significant fraction of the suspended solids only during low flow rates. For medium to high flow rates, it was assumed that the incoming suspended load consisted primarily of medium and coarse particles whose fractions of the total load depended on the flow rate. For flow rates less than 150 m^3/s, all the incoming solids were medium size. For flow rates between 150 and 600 m^3/s, the medium-size fraction was a function of flow rate Q and was given by $1.0 - 0.00156 (Q - 150)$, whereas the coarse-size fraction was the remainder. For flows greater than 600 m^3/s, the medium-size fraction was assumed constant at 0.3, whereas the coarse-size fraction was 0.7.

In addition to suspended solids transport, there is bedload transport. Bedload was assumed to act independently of suspended load. To predict bedload, the method and equations developed by Van Rijn (1993) were used.

6.4.5.1 Sediment Transport during Spring Runoff

The first results illustrated here are for changes in bathymetry during and after the spring runoff, when the maximum erosion and transport occurs. Figure 6.24(a) shows the changes in bathymetry due to sediment resuspension/deposition from April 1, the beginning of spring runoff, to April 19, the peak of spring runoff. Resuspension occurs in the channel, whereas deposition occurs in the nearshore areas. Large amounts of resuspension occur in the middle of the narrow parts of the river, whereas deposition is maximum (although relatively small) in a few nearshore areas mostly in the downstream part of the river. The suspended sediment concentration is relatively constant in the upper part of the river and decreases by only a few milligrams per liter (mg/L) in the lower part of the river. The reason for this is that, at these high flow rates, sediments generally cannot

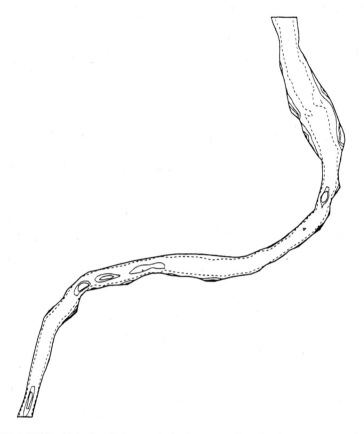

FIGURE 6.24(a) Calculated changes in bathymetry. Changes due to resuspension and deposition from April 1, the beginning of spring runoff, to April 19, the peak of spring runoff. The dashed line is no change. Contour intervals for resuspension (in the middle of the river) are 5 g/cm². Contour intervals for deposition (in the nearshore) are 1 g/cm². (*Source:* From Cardenas et al., 1995. With permission.)

deposit in most areas except for small, nearshore areas where the flows and shear stresses are small.

After the peak of spring runoff, the flows decrease until the end of the period on May 13. During this time, deposition occurs but is limited to the nearshore areas and the broad downstream part of the river. More deposition is present compared with that on April 19, but the general pattern of resuspension/deposition is similar.

The changes in bathymetry from April 1 to May 13 due to bedload are shown in Figure 6.24(b). The maximum amounts of erosion and deposition are about 25 g/cm². The effects of bedload occur predominately in small, isolated areas with alternating erosion and accumulation: erosion where the bedload is increasing with distance downstream because of increasing velocities and shear stresses (decreasing cross-sectional areas), and accumulation where the bedload is decreasing with distance downstream because of the decreasing velocities and shear

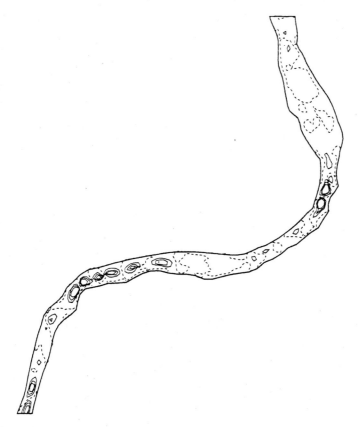

FIGURE 6.24(b)   Calculated changes in bathymetry. Changes due to bedload. Contours are from −15 to +15 g/cm², in intervals of 5 g/cm². The dashed line is zero change. (*Source:* From Cardenas et al., 1995. With permission.)

stresses (increasing cross-sectional areas, similar to that described in Section 6.3). Although there are large local changes in bed thickness due to bedload, the net transport of sediments is due primarily to suspended load rather than bedload.

For verification of the model, calculations of the changes in bathymetry from August 28, 1991, to May 13, 1992, were compared with the measured changes in the average channel depth at the nine transects on the river (Figure 6.22). In general, the results agreed reasonably well with the measured changes. The general trends of erosion/deposition along the river were correct, and the calculated results were usually within a few centimeters of the measured changes.

6.4.5.2 Long-Term Sediment Transport Predictions

Because sediment resuspension is a highly nonlinear function of flow rate and flow rates in the river are highly variable, sediment transport is also highly variable, with long periods of low transport interspersed with short periods with very high sediment transport. Because of this variability, the proper procedure for the prediction of sediment erosion and transport over a long time period is not obvious. For example, for an average flow rate, very little erosion occurs, mainly deposition, and therefore an average flow rate cannot be used to predict sediment erosion and transport over a long period of time. Similarly, the approach of choosing a "representative" year and repeating it 25 times to construct a 25-year scenario is also incorrect. In this case, choosing the year with the maximum flow would not be representative because the resulting simulation would show a very large amount of sediment transport, an amount much higher than the average. Choosing a year without the maximum flow would also not be representative because much of the sediment and contaminant transport occurs during this maximum flow.

For the purpose of predicting flow rates and sediment and contaminant transport in the future, the most reasonable assumption is that flows in the future will be statistically similar to flows in the past; that is, flow events with a certain magnitude will have the same frequency in the future as in the past. This assumes, of course, that no significant changes in the drainage area or controls on the river have been or will be made.

Based on this idea, the approach used to determine the long-term sediment transport in the Saginaw River was first to determine the frequency of representative flow events of different magnitudes from the historical record — that is, from the 48 years of flow data indicated above. The magnitudes of the representative flow events were chosen to be 500, 1000, 1500, and 1900 m^3/s (the largest flow during the 48-year period). Except for the largest event, these events were then distributed evenly over a 25-year period at the frequency determined. The resulting distribution of flow events was that twenty-five 500-m^3/s, eight 1000-m^3/s, and two 1500-m^3/s flow events were placed evenly over the 25-year period. Because of the significance of the largest flow event (1900 m^3/s), calculations were made for comparative purposes with this flow at the beginning, middle, or end of the time period. The actual flows during the events that were used to model the 25-year long scenarios resulted from composites of many different events that occurred in the past.

The amount of resuspension/deposition that was calculated during this 25-year period is highly variable in space and time. To describe the results concisely but still present results that are characteristic of most locations in the river, changes in sediment bed thickness with time due to resuspension and deposition are presented for three different but representative locations. These locations are all in shallow, nearshore waters and are shown in Figure 6.22. They include a downstream location (70, 9), which is always depositional; an upstream location, (28, 3), where both resuspension and deposition occur but little net resuspension occurs over the 25-year period; and a location in the middle of the intensive study area (35, 2), which is generally depositional.

The deeper parts of the river (Figure 6.22) are subject to stronger shear stresses than the shallower parts of the river. Because of this, they are generally erosional in moderate to strong flows. Contaminated sediments (which tend to be finer-grained) are generally not found in these deeper parts of the river because any contaminated sediments that might settle to the bottom during a low-flow period are swept away in the next moderate to strong flow.

Figure 6.25(a) shows the thickness of the sediment bed as a function of time at location (70, 9) for the three cases where the largest event occurs at the beginning, middle, and end of the time period. It can be seen that the flow is always depositional. At this location, the largest flow event causes only a modest amount of deposition. If the largest event occurs at the beginning of the period, the amount of deposited sediments at any time is greater than if the largest event had not yet occurred. The final change in the thickness of the sediment bed is the same for all three cases.

Figure 6.25(b) shows the sediment bed thickness as a function of time at location (28, 3), which is upstream and in a narrower part of the river compared to location (70, 9). This area shows both resuspension and deposition for each of the three simulations, with deposition occurring during low flow periods and erosion occurring during increased flows. The largest erosion occurs during the largest flow event. The time of the event greatly affects the maximum negative net change in bed thickness at this location in the river. When the largest flow occurs at the beginning of the period, the maximum negative net change is about 28 g/cm^2. When the largest flow occurs in the middle of the period, this change is about 14 g/cm^2. When the largest flow occurs at the end of the period, it is about 1 g/cm^2. The significance of this is that if the contaminant is located at a certain depth in the sediment (and possibly under clean sediments), there is more likelihood that a large flow at the beginning of the period can reach that contaminated material. In the cases of the second and third simulations, significant deposition has occurred before the large flow, and the erosion of the bed relative to the initial sediment-water interaction is therefore less.

Location (35, 2) is near shore in the middle of the intensive study area. The calculations show that this area is generally depositional (Figure 6.25(c)); the largest flow event causes a small amount of erosion (less than 1 g/cm^2), whereas the other flow events are depositional. Because the largest event causes little change in the sediment bed thickness at this location, the magnitude of the sediment bed thickness for all three simulations is nearly the same throughout the entire time period.

As a first approximation, the net effect on sediment transport and the change in sediment bed thickness at any river location over a 25-year period are independent of the time at which the largest flow event occurred (as long as it does occur within the period). However, both the transport and the changes in sediment bed thickness at any particular time during this period generally depend greatly on whether the largest flow event has already occurred. These calculations indicate the potential and inherent variability (uncertainty) in any 25-year prediction because the time of occurrence of a 25-year flow cannot be predicted. All that is known is the probability of its occurrence. The effects of this inherent variability are different at different sites, as has been discussed above. In any long-term prediction, this variability is significant and must be considered.

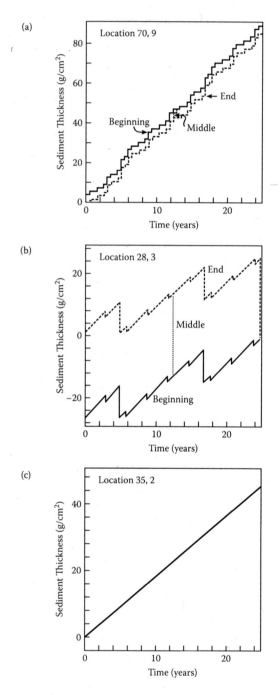

FIGURE 6.25 Change in sediment bed thickness as a function of time: (a) location (70, 9); (b) location (28, 3); and (c) location (35, 2). (*Source:* From Cardenas et al., 1995. With permission.)

6.5 LAKES AND BAYS

In rivers, bottom shear stresses are primarily due to currents; wave action is generally not significant except near the river mouth, where waves may enter from open waters. In contrast, in lakes and nontidal bays, bottom shear stresses are primarily due to wave action, with wind-driven currents generally of secondary importance. Because of these differences, patterns of bottom shear stress, erosion/deposition, and particle size variations are quite different in the two types of systems.

In the present section, the sediment transports in Lake Erie (Lick et al., 1994a) and in Green Bay (Wang et al., 1996) are examined by means of numerical models. The differences between rivers and enclosed (or semi-enclosed) systems such as lakes and bays will be evident. The calculations for Lake Erie emphasize and illustrate the effects of big events on sediment transport and are compared with geochronological data. Calculations for Green Bay were made by means of (1) a three-dimensional, nonstratified (constant-density) model; (2) a two-dimensional, vertically integrated model; and (3) a three-dimensional, thermally stratified (variable density) model. The results of these models are then compared under the same conditions and illustrate the effects (and non-effects) of thermal stratification; there are strong similarities between the results of the three models.

6.5.1 MODELING BIG EVENTS IN LAKE ERIE

In the calculations for Lake Erie, the SEDZL model with a 5-km grid was used. The sediment bed was assumed to consist of 11 layers in the vertical direction, with the properties of each layer depending on time after deposition; in each layer, τ_c varied from 0.01 N/m² for freshly deposited sediments to 0.1 N/m² for deposition times greater than 1 day. The amplitudes and periods of the wind-driven surface gravity waves were calculated by means of the SMB hindcasting method (U.S. Army CERC, 1973; Section 5.5). Although waves are the major cause of the bottom shear stress, currents also contribute significantly to this stress under certain conditions. Because of this, the combined effects of currents and waves on the bottom stress were considered, and the theoretical results due to Christoffersen and Jonsson were used (Section 5.1).

6.5.1.1 Transport due to Uniform Winds

Winds in the vicinity of Lake Erie vary widely in magnitude and direction and as a function of space and time. Because of the later emphasis on the geochronology of the Eastern Basin, because of the reasonably extensive records of wind speed and direction at the Buffalo Airport, and in an effort to simplify the calculations, all wind conditions were referenced to those at the Buffalo Airport. At this airport, the average wind speed is approximately 11 mph, whereas the maximum sustainable winds during the year are about 45 mph (National Climatic Data Center, 1990, and airport data from NOAA). Calculations of sediment transport were therefore made for winds of 11 mph, 22.5 mph (half of 45 mph and approximately twice 11 mph), and 45 mph. For each of these wind speeds, calculations were done for a southwest wind (the dominant wind throughout the year), a northeast wind (the dominant direction during large storms), a southeast wind, and a northwest wind.

For each calculation, it was assumed that the wind direction was constant throughout the period. However, the wind speed was assumed to vary as follows: constant at 5 mph for 2 days, constant at the specified stormy wind speed for 1 day, and then constant at 5 mph for another 10 days. This latter period was needed to allow the sediments to settle out of the water column so that net deposition/ erosion for the entire period could be calculated.

Results of the hydrodynamic calculations for a 22.5-mph wind from the southwest were summarized in Section 5.2; wind-driven currents are shown in Figure 5.6, whereas the bottom shear stress is shown in Figure 5.7. Figure 6.26(a) shows the suspended solids concentrations at the end of the stormy period. Concentrations in large central regions of the Western, Central, and Eastern basins are less than 1 mg/L. However, the concentrations increase rapidly toward shore (where the depths are less and wave action and hence resuspension are greater) and toward the northeast (because wave action increases in the direction of the wind).

Figure 6.26(b) shows the net deposition/erosion at the end of 13 days. At this time, almost all the sediments resuspended by the storm have settled out of the water. The nearshore sediments eroded during the storm have been transported offshore by the currents, where they are then deposited. The amount of deposition generally reaches a maximum just offshore and then decreases rapidly as the center of

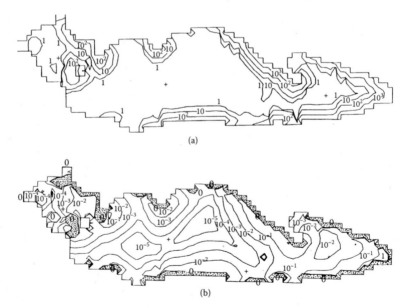

(a)

(b)

FIGURE 6.26   Lake Erie. Winds are from the southwest at 22.5 mph for a period of 1 day. (a) Suspended solids concentration (mg/L) at end of storm; and (b) deposition/erosion (g/cm^2) at end of 13-day period, 10 days after end of storm. Areas of erosion are shown as dotted. (*Source:* From Lick et al., 1994a. With permission.)

TABLE 6.5
Deposition in Eastern Basin of Lake Erie (g/cm²)

Wind direction	Wind magnitude (mi/hr)		
	11	22.5	45
Southwest	<10^{-6}	0.01	2.4
Northeast	<10^{-6}	0.0008	0.7
Northwest	<10^{-6}	0.005	1.6
Southeast	<10^{-6}	0.0003	1.0

the basin is approached. In the Central Basin, maximum depositions are on the order of 10^{-2} to 10^{-1} g/cm², whereas the amounts deposited in the center are as low as 10^{-5} g/cm². In the Eastern Basin, the maximum deposition (in a small northeastern area) is on the order of 1 g/cm²; the deposition decreases to 10^{-2} g/cm² in the middle of the basin. In the Western Basin, the net deposition is fairly small everywhere.

For purposes of comparison of the effects of different wind speeds and directions, the net deposition at the core location in the Eastern Basin, denoted by a plus sign in Figure 6.26(b), is shown in Table 6.5 for each wind speed and wind direction. For the average 11-mph wind, net deposition at this location is less than 10^{-6} g/cm² for all wind directions. For the 22.5-mph wind, the net deposition is largest for the southwest wind (0.01 g/cm²) and decreases to 0.0003 g/cm² for the southeast wind. For the 45-mph wind, net deposition is relatively large for all wind directions, is largest for the southwest wind (2.4 g/cm²), and is least for the northeast wind (0.7 g/cm²), but is generally on the order of 1 g/cm². The very nonlinear effect of increasing wind speed is quite evident, as is the lesser effect of wind direction.

6.5.1.2 The 1940 Armistice Day Storm

On November 10 and 11, 1940, probably the largest storm of the century over Lake Erie occurred (U.S. Weather Service, 1976; Murty and Polavarapu, 1975). Meteorological data for this period of time are quite sparse; however, wind magnitude and direction for the Buffalo Airport were available from NOAA and were used in a calculation as the wind forcing the waves and currents on Lake Erie. In general, this period was very stormy, with large storms on November 6, 15, 19, and 21 as well as the Armistice Day storm on November 10 and 11. During this period, the maximum sustainable winds were 22.4 m/s, or 50.2 mph. The winds varied greatly in direction but, when the winds were large, were generally from the northeast. The hydrodynamic and sediment transport calculations were begun on November 5 and were continued through November 23 with the winds as measured. The calculations then were continued for 20 more days with winds of 5 mph; this was to allow the sediments to settle out of the water so that net erosion/deposition could be determined.

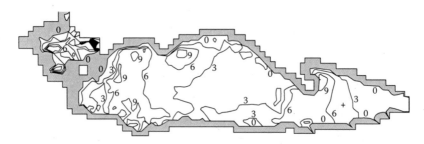

FIGURE 6.27 Deposition/erosion (g/cm^2) at end of 1940 storm on November 23. Areas of erosion are shown as dotted. (*Source:* From Lick et al., 1994a. With permission.)

Net erosion/deposition at the end of this period is shown in Figure 6.27. Net erosion occurred in almost all the nearshore areas. Because the winds were predominantly from the northeast, the major wave activity and resulting erosion occurred near the western shores of the Western Basin and the shallow ridge between the Western and Central basins. In these regions, up to 100 g/cm^2, or 1 to 2 m, of erosion occurred. In the Western Basin, this eroded material was deposited primarily in the middle of the basin. In the Central Basin, the sediments eroded from the western nearshore areas were transported and deposited further offshore, generally in an easterly direction. In the Eastern Basin, the major erosion occurred near Long Point; much of this eroded sediment was then transported a short distance offshore, with a maximum deposited area just offshore of Long Point with a deposition of more than 9 g/cm^2. At the core location in the Eastern Basin, approximately 4.5 g/cm^2 of sediment was deposited.

6.5.1.3 Geochronology

For the site in the Eastern Basin of Lake Erie denoted by a plus sign in Figure 6.26(b), measurements of excess ^{210}Pb and ^{137}Cs were made by Robbins et al. (1978) and are shown as a function of depth in Figure 6.28. Consider the ^{210}Pb data first. The half-life for ^{210}Pb decay is 19.4 years. If the sedimentation rate were uniform, the age of the sediments would be proportional to the depth, the plot of log ^{210}Pb as a function of depth should be a straight line, and a (uniform) deposition rate could then be uniquely defined. To a first approximation, the data can be approximated by a straight line. However, closer examination indicates significant deviations from a straight line; in fact, a better approximation would be a series of steps. As noted by Robbins et al. (1978), these deviations are common in other cores and are reproducible not only in cores from the same location but also from cores over a wide area. Robbins et al. have suggested that these deviations are related to major storms.

Based on storm records by Lewis (1987), the discussion of early storms by Murty and Polavarapu (1975), and previous calculations, an interpretation of the ^{210}Pb data as a series of steps caused by major identifiable storms is as shown in Figure 6.28. The justification for this interpretation is as follows. At the site in the Eastern Basin being considered here, the depth is 58 m; even in major storms, the

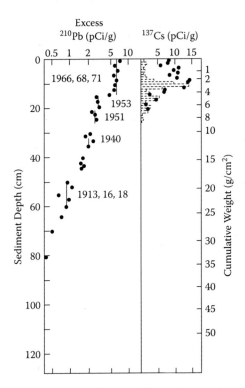

FIGURE 6.28   Distribution of excess ^{210}Pb and ^{137}Cs with depth in cores from the Eastern Basin of Lake Erie. Also shown is the interpretation of the ^{210}Pb data as a series of depositions due to major identifiable storms. Horizontal dashed lines indicate cesium production. (*Source:* Modified from Robbins et al., 1978. From Lick et al., 1994a. With permission.)

amount of resuspension during a storm is very small, and hence net deposition generally occurs over the period of an entire storm at this site. Because of this, each of the steps shown in Figure 6.28 is due to deposition by one or more (closely spaced) storms. It can be seen that at the surface, the top 15 cm of sediment has a reasonably constant ^{210}Pb concentration. This is attributed to deposition due to a series of large storms that occurred in 1966, 1968, and 1971. Because of their proximity, these storms cannot be separately identified in the ^{210}Pb record. In contrast, between 1953 and 1966, the winds over Lake Erie were remarkably calm, with no major storms and hence no significant deposition at the site during this period. This lack of storm activity is responsible for the break in the ^{210}Pb data at about 15 cm. A large storm did occur in 1953 and another in 1951, and these two are responsible for the much older sediments found at depths of 15 to 25 cm. The Armistice Day storm and the other storms in November 1940 and possibly a storm in 1947 are responsible for the deposition at 30 to 40 cm. At depths of 40 to 50 cm, another major storm (or series of closely spaced storms) is indicated, but meteorological records are not sufficient to identify the storms. Storms during

1913, 1916, and 1918 have been qualitatively recorded (Murty and Polavarapu, 1975) and can be identified in the ^{210}Pb data at depths of 50 to 65 cm as shown.

Depositions deduced from the ^{210}Pb data are consistent with the calculated depositions. For example, the calculated deposition during the 1940 storm was 4.6 g/cm^2, whereas the ^{210}Pb data indicated a deposition on the order of 5 g/cm^2. Calculations of depositions for more recent storms since 1940 are also in general agreement with the ^{210}Pb data. The theoretical excess ^{210}Pb as determined from radioactive decay with a half-life of 19.4 years is shown in Figure 6.28 for each storm event and is, of course, consistent with the dates of the storms shown.

Also shown in Figure 6.28 are the observed ^{137}Cs concentrations as a function of depth as well as the rate of ^{137}Cs production (and presumably deposition) as a function of time for each year since 1950. The ^{137}Cs production began about 1950, peaked in 1960, and then decreased rapidly as shown. If sediment deposition was uniform with time, the ^{137}Cs concentration would show the same pattern as a function of depth as the ^{137}Cs production; it is quite clear that they are significantly different. However, the ^{137}Cs data are consistent with the storm activity and deposition deduced from the ^{210}Pb data as described above. The top 12 to 15 cm of sediment shows relatively uniform ^{137}Cs concentrations, as would be expected from strong mixing and deposition due to the 1966 through 1971 storms. The 1951 and 1953 storms deposited sediments with nonzero but relatively small ^{137}Cs concentrations at depths of 15 to 25 cm. No ^{137}Cs appears in the sediments before this time.

A layer of constant properties near the sediment-water interface is often attributed to activity by benthic organisms. In the present case, because of the depth at the Eastern Basin coring site (58 m), this effect of benthic organisms is probably negligible. As indicated by the present calculations, a layer of constant properties occurs at this location, can occur almost anywhere in the lake, and is due to resuspension of sediments during a storm and deposition after the storm. After this deposition, no further movement or mixing of the sediments occurs. This is discussed further in Section 8.5.

In these calculations, sediment properties (except for those for newly deposited sediments consolidating with time) were assumed to be uniform throughout. As emphasized previously, particle sizes, bulk densities, and erosion rates vary markedly throughout a system. Because of this, the above results are valid to describe the major characteristics of the system (i.e., the importance of big events) but are not sufficient for an accurate description of sediment transport throughout the system. For this purpose, SEDZLJ or a similar model, along with Sedflume data for erosion rates, are necessary.

6.5.2 COMPARISON OF SEDIMENT TRANSPORT MODELS FOR GREEN BAY

Green Bay is a narrow body of water approximately 188 km long and 37 km wide that opens into Lake Michigan (Figure 6.29). The bay consists of a northern, or outer, bay and a southern, or inner, bay separated by a shallow sill and Chambers Island. The inner bay is relatively shallow, with an average depth of approximately

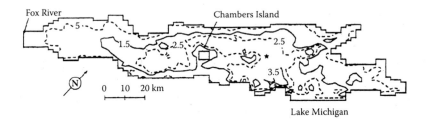

FIGURE 6.29   Green Bay bathymetry. Depth contours of 5, 15, 25, and 35 m are shown. (*Source:* From Wang et al., 1996. With permission.)

10 m and a maximum depth of 33 m, whereas the outer bay is deeper, with an average depth of 17 m and a maximum depth of almost 45 m.

During much of the summer, the bay is thermally stratified. In general, thermal stratification can cause significant changes in the wind-driven currents of a lake or bay. Because of these changes in the currents, the transport and fate of sediments and contaminants also are modified. To investigate these effects, the transport and fate of sediments in a thermally stratified Green Bay were investigated by means of numerical models (Wang et al., 1996).

In this modeling, the process models used were those describing (1) currents; (2) wave dynamics; (3) the bottom shear stresses due to waves and currents; and (4) sediment resuspension, transport, and deposition. For purposes of comparison of different transport model approximations, results were calculated and are shown here for (1) a three-dimensional, nonstratified (constant-density) model; (2) a two-dimensional, vertically integrated model; and (3) a three-dimensional, thermally stratified (variable density) model. In all calculations, wave action was described by the SMB method, shear stresses due to both wave action and currents were calculated by the procedure due to Christofferson and Jonsson (1985), and sediment bed dynamics were described as in SEDZL. The emphasis of the modeling was on the effects of large storms.

The three-dimensional, time-dependent model of the wind-driven currents used here (both constant- and variable-density versions) was developed by Blumberg and Mellor (1980, 1987) and is called ECOM-3D. To describe the transport of sediments more economically, a two-dimensional, vertically integrated, time-dependent model of the wind-driven currents and sediment transport, as in SEDZL, was used. This type of model is valid when the water is nonstratified and when the suspended sediment concentrations are approximately independent of depth. It will be seen that this model is also a very good approximation for the prediction of the resuspension and deposition of bottom sediments in many situations, even when the flow is stratified.

Using these models, calculations of sediment transport in Green Bay were made for a variety of wind conditions. The numerical elements used were rectangular and 2.5 × 1.2 km in size. In the calculations illustrated here, it was assumed that initially there was a short stormy period of 2 days with high winds of 20 m/s

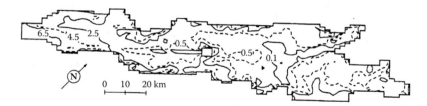

FIGURE 6.30　Shear stresses due to combined wave action and currents at end of the stormy period. Contours of 0.1, 0.5, 2.5, 4.5, and 6.5 N/m² are shown. Winds are from the northeast at 20 m/s. (*Source:* From Wang et al., 1996. With permission.)

from the northeast; this was followed by a 4-day period with low winds of 2 m/s from the northeast.

Results of calculations by means of ECOM-3D for nonstratified conditions are shown in Figures 6.30 and 6.31(a), (b), (c), and (d). Figure 6.30 shows the bottom shear stresses at the end of the stormy period. These shear stresses are almost entirely due to wave action. They generally increase from right to left (in the downwind direction) as the wave action increases and are below 0.1 N/m² in the deeper parts of the inner and outer bay. Due to increasing wave activity and the shallow waters in the southwestern part of the bay, the shear stresses reach a maximum of almost 6.5 N/m² near the Fox River.

The vertically averaged currents at the end of the 2-day stormy period are shown in Figure 6.31(a). In the nearshore, the currents are primarily in the direction of the wind, whereas in the middle of the bay, a return flow in the opposite direction to the wind occurs. The vertically averaged suspended sediment concentrations at the end of the stormy period are shown in Figure 6.31(b). Concentrations increase from right to left as the wave activity and shear stresses increase and are highest in the high-shear-stress, shallow area near the Fox. Figure 6.31(c) shows the net amount of sediment eroded at the end of the 2-day stormy period. At this time, little deposition has occurred. The deeper parts of the bay show a small amount of net deposition (shaded area), whereas net erosion occurs in most of the bay, especially near the Fox River. After the stormy period, sediments are transported by the relatively slowly moving currents after the storm and also deposit on the bottom. The net amount of deposition at the end of the 6-day event is shown in Figure 6.31(d). Net deposition occurs over most of the bay except in small nearshore areas. The maximum deposition occurs in the shallow part of the bay a short distance from the Fox. Little sediment is transported and deposited in the outer bay.

For comparative purposes, this same example was calculated using a two-dimensional, vertically-integrated model of the wind-driven currents and sediment transport. All other process models and parameters remained the same. Because the bottom shear stresses were primarily caused by wave action, the shear stresses as predicted by wave action and the two-dimensional model of the currents are essentially the same as those produced by wave action and the three-dimensional model of currents. Suspended sediment concentrations were also similar to those in the previous case. At the end of the 2-day stormy period, the net erosion of the bottom

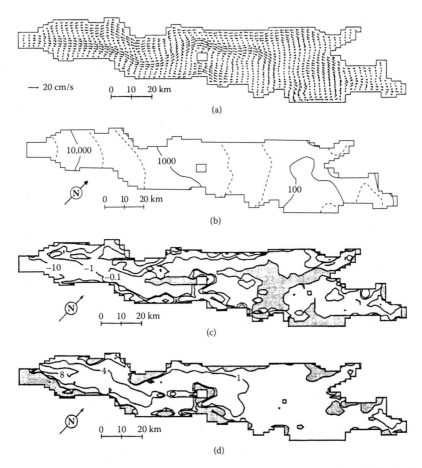

— 20 cm/s

(a)

(b)

(c)

(d)

FIGURE 6.31 Calculations by means of ECOM-3D. Constant density. (a) Vertically averaged currents at the end of the stormy period; (b) vertically averaged suspended sediment concentrations (mg/L) at the end of the stormy period; (c) net erosion (g/cm²) of bottom sediments at end of stormy period. Net deposition areas are shown as shaded; and (d) net deposition (g/cm²) at 4 days after the end of the stormy period. Net erosional areas are shown as shaded. (*Source:* From Wang et al., 1996. With permission.)

sediments is shown in Figure 6.32(a), whereas the net deposition at the end of the 6-day event is shown in Figure 6.32(b). The net erosion during the storm is almost identical to that predicted by the three-dimensional model (Figure 6.31(c)), whereas small (but generally insignificant) differences between the two models exist for the net deposition at the end of the event (Figures 6.32(b) and 6.31(d)).

Three-dimensional, time-dependent calculations were also made by means of ECOM-3D for a thermally stratified flow for these same wind conditions. It was assumed that the initial temperature in the epilimnion (the top 20 m of the water) was 18°C, whereas the temperature in the hypolimnion (the water below 20 m) was 12°C. The bottom shear stresses, once again, were dominated by wave action

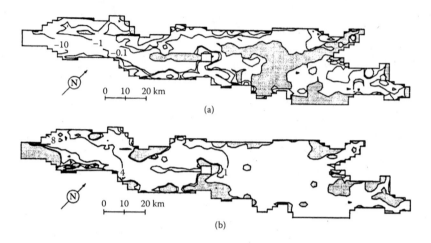

FIGURE 6.32   Calculations by means of two-dimensional, vertically integrated model. (a) Net erosion (g/cm^2) at end of the stormy period. Depositional areas are shown as shaded. (b) Net deposition (g/cm^2) at 4 days after the end of the stormy period. Net erosional areas are shown as shaded. (*Source:* From Wang et al., 1996. With permission.)

so that the bottom shear stresses were essentially the same as those in the previous two cases. During the event, because of the stratification, the currents were quite different in the epilimnion and hypolimnion and quite different from the nonstratified case. Because of these differences and because of the settling of the particles, there are differences in the sediment concentrations at the surface and at the bottom of the bay at the end of the stormy period and generally throughout the event. The vertically averaged concentrations also are somewhat different from those for the nonstratified case, especially near the end of the event, where the effects of transport are more important.

Despite these differences in the currents and sediment concentrations, it can be seen by comparing Figures 6.33(a) (stratified flow) and 6.31(c) (nonstratified flow) that the net erosion of bottom sediments at the end of the stormy period is essentially the same for the stratified and nonstratified cases. The reasons for this are as follows: (1) during the storm, the suspended sediment concentration is primarily governed by the erosion of sediments, whereas deposition and even transport are small effects, and (2) erosion is dominated by shear stresses due to wave action, which is not strongly influenced by thermal stratification.

For the stratified flow, the net deposition at the end of the event is shown in Figure 6.33(b). Because of the differences in currents in the stratified and nonstratified cases, the net deposition for the stratified case is noticeably different from that for the nonstratified case (Figure 6.31(d)). However, for practical purposes and considering our limited knowledge of sediment resuspension and deposition parameters, the results for net sediment erosion and deposition as predicted by (1) the three-dimensional, nonstratified case; (2) the three-dimensional, stratified case; and (3) the two-dimensional, vertically integrated case are all essentially the

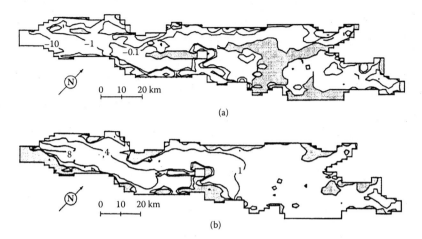

(a)

(b)

FIGURE 6.33   Calculations by means of ECOM-3D. Thermally stratified, variable density. (a) Net erosion (g/cm²) of bottom sediments at end of stormy period. Net deposition areas are shown as shaded. (b) Net deposition (g/cm²) at 4 days after the end of the stormy period. Net erosional areas are shown as shaded. (*Source:* From Wang et al., 1996. With permission.)

same. For longer storms, and especially for longer periods of moderate to strong winds after the storm, the effects of transport will be more significant, and the differences between the results from the different models may be more obvious.

Calculations also were made for different wind directions and different wind speeds (20, 10, and 5 m/s). For specific conditions, the results from all three model calculations were essentially the same. At low wind speeds, the three-dimensional and stratification effects on the currents were more significant; however, the transport due to these currents again had a relatively small effect on the overall pattern of resuspension and deposition of bottom sediments.

6.6 FORMATION OF A TURBIDITY MAXIMUM IN AN ESTUARY

The turbidity maximum is a region in an estuary where a maximum in surface sediment concentration occurs relative to sediment concentrations both upstream and downstream. A qualitative explanation of this maximum is as follows. From field and laboratory studies, it is well known that in estuaries, a two-layer circulation pattern tends to form, with the upper layer of fresh river water flowing downstream and a bottom layer of denser sea water flowing slowly upstream and mixing with the fresh water. Sediments are carried downstream with the river water and tend to settle out as they reach deeper, slower-moving, and less turbulent areas of the river and bay. As they settle toward the bottom, the upstream-flowing sea water near the bottom can transport these sediments back up the river. Where the upstream flow meets the downstream flow in a strongly stratified estuary, vertical fluid velocities can occur that are relatively large compared to those present in nonstratified flows. These vertical velocities carry sediments back toward the surface, where they are then transported downstream and tend to settle out again. This continuous process causes a recirculation of sediments in the vicinity of the fresh water/salt water interface with an associated increase in local suspended solids concentration — that is, the turbidity maximum. The recirculation and accumulation of sediments are complicated further by the bathymetry and the effects of tides, that is, by the increased vertical velocities near the salt water wedge that are caused by the tidal oscillations and also by the deposition of sediments on the bottom during slack periods and the subsequent resuspension of these recently deposited, low-density, and easily resuspendable sediments during high flow periods.

6.6.1 Numerical Model and Transport Parameters

Calculations have been made to investigate more quantitatively the processes and parameters that affect the turbidity maximum (Pickens et al., 1994). To be general but fairly realistic, a simplified geometry and flow parameters representative of the Passaic River–Newark Bay system were chosen for the calculations. A plan view of the assumed geometry is shown in Figure 6.34. A dam is located 26 km upstream of the river mouth, and the fresh water flow rate and sediment concentration are specified at this dam. The water in the river then flows into the bay; mixing of the river water with sea water occurs partially in the river near the mouth as well as in the bay. Specified boundary conditions in the bay are such that flow into and out of the bay can occur at the three open boundaries of the bay. The actual computational region extends for 64 km beyond the river mouth so that the total length of the computational region is 90 km. This length is needed so that assumed conditions at the boundaries of the bay do not significantly affect the flow near the river mouth. For clarity of presentation, results are presented here for only 60 km of the computational region.

The hydrodynamic model employed for this study was three-dimensional and time dependent (Paul and Lick, 1985). A sigma transformation in the vertical direction was used so as to include irregular bottom topography. To close the three-dimensional conservation equations for mass, momentum, and salinity, values for the horizontal and vertical eddy viscosity and diffusivity coefficients

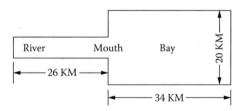

FIGURE 6.34   Plan view of estuary. Actual computation region extends for 90 km, 64 km beyond the mouth of the river; only a region of 60 km is shown here and in the following figures for clarity. (*Source:* From Pickens et al., 1994. With permission.)

must be determined. As in Section 5.1, these were assumed to be functions of the Richardson number, the ratio of buoyancy to inertia forces. With the inclusion of salinity effects on the buoyancy, the appropriate formulas are:

$$A_v = \frac{A_{v0}}{\left[1 + \delta_1 \left(\frac{H_0}{U_0}\right)^2 \frac{\partial S}{\partial z}\right]^{1/2}} \tag{6.24}$$

$$B_v = \frac{B_{v0}}{\left[1 + \delta_2 \left(\frac{H_0}{U_0}\right)^2 \frac{\partial S}{\partial z}\right]^{3/2}} \tag{6.25}$$

where A_{v0} and B_{v0} are the eddy viscosity and eddy diffusivity in the absence of stratification; H_0 and U_0 are a reference depth and velocity, respectively; δ_1 and δ_2 are empirical parameters that determine the magnitude of the effects of salinity gradients on turbulent mixing in the vertical direction; and $\partial S/\partial z$ is the vertical salinity gradient. It was assumed that vertical mixing for salinity and sediment was governed by the same turbulent eddies. Therefore, B_{v0} and δ_2 for salinity were the same as those for sediments. The horizontal viscosity and diffusivities were all assumed to be the same and constant.

The upstream boundary conditions are that the incoming fresh water velocity and suspended solids concentration are specified at the dam. The suspended sediments at the dam are assumed to be in a quasi-steady state such that the vertical distribution of sediments is determined by a balance between settling and vertical diffusion. The suspended sediment concentration C is therefore given by

$$C = C_S \exp(-w_s(z-h)/A_{v0}) \tag{6.26}$$

where C_S is the incoming suspended sediment concentration at the air-water interface, z is the vertical coordinate and is positive in the upward direction from the sediment-water interface, and h is the water depth. Due to changing hydrodynamic conditions, the sediment distribution will change as a function of distance downstream as well as across stream. The dynamics of the sediment bed, including resuspension, deposition, and compaction with time, were described by SEDZL.

6.6.2 NUMERICAL CALCULATIONS

6.6.2.1 A Constant-Depth, Steady-State Flow

To examine the effects of various flow parameters, the first case studied was for a constant-depth, steady-state flow; a water depth of 7.2 m was assumed. Values for other parameters were as follows: $C_S = 14.8$ mg/L, $A_{v0} = B_{v0} = 1.0$ cm^2/s,

$\delta_1 = \delta_2 = 0.1$, $A_H = B_H = 1.5 \times 10^6$ cm^2/s, and the salinity of the incoming sea water = 26 ppt. The formation of a turbidity maximum strongly depends on the settling speeds, w_s, of the sedimentary particles. In the calculations shown here, w_s was specified as 0.008 cm/s, a typical value for flocs of fine-grained sediments. The effects of different values of w_s are discussed below.

For a constant-depth estuary, the intrusion length of the salt wedge and turbidity maximum are sensitive to the river flow velocity. At low flows (river velocity of 6 cm/s), the salt wedge intrudes upstream to the dam. For higher flow rates of 10 to 20 cm/s, the salt wedge intrusion length is limited to within 10 km of the river mouth. However, formation of a distinct turbidity maximum does not occur for these velocities. The formation of a distinct turbidity maximum was found to require a strongly stratified estuary, with vertical fluid velocities significantly exceeding the sediment-settling velocity. For this to occur, the inlet flow velocity for a constant-depth system must be on the order of 40 cm/s or higher.

For a river velocity of 40 cm/s and no tidal effects, the calculated flow velocities in a vertical cross-section of the estuary are shown in Figure 6.35(a). The sea water intrusion and the general circulation pattern of fresh river water flowing over the slowly inflowing sea water are evident. The maximum vertical fluid velocities at mid-depth are approximately 0.04 cm/s. As seen in Figure 6.35(b), there is a large gradient in the salinity near the river mouth. Due to the strength of the inertial forces of the river flow, the sea water intrusion (defined as the 1-ppt salinity contour) is limited to 3 km upstream of the river mouth. In Figure 6.35(c), the vertical transport of sediment near the tip of the salt wedge and the resulting circulation of the sediment in this region can be readily seen. The turbidity maximum is quite pronounced, with sediment concentrations at the surface increasing from 14 mg/L in the upstream direction to more than 140 mg/L just 2 km downstream of the tip of the salt wedge.

The strength of the turbidity maximum depends on the magnitude of the settling speed. At values of w_s below 0.008 cm/s, the concentration profiles exhibit a more well-mixed character, with relatively small changes in the concentration from the surface to the mid-depth fluid layers. Increasing the sediment-settling speed to values equal to or greater than 0.01 cm/s also results in a weaker turbidity maximum. In this case, the heavier sediments settle into the bottom layers more rapidly, producing relatively low sediment concentrations at the surface and at mid-depth. Although a significant upward flux of sediments is observed from the bottom to the mid-depth layers, this transport does not extend to the surface, and a high sediment concentration is therefore not visible at the surface. Hence, the turbidity maximum is weakened for high as well as for low values of w_s.

6.6.2.2 A Variable-Depth, Steady-State Flow

To investigate the effects of variable depth, calculations were also performed for more realistic bottom topographies. In the case illustrated here (Figure 6.36), the

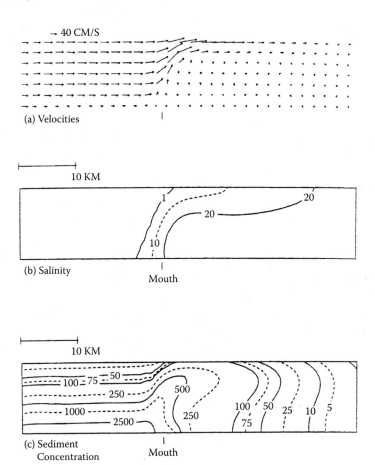

FIGURE 6.35 Velocities, salinity concentrations (ppt), and sediment concentrations (mg/L) in vertical cross-section along centerline of region for a steady-state flow in a constant-depth estuary. (*Source:* From Pickens et al., 1994. With permission.)

depth of the river at the dam was specified as 2.4 m. The depth then increases gradually to 4.8 m at a point 13 km downstream, where it remains constant for another 7 km. After this, the depth gradually increases to 7.2 m just before the mouth of the river and remains constant for 15 km out into the bay. At this point, the depth of the bay gradually increases in the seaward direction such that, at a distance of 64 km seaward of the river mouth, the bay has a depth of 21.6 m. As noted above, to increase the clarity of the illustrations, only the first 60 km (34 km seaward of the river mouth) of the system are shown in the figures.

To ensure an identical fresh water flow to the estuary, the volumetric flow rate was the same in the present case as in the constant-depth case. All other parameters were also the same. From Figure 6.36 it can be seen that the location of the

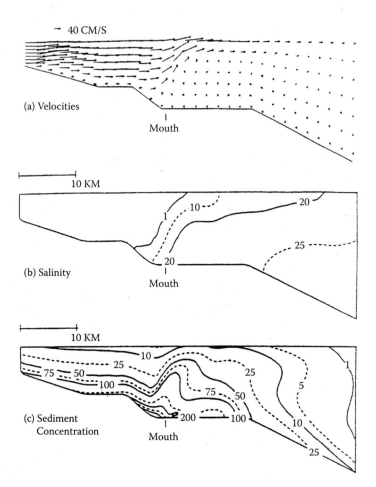

FIGURE 6.36 Velocities, salinity concentrations (ppt), and sediment concentrations (mg/L) in vertical cross-section along centerline of region for a steady-state flow in a variable-depth estuary. (*Source:* From Pickens et al., 1994. With permission.)

salt wedge has changed only slightly from the previous case. However, the turbidity maximum is significantly weaker than before, with the sediment concentration increasing from a minimum of 5 mg/L in the upstream direction to only 12 mg/L at a point 2 km downstream of the salt wedge. The weakening of the turbidity maximum is due to the reduction in suspended solids concentration at mid-depths as the river deepens just before the mouth.

In general, although the results for variable-depth systems are qualitatively similar to those derived for constant depth, the differences are significant. The most obvious difference is the location of the salt wedge at low flows. Because it is the salt wedge that governs the two-layer circulation pattern, it is also the salt wedge that governs the sediment deposition in the estuary. The deposition is heaviest just upstream of the mouth, coincident with the sloping bottom present there. As the flow rate increases, the depositional area moves further out toward the bay. At lower flow rates, it moves upstream but is limited in its upstream extent by the bottom slope and the higher velocities at smaller depths. From this it can be deduced that the bottom contours influence the deposition but, over a long period of time, it is clear that the deposition also will affect the depth. Calculations for a constant-depth system also showed heavy deposition in the vicinity of the river mouth. The bottom slope upstream of the mouth will thus develop over a number of years, even if the system is initially at constant depth; that is, the general features of the topography shown in Figure 6.36 are a natural consequence of sediment resuspension and deposition in a salinity-stratified estuary.

6.6.2.3 A Variable-Depth, Time-Dependent Tidal Flow

The amplitudes of tides in estuaries are highly variable but in this area are often on the order of 50 cm. To study the effects of tides, a series of calculations was made with different tidal amplitudes in the range from 0 to 50 cm. An amplitude of 25 cm was chosen to illustrate the results. A tidal period of 12 hr was specified. The bathymetry and other parameters were assumed to be the same as above.

Figure 6.37 shows the sediment concentration along the centerline during the tidal cycle. The general features of the sediment concentration distribution are the same as in the steady-state case; however, the sediment concentrations throughout the tidal cycle are generally greater than in the steady-state case, and the turbidity maximum is now more pronounced than before, with the sediment concentration reaching a low of 5 mg/L in the river, then increasing to 25 mg/L at its maximum further downstream, and then decreasing to almost zero far out in the bay. From the calculations and the results shown, it also can be demonstrated that (1) at the surface, concentrations are greatest during slack flow periods (at 3 and 9 hr); (2) at the sediment-water interface, resuspension is greatest during ebb tide (at zero hours), with a secondary maximum during flood tide (at 6 hr); and (3) deposition is greatest during slack periods. The sediment concentration at the turbidity maximum does not reach its maximum value at the same time that the tidal velocities (both horizontal and vertical) and the resuspension reach their maximum values.

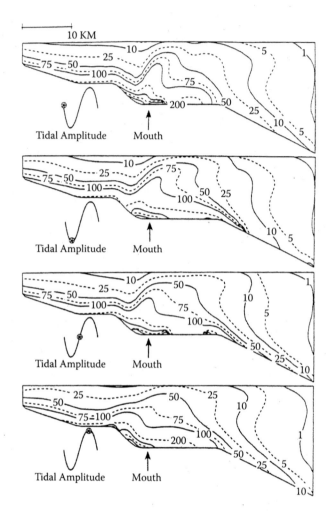

FIGURE 6.37 Suspended sediment concentrations (mg/L) at ebb tide and 3, 6 (flood tide), and 9 hr later as forced by tides with a 25-cm amplitude and 12-hour period. (*Source:* From Pickens et al., 1994. With permission.)

7 The Sorption and Partitioning of Hydrophobic Organic Chemicals

To understand and predict the transport and fate of hydrophobic organic chemicals (HOCs) in surface waters and bottom sediments, knowledge of the rates of sorption to and from solid sedimentary particles and the partitioning of these chemicals between solid particles and water is necessary. In early work, it was often assumed that adsorption and desorption occurred rapidly and that chemical equilibrium between the solids and water was attained in a very short time (e.g., see reviews by Sawhney and Brown, 1989; DiToro et al., 1991; Baker, 1991). This equilibrium was quantified by means of a partition coefficient, K_p(L/kg), defined as

$$K_p = \frac{C_s}{C_w} \qquad (7.1)$$

where C_s (kg/kg) is the mass of HOC sorbed to the sediment divided by the mass of the sediment, and C_w (kg/L) is the mass of HOC dissolved in the water divided by the volume of water. However, early sorption experiments were generally short term, hours to a few days, and were misleading; long-term experiments later demonstrated that both adsorption and desorption processes are often quite slow, with time scales of days to months or even longer before equilibrium is attained (e.g., Karickhoff and Morris, 1985; Coates and Elzerman, 1986). By comparison, the time of transport of a sediment particle in a river or lake may be as short as minutes to a few hours. Because of this, the assumption of chemical equilibrium in surface waters may not be a good approximation in many real situations, and therefore time-dependent sorption processes must be considered in detail.

In the first section of this chapter, experiments that illustrate basic and important characteristics of this time-dependent sorption as well as steady-state partitioning are presented and qualitatively analyzed. These experiments, as well as others, demonstrate that sorption times are long, that sorption processes depend on the HOC, and that these processes are significantly modified by colloids from the water, colloids from the sediments, organic content of the sediments, and particle and floc size and density distributions. The effects of these parameters on the steady-state partitioning of an HOC between sedimentary particles and water are discussed subsequently. In this first set of experiments, linear isotherms (i.e., K_p values, which are constant and independent of dissolved HOC concentration at constant temperature) were obtained. However, nonlinear and interactive effects on isotherms are often observed and have been reported in the literature; experiments and analyses that delineate these processes are also presented.

For a quantitative understanding of sorption dynamics and also an accurate ability to predict the sorption, transport, and fate of HOCs in aquatic systems, quantitative models of the time-dependent sorption processes are needed. In Section 7.2, a quite general but complex model of time-dependent sorption is described first; this model includes effects of particle and floc size and density distributions. A simpler, less accurate, but computationally efficient model is then presented. This model is sufficient to describe major characteristics of the sorption experiments. However, for more accurate descriptions of the experimental results, the general model is needed. Results with this latter model are then compared with experimental results.

The discussion in this chapter primarily concerns the sorption of HOCs to suspended particles. The sorption and flux of HOCs in bottom sediments are discussed in Chapter 8.

7.1 EXPERIMENTAL RESULTS AND ANALYSES

7.1.1 BASIC EXPERIMENTS

Results of several experiments are presented here that illustrate the basic characteristics of time-dependent adsorption, desorption, and short-term adsorption followed by desorption processes for one HOC, and the effects of different HOCs on adsorption and desorption (Jepsen et al., 1995; Tye et al. 1996; Borglin et al., 1996). From these experiments, sorption rates as well as partitioning of the HOC to suspended sedimentary particles can be determined as a function of time. All experiments were long term and were usually continued until chemical equilibrium was attained. All sediments used in this set of experiments were subsamples of the same batch of natural sediments from the Detroit River. The median particle size was 7 µm, and the organic carbon content was 1.42%. All HOCs used were carbon-14 labeled in order to simplify the analytical procedures and to enhance sensitivity. Filtration of the sediment-water mixtures separated operationally defined sediment-sorbed and dissolved fractions of the HOC at a particle size of 1 µm. Details of the experimental procedures and additional results can be found in the articles referenced.

The adsorption experiments were batch-mixing experiments. To initiate the experiments, a dissolved HOC and clean sediments (i.e., no sorbed HOC) at concentrations from 2 to 10,000 mg/L were mixed together with water in amber Qorpak glass jars. The jars were then rotated on a rolling table to ensure continuous mixing of the contents until they were sampled. The experiments were conducted for different periods of time up to 6 months. One sample jar was prepared for each sample point. Results of the experiments were reported as the logarithm of the partition coefficient as a function of time.

Typical results of experiments with hexachlorobenzene (HCB), sediments, and Optima pure water are shown in Figure 7.1, where log K_p is plotted as a function of time for sediment concentrations of 2, 10, 100, 500, 2000, and 10,000 mg/L. As the HOC is adsorbed to the sediment, log K_p varies relatively rapidly at small time but more slowly as time increases and a steady state is approached. The time to steady state varies from less than 1 day (for a sediment concentration of 2 mg/L) to about 30 days (for a sediment concentration of 10,000 mg/L). The measured steady-state partition coefficient depends somewhat on sediment concentration; that is, log K_p = 4.0 for a sediment concentration of 2 mg/L and decreases monotonically to 3.78 at a sediment concentration of 10,000 mg/L, a factor of 1.66 for K_p. Experiments of this type were performed with Optima pure water as well as with filtered tap water, with different-size fractions of the original sediments, with organically stripped sediments, and with three PCB congeners. All gave results qualitatively similar to those shown in Figure 7.1.

The desorption experiments used a purge-and-trap procedure. Sediments to which an HOC was sorbed, typically for 3 to 12 months (a time sufficient for sorption equilibrium to be attained), were mixed with water in a flask. The mixture was kept suspended and continuously purged with water-saturated compressed air. The air bubbled through the suspension and exited through a resin column that trapped the HOC. The HOC then was extracted from this column using a methanol solution and sonication. The experiments continued until essentially all the HOC had desorbed from the sediments.

Typical results of desorption experiments with HCB and pure water at different sediment concentrations are presented in Figure 7.2. The percent of

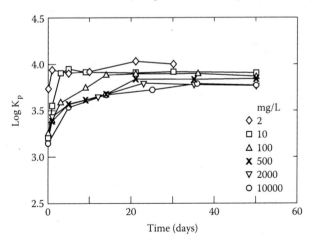

FIGURE 7.1   Adsorption experiments with HCB and pure water. Log K_p as a function of time with sediment concentration as a parameter. (*Source:* From Jepsen et al., 1995. With permission.)

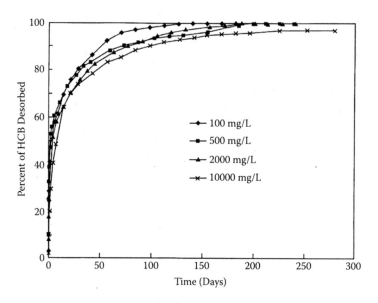

FIGURE 7.2 Desorption experiments with HCB and pure water. Percent of the initially sorbed HCB that has desorbed as a function of time. Sediment concentration as a parameter. (*Source:* From Borglin et al., 1996. With permission.)

the HCB initially adsorbed to the particles that has subsequently desorbed is shown as a function of time. Within experimental error (a few percent), all of the HCB initially sorbed to the sediments is desorbed with time, indicating that the adsorption and desorption processes are reversible. The desorption rate is greatest at the beginning and then decreases as the HCB is desorbed and its concentration goes to zero. The rate is slightly dependent on sediment concentration; the time for 90% desorption is on the order of 50 days (at 100 mg/L) to 100 days (at 10,000 mg/L). The desorption times for this type of experiment are significantly longer than the adsorption times for the adsorption experiments described above (compare Figures 7.1 and 7.2). However, it should be noted that the desorption experiments (purge-and-trap) are inherently different from the adsorption experiments (batch-mixing). Hence, there is no direct and/ or simple relation between adsorption and desorption times for these experiments. This is discussed further below and in the next section. Experiments of this type have been performed with pure water as well as with filtered tap water, with different-size fractions of the original sediment, with organically stripped sediments, and with two PCB congeners. Qualitatively similar results were obtained in all cases.

In addition to adsorption and desorption experiments, several short-term adsorption followed by long-term desorption experiments were performed. In these experiments, the HOC was adsorbed to sediments for either 2 days or 5 days (batch-mixing experiments); in this short period of time, sorption equilibrium was not attained. This was followed by desorption, which lasted until essentially

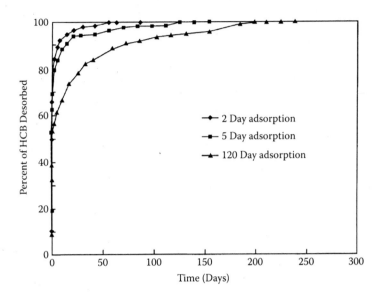

FIGURE 7.3   Short-term adsorption followed by desorption experiments. Sediment concentration is 500 mg/L. Percent of initially sorbed HCB that has desorbed as a function of time. (*Source:* From Borglin et al., 1996. With permission.)

all the HOC had desorbed (purge-and-trap experiments). For experiments with HCB in pure water and at a sediment concentration of 500 mg/L, results for desorption are shown in Figure 7.3 and are there compared with the standard desorption experiment (where HCB had been adsorbed for 120 days and therefore equilibrated before desorption began). For 2- and 5-day adsorption times, the desorption times are proportional to the adsorption times but are longer. For each experiment, essentially all the HCB that was sorbed to the sediment during the adsorption phase of the experiment is desorbed, again indicating reversibility of the processes. Experiments also were performed at sediment concentrations of 100 and 10,000 mg/L; the results were similar in character.

The results shown in Figures 7.1 through 7.3 are all for HCB. For other HOCs, the results are qualitatively the same but depend quantitatively on the partition coefficient of the HOC. To illustrate this, results of adsorption experiments are shown here for three PCB congeners: a monochlorobiphenyl (MCB), a dichlorobiphenyl (DCB), and a hexachlorobiphenyl (HPCB). For each of these HOCs and for HCB, log K_p is shown in Figure 7.4 as a function of time at a sediment concentration of 2000 mg/L. The times to steady state increase as the steady-state value of K_p increases. For the PCBs, it was shown that the times to steady state depend on the sediment concentration in a similar manner as for HCB.

Results of desorption experiments (percent desorbed as a function of time) are shown in Figure 7.5 for HCB, MCB, and HPCB at a sediment concentration of 2000 mg/L. Desorption times increase as K_p increases; for each HOC, they are proportional to adsorption times (Figure 7.4) but are longer. All the MCB

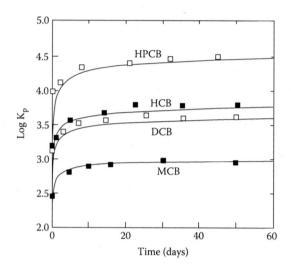

FIGURE 7.4   Partition coefficients for the adsorption of HCB and three PCB congeners (MCB, DCB, and HPCB). Log K_p as a function of time. Experimental data are shown as open and closed symbols, whereas the modeling results are shown as solid lines. (*Source:* From Lick et al., 1997. With permission.)

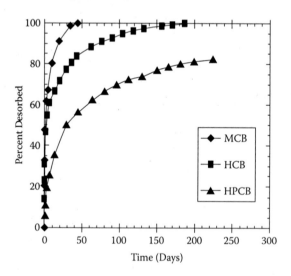

FIGURE 7.5   Desorption experiments with HCB, MCB, and HPCB. Percent desorbed as a function of time. (*Source:* From Borglin et al., 1996. With permission.)

and HCB desorbed completely during the experiments; HPCB was at 80% desorption at 200 days and was still desorbing when the experiment concluded at 230 days.

7.1.2 Parameters That Affect Steady-State Sorption and Partitioning

The experiments described above, as well as similar ones, demonstrate that (1) sorption times are long (days to months or even longer), (2) desorption times are longer than adsorption times (but adsorption and desorption experiments are inherently different), (3) adsorption and desorption are reversible processes, (4) sorption times and the measured partition coefficients depend on the sediment concentration, and (5) sorption times depend on the partition coefficient. These statements have been difficult to quantify and interpret because of seemingly contradictory experimental results and analyses reported in the literature. To assist in the clarification and quantification of these statements, various factors that affect the steady-state sorption and partitioning processes are reviewed and discussed here. The most significant of these factors are colloids from the sediments, colloids from the water, and the organic content of the sediments (Lick and Rapaka, 1996). The dynamics of sorption, including the effects of particle and floc size and density distributions, are discussed in Section 7.2.

7.1.2.1 Colloids from the Sediments

Colloids are here operationally defined as particles or flocs less than 1 μm in diameter. In pure water, no colloids are present. However, they are always present in natural waters but vary widely in amount and character. In addition, because there is a wide distribution of particle sizes in natural sediments (inevitably including some particles less than 1 μm in diameter), colloids are inherently present in any sample of natural sediments; they are a natural part of the sediments, and their amount in the water is more or less proportional to the amount of sediments in suspension.

In the adsorption experiments with results as shown in Figure 7.1, pure water was used and there were therefore no colloids from the water. However, because natural sediments were used, colloidal particles from the sediments were present. In these experiments, HCB was truly dissolved in the water and also adsorbed to the solid sedimentary particles greater than 1 μm in diameter, to the colloidal particles, and to colloidal particles that had flocculated such that the floc diameter was greater than 1 μm (i.e., no longer colloidal). For these experiments, it was demonstrated that the mass of HCB adsorbed to the flocculated colloidal matter was generally small in comparison with the HCB adsorbed to the solid particles, and it was therefore ignored.

To interpret and quantify the steady-state partition coefficients as shown in Figure 7.1, especially the dependence of K_p on sediment concentration, consider the following. During filtration to separate C_s and C_w and hence to determine K_p from Equation 7.1, the amount of HCB retained on the filter consists of the HCB sorbed to the sediment particles greater than 1 μm, m_{Hs}, whereas the amount of HCB in the filtrate consists of the truly dissolved HCB, m_{Hd}, plus the amount of

HCB sorbed to the colloidal matter from the sediments, m_{Hdc}. It follows that the measured partition coefficient in this case is given by

$$K_{pm} = \frac{\dfrac{m_{Hs}}{m_{sed}}}{\dfrac{m_{Hd} + m_{Hdc}}{V}} = \frac{\dfrac{m_{Hs}}{m_{sed}}}{\left(\dfrac{m_{Hd}}{V}\right)\left(1 + \dfrac{m_{Hdc}}{m_{Hd}}\right)}$$

$$= \frac{C_s}{C_w\left(1 + \dfrac{m_{Hdc}}{m_{Hd}}\right)} = \frac{K_p}{1 + \dfrac{m_{Hdc}}{m_{Hd}}} \tag{7.2}$$

where m_{sed} is the mass of sediments, V is the volume of water, and K_p is the true partition coefficient as defined in Equation 7.1.

A partition coefficient for the colloidal matter can be defined as

$$K_c = \frac{\dfrac{m_{Hdc}}{m_{dc}}}{\dfrac{m_{Hd}}{V}} \tag{7.3}$$

where m_{dc} is the mass of colloidal particles from the sediments. In general, m_{dc} should be proportional to the mass of the sediments; that is, $m_{dc} = \alpha m_{sed}$, where α is the fraction of colloidal particles in the sediments. It follows from the above that

$$\frac{m_{Hdc}}{m_{Hd}} = K_c \frac{m_{dc}}{V} = K_c \alpha \frac{m_{sed}}{V}$$

$$= K_c \alpha C \tag{7.4}$$

where C is the sediment concentration. By substituting this expression into Equation 7.2, one obtains

$$K_{pm} = \frac{K_p}{1 + K_c \alpha C} \tag{7.5}$$

It can be seen that K_{pm} depends on the sediment concentration and reduces to K_p as the sediment concentration decreases to zero. From this and Figure 7.1, it follows that log K_p is approximately equal to 4.0 ± 0.1, or $K_p = 10,000$ L/kg.

If it is further assumed that the partition coefficient for the colloidal matter is approximately the same as that for the sediments, the above equation reduces to

$$K_{pm} = \frac{K_p}{1 + K_p \alpha C} \tag{7.6}$$

This expression is similar to that derived by Wu and Gschwend (1986). From this equation and the data in Figure 7.1, α can be estimated and is approximately 0.005, a reasonable number for the fractional mass of colloidal particles in natural sediments. With this, the above equation is consistent with the results in Figure 7.1. By comparison of K_p for 2 mg/L and 10,000 mg/L, the maximum effect of colloids from the sediments on K_p in these experiments is a factor of about 1.66.

As with these experiments, most sorption experiments have been performed with suspended sediments at relatively low concentrations of 10^4 mg/L or less. The extension of these results to high concentrations, such as may occur in surface waters during large floods or storms but especially in consolidated bottom sediments, was questionable. Because of this difficulty, adsorption experiments with HCB and pure water were done at sediment concentrations from 10^2 mg/L up to 6.25×10^5 mg/L, concentrations approaching those of consolidated sediments (Deane et al., 1999).

Measured partition coefficients for HCB and Detroit River sediments (with 3.2% organic carbon and different from that above) are shown in Figure 7.6(a) as a function of time and for sediment concentrations of 10^2, 10^3, 10^4, 5×10^4, 10^5, and 6.25×10^5 mg/L. At the largest sediment concentration, the effect on K_p is a factor of about 5 compared with K_p at 100 mg/L. For each of these concentrations, the colloidal fraction, α, was determined by means of a submicron particle sizer and also by the difference in mass between the filtrates from a 1-μm and a 0.1-μm filter (Table 7.1). With this data, Equation 7.6 then was used to determine

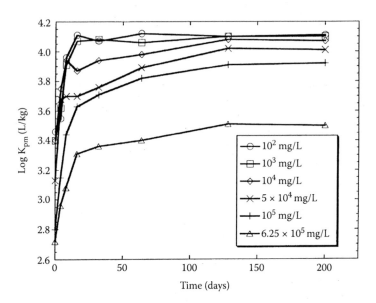

FIGURE 7.6(a) Partition coefficients as a function of time during adsorption at different sediment concentrations: measured partition coefficients. (*Source:* From Deane et al., 1999. With permission.)

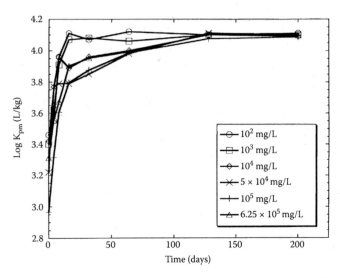

FIGURE 7.6(b)   Partition coefficients as a function of time during adsorption at different sediment concentrations: partition coefficients corrected for colloidal effects. (*Source: From Deane et al., 1999. With permission.*)

TABLE 7.1
Colloidal Fractions in Highly Suspended Sediment Concentration Experiments

Sediment	Concentration (mg/L)	Colloidal Fraction Measured by Submicron Particle Sizer	Colloidal Fraction Measured by 0.1 μm Filtration
Detroit River	10^2	0.00057	0.00062
	10^3	0.00046	0.00039
	10^4	0.00039	0.00035
	5×10^4	0.00008	0.00012
	10^5	0.00011	0.00017
	6.25×10^5	0.00006	0.00016
Stripped Detroit River	10^3	0.00572	0.00613
	10^4	0.00540	0.00500
	5×10^4	0.00470	0.00430
	10^5	0.00325	0.00415
	6.25×10^5	0.00070	0.00150
Santa Barbara Mountain	10^2	0.00232	0.00236
	10^3	0.00221	0.00248
	10^4	0.00190	0.00215
	5×10^4	0.00057	0.00071
	10^5	0.00040	0.00036
	6.26×10^5	0.00027	0.00037

the true partition coefficient. These corrected partition coefficients are shown in Figure 7.6(b). The general trends of K_p with time are the same as in Figure 7.6(a) except that now all partition coefficients at the different sediment concentrations approach the same value of log $K_p = 4.1$, or $K_p = 12,600$ L/kg. This equilibrium value of K_p is independent of sediment concentration.

These results, along with previous experimental results, demonstrate that the steady-state measured partition coefficient after correction for effects of colloids from the sediments is independent of sediment concentration, even at very high sediment concentrations approaching those of consolidated sediments. This true equilibrium partition coefficient therefore can be used as the appropriate partition coefficient in studies of consolidated sediments and saturated soils.

7.1.2.2 Colloids from the Water

Colloids are always present in natural waters, even without resuspension of bottom sediments — for example, in slowly moving streams or even in tap water. HOCs naturally sorb to these colloids as well as to colloids from resuspended sediments. Because of this sorption capability, colloids from the water can significantly affect the adsorption, desorption, and partitioning processes. They can do this in two different ways. First, HOCs sorbed to colloids are, in general, experimentally considered part of the dissolved chemical component, C_w. This increases the value of C_w and hence reduces the value of the partition coefficient from its value in pure water (i.e., water without colloids). Second, colloids can affect the sorption and partitioning processes through aggregation of the colloids into flocs that are greater than 1 μm in diameter. As this happens, the chemical adsorbed to these flocs is then generally considered part of the solid component, C_s. This increases the value of the partition coefficient from its value in the absence of colloid flocculation.

To illustrate the effects of colloids from the water, consider experiments by Jepsen and Lick (1996) where Santa Barbara tap water was filtered with a 0.2-μm pore-size filter and then mixed with HCB. After 5 days, the HCB-water mixture was filtered with a 1-μm filter just prior to the addition of sediments. Results for log K_p as a function of time with sediment concentration as a parameter are shown in Figure 7.7.

As in other experiments, the measured K_p at steady state is a function of the sediment concentration but is a stronger function of C than that shown in Figure 7.1. For this set of experiments, it was demonstrated that the main reason for this dependence was the inclusion, in the determination of K_p, of the HCB adsorbed to the flocculated colloidal matter from the water. A partition coefficient independent of this flocculated colloidal matter can be determined as follows. The total amount of HCB retained on the filter at steady state consists of the sum of the HCB on the sediments and the HCB sorbed to the flocculated colloidal matter, m_{Hf}. It follows that the measured K_p is given by

$$K_{pm} = \frac{\dfrac{m_{Hs} + m_{Hf}}{m_{sed}}}{C_w} = \frac{m_{Hs}}{m_{sed}C_w} + \frac{m_{Hf}}{m_{sed}C_w} \tag{7.7}$$

The last term can be rewritten as

$$\frac{m_{Hf}}{m_{sed}C_w} = \frac{m_{Hf}V}{m_{sed}m_{Hd}} = \frac{m_{Hf}}{m_{Hd}C} \tag{7.8}$$

From these two equations, K_{pm} can be written as

$$K_{pm} = K_p + \frac{m_{Hf}}{m_{Hd}C} \tag{7.9}$$

The first term on the right is the partition coefficient in the absence of flocculated colloidal matter from the water. The second term is due to this flocculated matter. The ratio of m_{Hf} to m_{Hd} was determined independently by means of mixing experiments with just the chemical, water, and the colloids inherent in the water, but with no sediments; it was shown to have a value of 0.087 for the filtered tap water used in the experiments. With this value of m_{Hf}/m_{Hd}, the above equation gives results that are in good agreement with those shown in Figure 7.7.

This equation shows the correction to K_p due to flocculated colloids from the water. In general, the effect on the measured partition coefficient of colloids

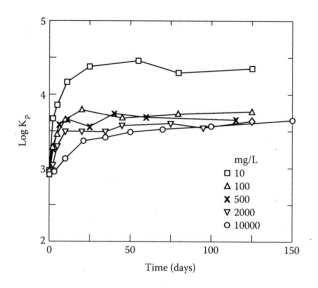

FIGURE 7.7 Adsorption experiments with HCB. Log K_p as a function of time with sediment concentration (mg/L) as a parameter. (*Source:* From Jepsen and Lick, 1996. With permission.)

from the sediments and of the flocculated colloidal matter from the water can be described by

$$K_{pm} = \frac{K_p}{1 + K_p \alpha C} + \frac{m_{Hf}}{m_{Hd}C(1 + K_p \alpha C)} \qquad (7.10)$$

With this correction, the partition coefficients shown in both Figures 7.1 and 7.7 all reduce to a single value, independent of sediment concentration. For these experiments, the effects of colloids from the tap water, Equation 7.9, were greater than the effects of colloids from the sediments, Equation 7.6. This is probably true for most natural waters at low sediment concentrations. At high sediment concentrations, as in Figure 7.6, the effects of colloids from the sediment are greater.

In addition to its effect on equilibrium partitioning, the flocculation of colloids may affect the sorption rates. Because of the small sizes of (unflocculated) colloids, the sorption of a chemical to colloids is probably very rapid, so this should not affect the overall rates of sorption to the suspended sediments. However, the sorption of a chemical to aggregated colloids may be relatively slow, and this will affect the measured sorption rates. The time-dependent flocculation of colloids and the subsequent operational shift of the sorbed chemical from the dissolved to the solid compartment may also significantly affect the observed sorption rates. These effects have been qualitatively observed but have only been partially quantified (Jepsen et al., 1995).

7.1.2.3 Organic Content of Sediments

The sorption of HOCs, in general, is considered to be primarily to the organic matter in the sediments. This has been demonstrated by many investigators and is illustrated by results from long-term adsorption (Jepsen et al., 1995; Tye et al., 1996) and desorption experiments (Borglin et al., 1996). These experiments were performed with natural Detroit River sediments (1.42% o.c.) and with these same sediments stripped of their organic matter. The equilibrium K_p values for the stripped sediments were lower by a factor of about 16 compared to the natural sediments. It could not be ascertained whether the remaining sorption after stripping was due to sorption to the mineral surfaces of the sediments or due to sorption of HCB to a small amount of organic matter still remaining in the sediments. The adsorption and desorption times also were dependent on the presence of organic matter; for the stripped sediments, they were smaller by factors of 5 to 10.

As with much of the literature on this subject, the above discussion has implicitly assumed that HOCs sorb to the organic matter in the sediments and that this organic matter has universal, homogeneous sorption properties; for example, partition coefficients are commonly normalized to total organic carbon (TOC). However, recent investigations (Ghosh et al., 2003; Accardi-Dey and Gschwend, 2002, 2003; Lohmann et al., 2005) have indicated that different types of organic matter may sorb HOCs in different amounts and at different rates. In particular, investigators have identified two types of organic matter with differing sorption

properties: (1) amorphous organic matter (AOM) and (2) carbonaceous geosorbents (CG) such as black carbon, coal, and kerogen. Partition coefficients for CG
may be one to two orders of magnitude greater than for AOM. This indicates that
partition coefficients and the rates of sorption are determined not only by the total
amount of organic matter but also by the amounts of each type of organic matter.
A recent review of this topic is given by Cornelissen et al. (2005).

7.1.2.4 Sorption to Benthic Organisms and Bacteria

In assessing sediment contamination and the effects of HOCs on organisms,
equilibrium partitioning (EqP) of HOCs among the solids, water, and benthic
organisms is often assumed. For example, EqP has been applied to set sediment quality criteria for benthic organisms in sediments with PAHs. However, it
has been demonstrated that HOC concentrations in benthic organisms are often
one to two orders of magnitude less than expected based on EqP (Hellou et al.,
2002; Kraaij et al., 2002; Guthrie-Nichols et al., 2003). Reasons for this are the
finite rates of sorption to organisms as well as to sediments — even slower for
consolidated bottom sediments than for suspended sediments. This finite rate of
sorption also has been demonstrated for bacteria (Lunsman and Lick, 2005); in
these experiments, the time-dependent sorption of three HOCs to *Rhodococcus
rhodochrous* was investigated. It was demonstrated that the sorption depended
on whether the bacteria were living (different depending on whether they were
growing or nongrowing) or dead as well as on the state of aggregation (flocculation) of the bacteria.

7.1.3 Nonlinear Isotherms

For the experiments with HCB, MCB, DCB, and HPCB described above, C_s was
proportional to C_w at constant temperature for HOC concentrations that varied
over four orders of magnitude; that is, the isotherms were linear and K_p was constant over this range. However, there are numerous reports of nonlinear isotherms
in the literature. Some of the reasons for this nonlinearity were investigated (Jepsen and Lick, 1999) and are discussed here.

   The HOCs mentioned above have a range of K_p values from approximately
10^3 to 10^5 L/kg, but all are characterized by having low solubilities, from 1 µg/L to
a few mg/L. Because of these low solubilities and because the maximum amount
of an HOC that can be sorbed to a sediment depends on the solubility as well as
the partition coefficient of the HOC, only a relatively small amount of any of these
HOCs can be sorbed to the sediment. This amount may be inadequate to cause
nonlinear effects.

   To investigate this hypothesis, long-term sorption experiments were done
with additional HOCs with higher solubilities (tetrachlorobiphenyl, TCB; tetrachloroethylene, PCE; pentachlorophenol, PCP; and octanol), whereas previous
sorption experiments with HCB, MCB, DCB, and HPCB were extended to higher
values of C_w so as to obtain values of C_w as close to the HOC's solubility limit

as possible. The results of these experiments are presented here in terms of C_{oc} and K_{oc}. C_{oc} is the mass of an HOC sorbed to the sediments divided by the mass of organic carbon in the sediments and is dimensionless (i.e., kg of chemical/kg of organic carbon); K_{oc}, a partition coefficient normalized to the organic carbon content of the sediments, is defined as

$$K_{oc} = \frac{C_{oc}}{C_w} \tag{7.11}$$

and has units of liters per kilogram (L/kg). The solubility, S (kg/L), is the maximum amount of the HOC that can be in solution (i.e., the maximum value of C_w); the maximum value of C_{oc} is therefore $K_{oc}S$. For $K_{oc}S \ll 1$, it follows that $C_{oc} \ll 1$; the maximum amount of chemical that can be sorbed is therefore much smaller than the amount of organic carbon that is associated with the sediments. For $K_{oc}S = 0(1)$, $C_{oc} = 0(1)$, and the amounts of sorbed chemical and organic carbon are comparable. Define K_{oc}^o as the value of K_{oc} at low values of C_w, where C_s is a linear function of C_w. Values of K_{oc}^o, S, and $K_{oc}^o S$ for all HOCs tested are given in Table 7.2. For the chemicals listed, $K_{oc}^o S$ varies over a large range, from much less than 1 (2×10^{-3} for HCB) to much greater than 1 (36.5 for octanol).

As a representative HOC that demonstrates nonlinear effects, consider PCE. For PCE, K_{oc}^o is 1.1×10^3 L/kg, the solubility is 1.5×10^{-4} kg/L, and $K_{oc}^o S$ is therefore 0.165; $K_{oc}^o S$ is smaller than 1 but of the same order of magnitude as 1. Experimental results for C_{oc} as a function of C_w are shown in Figure 7.8 for two organic carbon concentrations (1.95 and 3.3%) and two sediment concentrations (100 and 1000 mg/L). It can be seen that C_{oc} is independent of organic carbon and sediment concentrations. At low concentrations of C_w, C_{oc} is a linear function of C_w and $C_{oc}/C_w = K_{oc}^o$. As C_w increases, C_{oc} is no longer a linear function of C_w and continually decreases below its linear value; K_{oc} therefore also decreases below K_{oc}^o. These deviations of C_{oc} and K_{oc} begin when C_{oc} is approximately 10^{-2} (i.e., when

TABLE 7.2
Partition Coefficients and Solubilities of Chemicals

Chemical	K_{oc}^o (L/kg)	S (10^{-6} kg/L)	$K_{oc}^o S$
Hexachlorobenzene	4×10^5	0.005	0.002
Monochlorobiphenyl	1.25×10^5	5.1	0.64
Dichlorobiphenyl	4×10^5	0.055	0.022
Tetrachlorobiphenyl	1.8×10^6	0.045	0.081
Hexachlorobiphenyl	6.3×10^6	0.001	0.0063
Tetrachloroethylene	1.1×10^3	150	0.165
Pentachlorophenol	2.6×10^3	11	0.03
Octanol	8.1×10^4	450	36.5

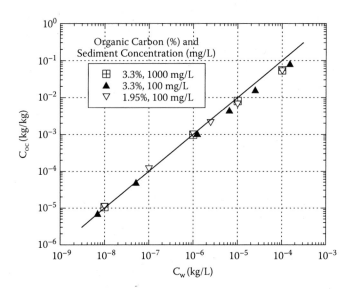

FIGURE 7.8　Tetrachloroethylene isotherm. C_{oc} as a function of C_w for different organic carbon contents and sediment concentrations. The solid line is the linear isotherm. (*Source:* From Jepsen and Lick, 1999. With permission.)

the mass of PCE sorbed to the sediment is approximately 10^{-2} of the mass of the organic carbon associated with the sediments) and continually increase until the solubility limit is reached. When $C_w = S$, C_{oc} is about 0.08 and K_{oc} has decreased from K_{oc}^o by a factor of about 2.

For all the HOCs tested, experimental results for C_{oc} as a function of C_w were similar in character to those for PCE. In particular, C_{oc} was a linear function of C_w for small values of C_{oc} but began to decrease below its linear value when C_{oc} was approximately 10^{-2}. The similarity of the experimental results suggests non-dimensionalizing, or normalizing, the values of C_w and plotting C_{oc} as a function of $K_{oc}^o C_w$. The results for C_{oc} as a function of $K_{oc}^o C_w$ for all the HOCs tested are shown in Figure 7.9(a). As with PCE, for each HOC and as long as C_w is less than S, C_{oc} is a linear function of $K_{oc}^o C_w$ until C_{oc} is approximately 10^{-2}; after this, C_{oc} deviates from this linear relation, with the deviation increasing as $K_{oc}^o C_w$ increases.

Once C_{oc} is determined as a function of C_w, K_{oc} can be calculated from Equation 7.11. It is informative to normalize K_{oc} by K_{oc}^o and to plot K_{oc}/K_{oc}^o as a function of $K_{oc}^o C_w$ (Figure 7.9(b)). The data for all HOCs essentially fall on the same curve, and therefore K_{oc}/K_{oc}^o is primarily a function of $K_{oc}^o C_w$. For each HOC, the extent of the reduction of K_{oc}/K_{oc}^o from one depends on $K_{oc}^o C_w$ but is limited by the solubility of the HOC. For HCB and HPCB, $K_{oc}^o S \ll 1$ and K_{oc}/K_{oc}^o is always essentially 1; for octanol, $K_{oc}^o S \gg 1$ and the maximum reduction in K_{oc}/K_{oc}^o is the largest of the HOCs tested, a factor of about 100; other HOCs with intermediate values of $K_{oc}^o S$ have intermediate reductions in K_{oc}/K_{oc}^o.

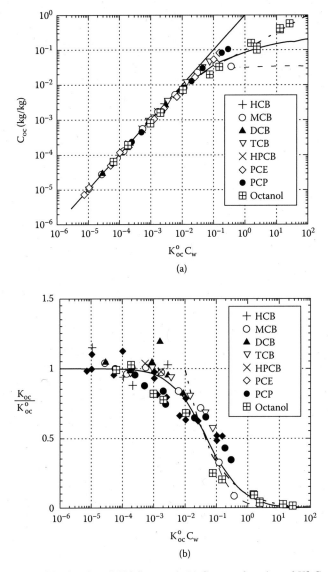

FIGURE 7.9   Partitioning for all HOCs tested: (a) C_{oc} as a function of $K_{oc}^0 C_w$ (the solid line is the linear isotherm), and (b) K_{oc}/K_{oc}^0 as a function of $K_{oc}^0 C_w$. (*Source:* From Jepsen and Lick, 1999. With permission.)

The reduction of K_{oc} from its value at low chemical concentrations, K_{oc}^0, can be attributed to the saturation by the HOC of the organic carbon in the sediments such that fewer adsorption sites are available to the HOC as C_{oc} increases. Analytic approximations for this reduction can be derived as follows. The desorption rate for an HOC, to a first approximation, is proportional to the concentration of the HOC on the solid and can be expressed as $k_2 C_{oc}$, where k_2 is a desorption coefficient. The adsorption rate is proportional to the dissolved HOC concentration and can be expressed as $k_1 C_w (1 - \varepsilon)$, where k_1 is an adsorption coefficient and ε is the fraction of the volume already occupied by the chemical and therefore not available for further sorption. At equilibrium, the rates of adsorption and desorption are equal and therefore

$$k_2 C_{oc} = k_1 C_w (1 - \varepsilon) \tag{7.12}$$

Rearranging gives

$$\frac{C_{oc}}{C_w} = \frac{k_1}{k_2}(1 - \varepsilon) = K_{oc}^o (1 - \varepsilon) = K_{oc} \tag{7.13}$$

where $K_{oc}^o = k_1 / k_2$.

Various assumptions can be made to relate ε to C_{oc}. For low HOC concentrations when $\varepsilon \ll 1$, the simplest and a quite reasonable approximation is that $\varepsilon = \alpha C_{oc}$, where α is a constant. The above expression then can be written as

$$\frac{C_{oc}}{C_w} = K_{oc}^o (1 - \alpha C_{oc}) \tag{7.14}$$

and is equivalent to the well-known Langmuir equation. From this it follows that

$$C_{oc} = \frac{K_{oc}^o C_w}{1 + \alpha K_{oc}^o C_w} \tag{7.15}$$

$$\frac{K_{oc}}{K_{oc}^o} = \frac{1}{1 + \alpha K_{oc}^o C_w} \tag{7.16}$$

and it follows that both C_{oc} and K_{oc}/K_{oc}^o are functions of $K_{oc}^o C_w$, as implied in Figures 7.9(a) and (b). For $\alpha = 30$, plots of Equations 7.15 and 7.16 are shown in Figures 7.9(a) and (b), respectively. Reasonably good agreement between theory and experiments is demonstrated.

A better assumption for ε is that

$$\varepsilon = 1 - e^{-\alpha C_{oc}} \tag{7.17}$$

For low values of αC_{oc}, this is equivalent to $\varepsilon = \alpha C_{oc}$. Substitution of Equation 7.17 into Equation 7.13 leads to

$$C_{oc} = K_{oc}^o C_w e^{-\alpha C_w} \tag{7.18}$$

$$\frac{K_{oc}}{K_{oc}^o} = e^{-\alpha C_{oc}} \tag{7.19}$$

These are both better approximations to the data at larger values of $K_{oc}^o C_w$ than Equations 7.15 and 7.16.

For octanol, neither of the above approximations for ε fits the data for C_{oc} well over the entire range of C_w (Figure 7.9(a)). The best fit to the data for octanol seems to be two straight lines, that is, $C_{oc} = K_{oc}C_w$ for low values of C_w (say, less than C_w^*) and, at higher values,

$$\frac{C_{oc}}{C_{oc}^*} = \left(\frac{C_w}{C_w^*}\right)^m \tag{7.20}$$

where $C_{oc}^* = K_{oc}^o C_w^*$ and m is a constant. It follows that

$$\frac{K_{oc}}{K_{oc}^o} = \left(\frac{C_w^*}{C_w}\right)^{1-m} \tag{7.21}$$

for $C_w > C_w^*$. For m = 0.5, these approximations are shown in Figures 7.9(a) and (b) as the dot-dash line. This approximation for C_{oc} as a function of C_w is reminiscent of a Freundlich isotherm, except, of course, that Equation 7.20 is only valid for $C_w > C_w^*$ and not for the entire range of C_w. For $C_w < C_w^*$, the isotherms are linear.

Nonlinear isotherms also can be caused by interactive effects between the HOC and a co-solvent (Jepsen and Lick, 1999). To investigate this, long-term experiments were first done with HOCs, all of which had low solubilities. For mixtures of these HOCs, no interactive effects were observed. Experiments then were performed with HCB–octanol, HCB–ethanol, octanol–ethanol, and HOC–methanol mixtures. For these mixtures, significant reductions in the partition coefficient occurred as the co-solvent concentration increased. This reduction was explained quantitatively by a partitioning of the primary HOC between the co-solvent, organic matter in the sediment, and water. The agreement between theory and experiment was quite good.

7.2 MODELING THE DYNAMICS OF SORPTION

To more quantitatively understand the laboratory experiments described and referenced above and also to be able to accurately predict the sorption, transport, and fate of HOCs in surface waters, quantitative models of sorption dynamics are needed. For this purpose, a description of a general model (Lick and Rapaka, 1996) is given in Section 7.2.1. This will later be shown to give quite accurate descriptions of the dynamics of the experiments described in the previous section. However, this model is quite complex and requires considerable auxiliary data and computer time. Because of this, a general calculation of the transport of contaminants in surface waters with this model is impracticable. A simplified model is therefore needed (Lick et al., 1997) and is described in Section 7.2.2. This simplified model is then used to analytically describe major characteristics of the experiments described above. In Chapter 8, it is used in transport calculations. In Section 7.2.3, results of numerical calculations with the general model are presented and compared with experimental results on sorption.

In the discussion, the emphasis is on the nonequilibrium dynamics of the sorption. The general model includes diffusion of the HOC through the pores within a particle/floc, quasi-equilibrium HOC partitioning locally within a particle/floc, approximate particle and floc size and density distributions, and effects of organic content and chemical properties. As discussed above, colloids from the water and from the sediments have significant effects on sorption rates and equilibrium partitioning. Although the effects of colloids on equilibrium partitioning are understood reasonably well and have been quantified, our knowledge of their effects on the dynamics of sorption is inadequate. Because of this, the effects of colloids on sorption dynamics are not considered here.

7.2.1 A DIFFUSION MODEL

In the present modeling, the transport of an HOC within a particle/floc (hereafter generally referred to as a particle for simplicity) is assumed to occur by diffusion of the dissolved chemical through the pores of the particle; this transport is then modified by adsorption of the HOC to organic substances within the particle and possibly to mineral surfaces of the particle. In general, this diffusion/reaction process must be described by partial differential equations for the concentrations of the chemical dissolved in the pore waters of the solid as well as in the solid. However, if it is assumed that the time for adsorption within the particle/floc is relatively fast by comparison with the time for diffusion, a local quasi-equilibrium assumption can be made (this is discussed in more detail in Section 8.2). In this case, the transport of the chemical within the particle can be described by a single time-dependent diffusion equation with no reaction term but with an effective diffusion coefficient given by (Berner, 1980; Wu and Gschwend, 1986)

$$D = \frac{D_m}{1 + \left(\frac{1-\phi}{\phi}\right)\rho_s K_p}$$

$$\cong \frac{D_m \phi}{(1-\phi)\rho_s K_p} \tag{7.22}$$

where D_m is the molecular diffusion coefficient in the fluid within the particle without consideration of any reaction but corrected for tortuosity, ϕ is the porosity of the particle, and ρ_s is the mass density of the solid in the particle. The second equality is valid for chemicals with large partition coefficients such that $(1 - \phi)$ $\rho_s K_p/\phi \gg 1$.

If it is assumed that the particle/floc is a homogeneous, porous sphere with radius R, the governing equation for the contaminant concentration within the sphere, $C_s(r,t)$, is

$$\frac{\partial C_s}{\partial t} = \frac{D}{r^2} \frac{\partial}{\partial r}\left(r^2 \frac{\partial C_s}{\partial r} \right) \tag{7.23}$$

where r is the distance in the radial direction, t is time, and D is assumed constant. For the small particles considered here, the flow around the particle is slow and the thickness of the fluid boundary layer through which mass transfer occurs is relatively small. As a result, the transfer of the chemical from the water to the surface of the particle is a relatively fast process. It follows that the concentration of the chemical at the surface of the particle, $C_s(R,t)$, is essentially in equilibrium with the chemical dissolved in the water, that is,

$$C_s(R,t) = K_p C_w(t) \tag{7.24}$$

For the desorption experiments, C_w is small and $C_s(R,t)$ is therefore approximately zero. For the adsorption experiments, C_w changes with time due to adsorption of the dissolved chemical by the particles. This variation with time can be calculated from the mass conservation equation for the sediment–contaminant–water mixture, which is

$$\frac{m}{V} = \overline{C}_s C + C_w \tag{7.25}$$

where m is the total mass of chemical in volume V and $\overline{C}_s(t)$ is the average concentration of the chemical in the particles.

In many cases, the effects of particle/floc size and density distributions on sorption dynamics are significant and must be included in the modeling for accurate results. The concentrations of the sediments in each size range, C_k, depend on the particle/floc size and density distributions and will vary with time because of flocculation. When different-size fractions are considered, it is convenient to introduce X_k, the fraction of sediments by mass in the k-th size range, that is,

$$X_k = \frac{C_k}{C} \tag{7.26}$$

In measurements of particle/floc size distributions, the quantity actually measured is the volume fraction, Y_k, where Y_k is the volume of the flocs in the k-th size range divided by the total volume of the flocs. X_k and Y_k are related by

$$X_k = \Phi_k Y_k \tag{7.27}$$

where $\Phi_k = (2.6 - 1.6\,\varphi_k)/(2.6 - 1.6\,\varphi)$ and it has been assumed that the density of the particles composing the floc is 2.6 g/cm^3. For the cases considered here, the size and density distributions are not wide, and the variation in Φ_k is less than 2. Because of this and because of our meager knowledge of the values of φ_k, it will be assumed that $\Phi_k = 1$ and therefore that $X_k = Y_k$.

For each size fraction, the appropriate generalization of Equation 7.24 is

$$C_{sk}(R_k,t) = K_p C_w(t) \qquad (7.28)$$

where C_{sk} and R_k are the chemical concentration and radius, respectively, for the sediments in the k-th size range. Equation 7.25 is still valid, except that now

$$\overline{C}_s(t) = \sum_k \overline{C}_{sk}(t) X_k \qquad (7.29)$$

where $\overline{C}_{sk}$ is the average concentration of the chemical in particles in the k-th size range.

7.2.2 A Simple and Computationally Efficient Model

The sorption model described above (hereafter referred to as the diffusion model) is quite complex, requires data on flocculation, and also requires extensive computer time when it is used as part of a general transport model. Some calculations with this model are illustrated in Section 7.2.3. For more practical purposes, a simplified and computationally efficient, but still reasonably accurate, model has been developed and is as follows. The major simplifications are that (1) the diffusion of C_s within a particle is approximated as a mass transfer process between the particle and the surrounding water, and (2) changes in particle and floc size and density distributions are not included. For sediments consisting of single-size particles, the time-dependent change of the average value of C_s within the particle, $\overline{C}_s$, is then approximated by

$$\frac{d\overline{C}_s}{dt} = -k(\overline{C}_s - K_p C_w) \qquad (7.30)$$

where k is a mass transfer coefficient (s^{-1}). The quantity $(\overline{C}_s - K_p C_w)$ is a forcing function that is proportional to the difference between $\overline{C}_s$ and its equilibrium value, $K_p C_w$, and is zero when the HOC sorbed to the particle/floc is in chemical equilibrium with the HOC dissolved in the water. This model will be referred to as the mass transfer model.

The coefficient k is not known from basic principles or directly from experimental results but is chosen so as to obtain good agreement between the solutions of the above equation and the diffusion equation, Equation 7.23. For this purpose, consider the case of desorption from a particle with a sorbed concentration of $\overline{C}_s$ to a dissolved HOC concentration in the water of zero. Integration of Equation 7.30 then gives

$$\overline{C}_s = C_o\, e^{-kt} \tag{7.31}$$

where C_o is the initial value of $\overline{C}_s$. The time for 75% desorption of C_o is given by

$$t^+ = \frac{1.386}{k} \tag{7.32}$$

For the diffusion equation (Equation 7.23), the time for 75% desorption is given approximately by (Carslaw and Jaeger, 1959)

$$t^+ = 0.0225\frac{d^2}{D} \tag{7.33}$$

where d is the diameter of the particle. For the results of the diffusion and mass transfer models to agree at t^+, Equations 7.32 and 7.33 must agree, and therefore k is determined by

$$k = \frac{D}{0.0165d^2}$$

$$= \frac{D_m}{0.0165\, d^2\left[1+\left(\dfrac{1-\phi}{\phi}\right)\rho\, K_p\right]} \tag{7.34}$$

with D from Equation 7.22.

To determine the accuracy of this approximation, desorption calculations were made with both the diffusion and the mass transfer models (Lick et al., 1997). Results are shown in Figure 7.10 for the desorption of HCB ($K_p = 10^4$ L/kg and $D = 2 \times 10^{-14}$ cm²/s, a reasonable value for D, as shown below) for a particle with a diameter of 15 μm. There is reasonable agreement between the results of the two models for all time, and they coincide for 75% desorption. The mass transfer model is much more computationally efficient than the diffusion model, usually by more than an order of magnitude, because it is not necessary to solve the diffusion equation (Equation 7.23) for a particle/floc as a function of r and t. For a mixture of particles of different size classes, Equations 7.30 and 7.34 are assumed to be valid for each size class.

For analytic purposes, an exponential desorption time is convenient and can be defined from Equation 7.31 as $kt_d^* = 1$ (t_d^* is the time for $\overline{C}_s$ to desorb to $e^{-1} = 0.368$ of its initial value). Equation 7.34 indicates that this time is given by

$$t_d^* = \frac{0.0165\, d^2\left[1+\left(\dfrac{1-\phi}{\phi}\right)\rho_s\, K_p\right]}{D_m} \tag{7.35}$$

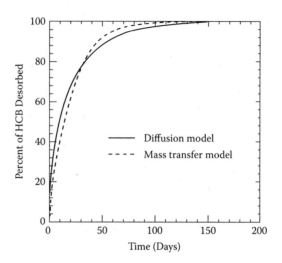

FIGURE 7.10   Desorption of HCB from suspended sediments as calculated by the diffusion and mass transfer models. Shown is the percent of sorbed HCB that has desorbed as a function of time. For this calculation, $D = 2 \times 10^{-14}$ cm²/s, log $K_p = 4.0$, and the particle diameter was 14.56 μm. (*Source:* From Lick et al., 1997. With permission.)

For large K_p, this states that t_d^* is proportional to K_p. This dependence of t_d^* on K_p is consistent with Figure 7.5, where desorption times for MCB, HCB, and HPCB are approximately in the ratio of 1:5:33, which is also the ratio of the K_p values. In contrast to the experimental results as shown in Figure 7.2, there is no obvious dependence of t_d^* on sediment concentration in the above equation. This sediment concentration effect depends on particle and floc size and density distributions and is discussed below.

For the adsorption problem, C_w is not zero but is finite and varies throughout the experiment, often by orders of magnitude; its time dependence is governed by Equation 7.25. Substitution of Equation 7.25 into Equation 7.30 gives

$$\frac{d\overline{C}_s}{dt} + k(1 + K_pC)\overline{C}_s = k\,K_p\frac{m}{V} \qquad (7.36)$$

At the beginning of the adsorption experiments, $\overline{C}_s = 0$. With this condition, the solution to the above equation is

$$\overline{C}_s = \overline{C}_{s\infty}\left[1 - e^{-k(1+K_pC)t}\right] \qquad (7.37)$$

where $\overline{C}_{s\infty} = K_pm/(1 + K_pC)V$ and is the value of $\overline{C}_s$ in the steady state as $t \rightarrow \infty$. This equation indicates an exponential adsorption time, t_a^*, given by

$$t_a^* = \frac{1}{k(1+K_pC)}$$

$$= \frac{0.0165\, d^2 \left[1 + \left(\frac{1-\phi}{\phi} \right) \rho_s K_p \right]}{D_m(1+K_pC)} \qquad (7.38)$$

This adsorption time is significantly different from the desorption time given by Equation 7.35 and, for the same conditions, is always smaller than t_d^*. The reason for the difference between t_a^* and t_d^* is, of course, that the adsorption and desorption experiments are inherently different from each other and, in particular, they are not mirror images of each other. In the desorption experiments, the particle is desorbing into a dissolved chemical concentration of zero. In the adsorption experiments, the particle is adsorbing from a dissolved chemical concentration that is not constant but is decreasing with time, often by an order of magnitude or more.

For small-enough sediment concentrations, $K_pC \ll 1$. In this case, the above equation indicates that t_a^* is proportional to K_p. However, for most adsorption experiments described above, K_pC is not small; t_a^*, therefore, is not directly proportional to K_p but instead is a weak function of K_p. This is consistent with the results of the adsorption experiments shown in Figure 7.4, where the ratios of the t_a^* for MCB, DCB, HCB, and HPCB are approximately 1:1.5:1.5:3.0, significantly different from the ratio of K_p values and the times for desorption.

Both of the above equations for t* indicate that t* is proportional to d^2, where d is an effective floc size for diffusion of the HOC within the floc. In general, this effective floc size depends on the floc size and density and is a complex function of the sediment concentration and fluid shear (Chapter 4); it is different in the adsorption and desorption experiments. This is discussed further below in numerical calculations with the general model.

Comparison of Equations 7.38 and 7.35 clearly indicates that adsorption times are less than desorption times, in agreement with experiments. However, the ratio of these two times, $1 + K_pC$, needs to be modified by the effects of differences in floc sizes between the adsorption and desorption experiments.

7.2.3 Calculations with the General Model and Comparisons with Experimental Results

Although the simple model presented above describes many of the general characteristics of the sorption experiments, for accurate representation of the experimental results, it is necessary to use the more general model to (1) describe the transport of C_s within the particle/floc by means of the diffusion equation, Equation 7.23, and (2) approximate the particle/floc size and density distributions.

For the latter purpose, flocculation experiments were performed under the same conditions as the sorption experiments. For Detroit River sediments (median disaggregated particle size of 7 μm and 1.42% o.c.) where the particles were initially disaggregated and then flocculated due to differential settling until a steady state was reached (as in the adsorption experiments), the steady-state median floc diameters, the times to steady state, and the settling speeds of the flocs were measured. From the settling speeds and Stokes law, the effective densities of the flocs can be calculated. The porosity, φ, of the flocs can then be determined from

$$\phi = \frac{\rho_s - \rho_f}{\rho_s - \rho_w} \quad (7.39)$$

where ρ_s is 2.6 g/cm³ and is the approximate density of the solid particles, ρ_f (g/cm³) is the density of the floc, and ρ_w is the density of water and is equal to 1.0 g/cm³. The median steady-state floc diameter, the time to steady state, the settling speed, the effective floc density, and the floc porosity are shown in Table 7.3 as functions of sediment concentration. As the sediment concentration increases from 2 to 10,000 mg/L, the floc density increases from 1.0009 to 2.2 g/cm³, the porosity decreases from 0.9994 to 0.25, and all the other quantities decrease by large amounts. Fluid shear, in general, also will affect these quantities; however, in the experiments described here, the fluid shear was essentially the same in all experiments and had no relative effect on the experiments.

During the adsorption and desorption experiments, particle size distributions also were measured. For convenience, these distributions are presented as the mass fraction of particles/flocs in four size ranges, with median sizes of 3, 7, 17, and 40 μm; the size ranges are 1–5, 5–10.5, 10.5–33.5, and >33.5 μm, respectively. The mass fraction in each size range is shown in Table 7.4 for (1) the initial

---

TABLE 7.3

Properties of Flocs as Functions of Sediment Concentration (Natural Sediments from the Detroit River)

Sediment Concentration (mg/L)	Median Steady-State Floc Diameter (μm)	Time of Flocculation to Steady State (hours)	Settling speed (μm/s)	Floc Density (g/cm³)	Porosity
2	4500	228	9700	1.0009	0.9994
10	1330	63	7900	1.008	0.995
100	244	8.5	1685	1.05	0.97
500	48	4.5	511	1.41	0.74
2,000	24	2.5	216	1.69	0.57
10,000	17	1.2	189	2.2	0.25

TABLE 7.4
Steady-State Floc Size Distributions: Initial Disaggregated Size Distribution and Size Distributions during Adsorption and at End of Desorption

Average diameter (µm)	3	7	17	40
Size range (µm)	1–5	5–10.5	10.5–33.5	>33.5
Initial disaggregated size	0.295	0.243	0.395	0.067
During adsorption:				
500 mg/L	0.037	0.076	0.282	0.605
2000 mg/L	0.125	0.143	0.357	0.375
10,000 mg/L	0.112	0.196	0.510	0.182
At end of desorption:				
100 mg/L	0.145	0.174	0.422	0.259
500 mg/L	0.295	0.278	0.328	0.099
2000 mg/L	0.375	0.288	0.252	0.085
10,000 mg/L	0.537	0.264	0.129	0.070

disaggregated size distribution, (2) the steady-state size distribution during the adsorption experiments as a function of sediment concentration, and (3) the size distribution at the end of the desorption experiments as a function of sediment concentration. The size distributions at the beginning of the desorption experiments are the same as those at the end of the adsorption experiments.

With this information on flocculation, the general model was then used for calculations of HCB desorption, adsorption, and short-term adsorption followed by desorption, as well as for the adsorption of three PCB congeners (Lick and Rapaka, 1996). Results of some of these calculations are reported here and are compared with the experimental results.

7.2.3.1 Desorption

Consider the desorption of HCB from sediments that have been adsorbing HCB for approximately 4 months in batch-mixing experiments and that are therefore in chemical equilibrium with the dissolved HCB at that time (Figure 7.2). Calculations are illustrated for sediment concentrations of 10,000 and 500 mg/L. The desorption at a sediment concentration of 10,000 mg/L is discussed first.

As a first approximation in the modeling, it was assumed that all particle/floc sizes were the same and that an average particle/floc size therefore could be used in the calculations. From Table 7.4, the average particle floc size is approximately 17 µm. A value for the diffusion coefficient was assumed to be 2.0×10^{-14} cm^2/s. Results of calculations with these values of the parameters are shown in Figure 7.11(a). The calculated results are in qualitative agreement with the experimental results, but the calculations generally do not approximate the experimental results well at small and large times.

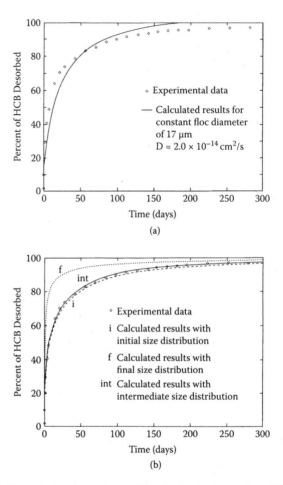

FIGURE 7.11   Numerical and experimental results for the desorption of HCB from sediments at a sediment concentration of 10,000 mg/L. Percent of sorbed HCB that has desorbed as a function of time. (a) Results for an average particle/floc diameter of 17 μm. (b) Results for different size distributions: initial (0.112, 0.196, 0.510, 0.182); final (0.537, 0.264, 0.129, 0.070); intermediate (0.137, 0.190, 0.528, 0.195). (*Source:* From Lick and Rapaka, 1996. With permission.)

The reason for this is that the sediments are a mixture of particles and flocs of widely different sizes, and the desorption rates and the times for desorption depend on the sizes of the particles and flocs. If a time for desorption, t_d, is defined as the time for 90% of the initial amount of chemical to be desorbed, then this time for a homogeneous sphere with diameter d is given approximately by (Carslaw and Jaeger, 1959)

$$t_d = \frac{d^2}{24D} \tag{7.40}$$

where d is in centimeters (cm). For a diffusion coefficient of 2×10^{-14} cm²/s and particle sizes of 3, 7, 17, and 40 μm, the desorption times are 1.9, 10, 60, and 333 days, a ratio of almost 200 between the 3-μm and 40-μm particles.

These differences in desorption times for different-size particles significantly affect the desorption rates for a mixture of particles of different sizes, especially at small and large times. Subsequent calculations were therefore made with four size fractions with average diameters of 3, 7, 17, and 40 μm. Another problem in the modeling is that the particle/floc size distributions change with time. For a sediment concentration of 10,000 mg/L, at the end of the long-term adsorption experiments and hence at the beginning of the desorption experiments, the average diameter of the particle/flocs was 17 μm and the size fractions were 0.112, 0.196, 0.510, and 0.182 (Table 7.4). During the desorption experiments, the particle/floc sizes decreased because of the disaggregation of the flocs due to fluid turbulence and because of grinding of the particles by the magnetic stirrer used in the mixing of the sediments. Much of the disaggregation of the flocs occurred during the first 24 hours of the experiment. At that time, the average particle/floc size was about 12 μm and was slowly approaching the disaggregated particle/floc size of the sediment of 7 μm. By the end of the experiment 280 days later, the average particle/floc size had decreased to about 5 μm and the size fractions were now 0.537, 0.264, 0.129, and 0.070.

Calculations were made with these initial and final size distributions. The results for the percent desorbed as a function of time are shown along with the experimental data in Figure 7.11(b). The initial size distribution gives good agreement with the experimental results at small and moderate times, whereas the final size distribution gives reasonable agreement with the experimental results only at large time. Also shown are calculated results for an intermediate size distribution of 0.137, 0.190, 0.528, and 0.145. Good agreement with the experimental results is shown for all times.

For 10,000 mg/L, the flocs were quite dense at all times, and the assumption that the floc densities did not change significantly with time gave reasonably good results. However, at 500 mg/L, the flocs were relatively porous at the end of adsorption (Table 7.3) and at the beginning of the desorption experiment but became smaller and denser as they disaggregated and compacted with time. For this calculation, to include the effect on the diffusion of the higher porosity at small time, the effective diffusion coefficient for 10,000 mg/L was modified, consistent with Table 7.3 and Equation 7.22, so that

$$D = 2.0 \times 10^{-14} \left(\frac{1-\phi_1}{\phi_1} \right) \left(\frac{\phi_2}{1-\phi_2} \right) \tag{7.41}$$

where $\phi_1 = 0.25$ and is the porosity of flocs at 10,000 mg/L, and $\phi_2 = 0.74$ and is the porosity of flocs at 500 mg/L. The modified diffusion coefficient is now 1.7×10^{-13} cm²/s, a factor of about 10 greater than that for 10,000 mg/L. For the calculated results with this value for D and with the initial size distribution for desorption of X_k = 0.037, 0.076, 0.282, and 0.605, there was good agreement with the experimental results for small and moderate times. For large time, the flocs have significantly disaggregated and decreased in size so that X_k = 0.295, 0.278, 0.328, and 0.099. For these flocs, which are denser, a value of D of 2.0×10^{-14} cm^2/s is more appropriate. For a calculation with these quantities, there was good agreement with the experimental results for moderate to large times.

For the two cases illustrated here (for small to moderate times when most desorption occurs), the ratio of the effective diameters in the 10,000 and 500 mg/L experiments was about 0.5; the ratio of the effective diffusion coefficients was approximately 0.1. From this and Equation 7.35, the ratio of the desorption times, t_d^*, for these two cases should then be about 2 to 2.5, consistent with the experimental results shown in Figure 7.2.

Similar calculations were made for 100 and 2000 mg/L. In both cases, the variations in particle/floc size and density distributions had to be taken into account to obtain good agreement between the calculated and experimental results. A general result of these calculations and experiments is the following. As the sediment concentration decreases, the floc size increases but the floc density decreases. The effects of these two quantities on the desorption rates are in opposite directions, but the net effect is that the desorption rate increases slowly as the sediment concentration decreases, in agreement with experiments.

7.2.3.2 Adsorption

For adsorption experiments with HCB, as shown in Figure 7.1 as well as for three PCBs (MCB, DCB, and HPCB), it was shown (Jepsen and Lick, 1996) that the time for adsorption to steady state, t_a^*, was relatively short (1 to 2 days) at the lowest sediment concentrations (typically about 2 mg/L) and increased significantly as the sediment concentration increased. Simple theories of the adsorption process (which do not include flocculation of the sedimentary particles and assume all parameters are constant as in the simple model above, Equation 7.38) indicate the reverse — that is, that t_a^* should decrease as the sediment concentration increases. The increase of t_a^* as C increases is due to flocculation and can be explained as follows.

In modeling the results of adsorption experiments, difficulties associated with changes in particle/floc size and density distributions with time are also present. However, as in the desorption case, the analyses are somewhat simpler at both high (10,000 mg/L) and low (10 mg/L) sediment concentrations. Calculations at a sediment concentration of 10,000 mg/L are described first. At this concentration, flocculation is quite rapid, and the steady-state floc size distribution is reached in approximately 1 hour (Table 7.3). Because of this, almost all adsorption occurs to flocculated particles; these have an average size of 17 μm; size fractions of 0.112, 0.196, 0.51, and 0.182; and are quite dense. With this size distribution, an equilibrium partition coefficient of 6310 (log K_p = 3.8), an initial dissolved HCB concentration of 1 μg/L, and D = 2.0×10^{-14} cm^2/s (the same as the D in the desorption calculations), adsorption calculations were made for a period of 900 days;

this was required so that even the largest-size fraction (d = 40 μm) could reach a steady state. Results are shown in Figure 7.12 for the first 60 days, when most of the adsorption occurred. The calculated results show that log K_p was 3.78 at 900 days (essentially steady state), was 3.70 at 60 days, and was 3.58 at 15 to 20 days. These values are in good agreement with the experimental results.

At a sediment concentration of 10 mg/L, flocculation is relatively slow; little flocculation occurs in the first 2 days, and a steady-state size distribution is not approached until almost 3 days. In this case, most adsorption occurs to individual particles before flocculation has occurred (Figure 7.1 and Table 7.3). Because of this, calculations were made with the initial disaggregated size distribution with size fractions given by 0.295, 0.243, 0.395, and 0.067. The same values for the equilibrium K_p and D as above were used here. Results for log K_p as a function of time for the first 60 days are shown in Figure 7.12 as the dashed curve. The calculations show that log K_p is 3.8 at 900 days and is 3.77 at 60 days. Log K_p reaches a value of 3.67 (a change of 0.1 from its value at 60 days) at approximately 5 days, a time somewhat longer than the experimental results show. However, this calculation is based on the initial size distribution before flocculation occurs. This calculation can be improved if the initial size distribution is used for the first 2.5 days (the time to steady-state flocculation), and an estimated steady-state floc size distribution (extrapolated from data in Table 7.4) is used thereafter. This is shown as a solid line in Figure 7.12. It can be seen that a deviation in log K_p of 0.1 from its value at 60 days now occurs at approximately 2 to 3 days. This final calculation is in excellent agreement with the experimental results. From this, it follows that the rapid adsorption at 10 mg/L is primarily due to sorption to individual disaggregated particles as compared to the slow adsorption to large, dense flocculated particles at 10,000 mg/L. This is consistent with experimental and calculated results at other sediment concentrations.

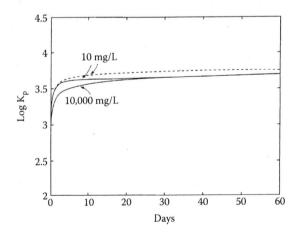

FIGURE 7.12 Numerical results for the adsorption of HCB to sediments at sediment concentrations of 10,000 and 10 mg/L. Log K_p as a function of time. (*Source:* From Lick and Rapaka, 1996. With permission.)

It should be noted that, because the emphasis here is on the dynamics of sorption, the effects of colloids have not been included in the calculations. The equilibrium partition coefficient therefore has been assumed to be the same for all sediment concentrations, although the experimental results for K_{pe} are somewhat dependent on sediment concentration because of the effects of colloids.

7.2.3.3 Short-Term Adsorption Followed by Desorption

For the short-term adsorption followed by desorption experiments as illustrated in Figure 7.3 (C = 500 mg/L), numerical calculations were performed for $K_p(t)$ and are as follows. During adsorption, because the time for flocculation (4.5 hours) was much less than the time for adsorption in the experiments considered here, the particle/floc size distribution was taken as the steady-state floc size distribution for adsorption at 500 mg/L (Table 7.4). For desorption, the particle/floc size distribution was taken to be the same as the distribution that gave the best result in the long-term desorption experiment at 500 mg/L described above. For adsorption, it was therefore assumed that $X_k = 0.037, 0.076, 0.282,$ and 0.605 (Table 7.4). For desorption, it was assumed that $X_k = 0.260, 0.190, 0.451,$ and 0.099. Results for a diffusion coefficient of 2.0×10^{-14} cm^2/s are shown in Figure 7.13 and are compared there with the experimental results.

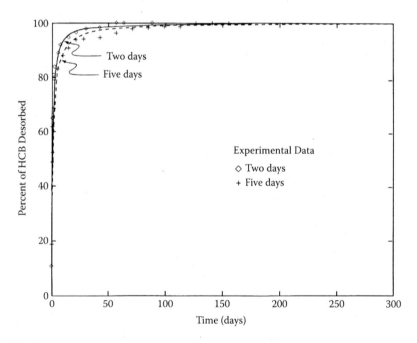

FIGURE 7.13 Numerical results for the short-term adsorption of HCB followed by desorption. Percent of initially sorbed that has desorbed as a function of time. (*Source:* From Lick and Rapaka, 1996. With permission.)

From the experimental results and the calculations, a general description of the adsorption/desorption process is as follows. During adsorption, the distance that an HOC diffuses into the interior of a particle increases with the length of the adsorption period. This causes the desorption time to increase as the adsorption time increases because the HOC has further to travel before it escapes the interior of the particle. In addition, during desorption after short-time adsorption, the HOC is still diffusing toward the interior of the particle as well as toward the surface of the particle. This causes the desorption time to be greater than the adsorption time.

7.2.3.4 Effects of Chemical Properties on Adsorption

For HCB, MCB, DCB, and HPCB, results of adsorption experiments for log K_p as a function of time at a sediment concentration of 2000 mg/L are shown in Figure 7.4. By means of the general model, numerical calculations were made for each HOC. In the calculations, the diffusion coefficients were modified to take into account the retardation effect due to chemical reaction as given by Equation 7.22. Because of the dependence of D on K_p, it was assumed for each PCB congener that

$$D = 2.0 \times 10^{-14} \left[\frac{K_p(\text{HCB})}{K_p(\text{PCB})} \right] \qquad (7.42)$$

where $K_p(\text{PCB})$ is the appropriate equilibrium partition coefficient for the particular PCB congener. For HPCB, DCB, and MCB, the logarithms of these quantities were assumed to be 4.8, 3.5, and 3.0, respectively. The experimentally determined steady-state size fractions for 2000 mg/L were used (Table 7.4).

The results of the numerical calculations are shown in Figure 7.4. The calculated times to steady state agree well with the experimental results. They also agree reasonably well with Equation 7.38 as far as the dependence on K_p is concerned. In particular, the times to steady state are largest for HPCB, intermediate for DCB, and smallest for MCB — in good agreement with the experimental results. The time for HCB is intermediate between those for HPCB and MCB and approximately the same as that for DCB — again in agreement with the experiments.

For sediments that contain both amorphous organic matter and carbonaceous geosorbents, partition coefficients and sorption rates must be determined for each component. The above type of analysis should then apply to each component.

8 Modeling the Transport and Fate of Hydrophobic Chemicals

Hydrophobic organic chemicals, such as PCBs and PAHs, often have large partition coefficients, on the order of 10^3 to 10^6 L/kg or even higher. In this case, much of the chemical is sorbed to particulate matter and is transported with it. This particulate matter, along with the sorbed HOC, usually settles onto the bottom of an aquatic system and forms deposits of contaminated sediments there that can be many meters thick. At most sites with contaminated sediments, approximate calculations of the amounts of contaminants in these bottom sediments and also in the overlying water lead to the conclusion that there are orders of magnitude more contaminant in the bottom sediments than in the overlying water. As a result, even after the cleanup of point sources of contamination, these bottom sediments can serve as a major and long-lasting source of contaminants to the overlying water. To predict sediment and water quality over long periods of time, the flux of these contaminants between the bottom sediments and the overlying water needs to be quantitatively understood and modeled.

The sediment-water flux of contaminants is primarily due to sediment erosion/deposition, molecular diffusion, bioturbation, and groundwater flow. Each of these processes acts in a different way, and hence each must be described and modeled in a different way. In general, they occur more or less simultaneously and there are interactions among them. All these processes are continuously and often significantly modified by the finite rates of adsorption and desorption of the HOC between the solid sedimentary particles and the surrounding waters. These rates of sorption and the resulting partitioning depend on the hydrophobicity of the chemical, that is, on K_p. Because of this, the transport of an HOC also strongly depends on K_p. This is especially true for HOCs with large partition coefficients and is a major emphasis in this chapter.

As bottom sediments erode, the contaminants associated with these sediments are transported into the water column, where they may adsorb or desorb, depending on conditions in the overlying water relative to conditions in the bottom sediments. Because erosion rates are highly variable in space and time, contaminant fluxes due to erosion/deposition are also highly variable in space and time. During calm periods and average winds, these fluxes are relatively small and are probably comparable with the fluxes due to molecular diffusion, bioturbation, and groundwater flow. However, major storms and floods can cause movement and mixing of bottom sediments by erosion/deposition more rapidly and to depths in the sediments much greater than that possible by these other processes. The contaminant flux due to the erosion of particles with their sorbed contaminants and the subsequent desorption of these contaminants into the surrounding water would also then be much greater than the contaminant fluxes due to these other processes.

The effects of bioturbation on sediment properties and the sediment-water flux are due to feeding and burrowing activities of benthic organisms, are quite diverse, and depend on the amounts and types of organisms. In fresh waters, benthic organisms disturb and/or mix the sediments down to depths of 2 to 10 cm. This does not occur instantaneously but over a period of time that depends on the number densities of the organisms and their activities; this can be months to years. For sea water, the depths of the disturbances due to benthic organisms are much greater, on the order of 10 cm to as much as 1 m.

The flux of contaminants from the bottom sediments due to molecular diffusion has often been considered negligible by comparison with other processes. However, rapid erosion and deposition (as caused by floods and storms) as well as chemical spills can cause sharp gradients in contaminant concentrations and hence large contaminant fluxes at the sediment-water interface. As will be seen below, finite sorption rates for HOCs with large partition coefficients will exacerbate this effect. In addition, molecular diffusion is ubiquitous and inherently modifies and is modified by all the other flux processes. As a result, the effects of molecular diffusion on fluxes can be quite large and must be considered in calculating and predicting sediment-water fluxes of HOCs.

In field studies, groundwater flow has been shown to be a major influence on the sediment-water flux of HOCs in certain areas. However, these fluxes are difficult to measure and, in addition, models of this flux as modified by finite sorption rates have not been extensively applied or verified. As a result, the effects of groundwater flow on the sediment-water flux of HOCs have not been well quantified. Nevertheless, the flux of HOCs due to groundwater flow can be significant and is a process that deserves careful consideration.

In water quality models, a common approach to modeling the contaminant flux between the bottom sediments and the overlying water is to use the equation

$$q = \frac{D}{h}(C_w - C_{wo})$$

$$= H\,(C_w - C_{wo}) \qquad\qquad (8.1)$$

where q is the flux; D is a diffusion coefficient; C_w and C_{wo} are representative contaminant concentrations in the pore waters of the sediment and in the overlying water, respectively; h is the thickness of an assumed "well-mixed" or "active" sediment layer; and $H = D/h$ and is a mass transfer coefficient with units of centimeters per second (cm/s). If local chemical equilibrium in the sediments is assumed, the above equation can be written as

$$q = \frac{D}{h}\left(\frac{C_s}{K_p} - C_{wo}\right) \tag{8.2}$$

where C_s is a representative contaminant concentration of the solids in the sediment. The parameters D, h, and H (only two of the three are independent) are generally determined by parameterization, that is, by comparison of results calculated from the water quality model with field measurements and then modifying these parameters until the model and field results agree (e.g., see the texts by Thomann and Mueller, 1987; Chapra, 1997). As a first approximation, typical recommended values for h are 10 to 15 cm, whereas a suggested value for D is the molecular diffusion coefficient for the HOC being considered. The effects of finite sorption rates on these fluxes are generally not explicitly considered.

The difficulty with Equations 8.1 and 8.2 is that they do not describe (especially in functional form) the time-dependent fluxes due to sediment erosion/deposition, molecular diffusion, bioturbation, or groundwater flow. Because these equations do not correspond to any real process, the parameters D, h, and H are purely empirical functions chosen to fit existing data; they generally are not valid for conditions for which they have not been calibrated, and the solutions to these equations do not have the proper functional dependence on time. Because of this, these equations have very little predictive capability. Equations 8.1 and 8.2 are often called diffusion equations, but they are not; a better term would be a mass transfer approximation.

The major processes that affect HOC transport and sediment-water flux as well as the modeling of these processes are described in this chapter. Erosion/deposition and the subsequent transport of HOCs in the overlying water are discussed in the following section. Section 8.2 describes the conventional one-dimensional, time-dependent diffusion (Fickian) approximation that is often used for the sediment-water flux of nonsorbing chemicals or for sorbing chemicals when chemical sorption equilibrium is a good approximation, that is, when sorption rates are relatively high. The mass transfer approximation as described by Equation 8.1 or 8.2 is also discussed and compared with the diffusion approximation.

For the diffusion of sorbing chemicals with finite sorption rates, the Fickian approximation is not valid. In this case, the basic conservation equations for the HOC must be supplemented by a rate equation for the transfer of the HOC between the solids and the pore water. This is done for the molecular diffusion of HOCs in Section 8.3, where experimental and theoretical results for HOCs with a wide range of partition coefficients are discussed; finite sorption rates are inherent in the results and analyses. The effects of bioturbation (including finite sorption rates) on the sediment-water flux are discussed in Section 8.4. Comparisons of the magnitudes and time dependencies of the different sediment-water fluxes, as well as a discussion of the approximation of a "well-mixed" layer, are given in Section 8.5.

A simple and idealized problem of contaminant release and transport during environmental dredging is discussed in Section 8.6; the purpose is to characterize and estimate the magnitudes of various processes that affect this release without the use of a complex model. Previously, in Section 1.2, an introduction to the problem of water quality modeling, parameterization, and the resulting non-unique solutions was given. In Section 8.7, this discussion is continued in the context of a more general model of PCB transport and water quality.

8.1 EFFECTS OF EROSION/DEPOSITION AND TRANSPORT

Two HOC transport problems are discussed in this section: (1) the transport of PCBs in the Saginaw River, including the assumption of equilibrium partitioning; and (2) the transport of PCBs in Green Bay as affected by finite sorption rates.

8.1.1 THE SAGINAW RIVER

The transport of sediments in the Saginaw River has been modeled, and results of these numerical calculations were presented in Section 6.4. Based on these sediment transport calculations, the transport of PCBs in the river has also been modeled (Cardenas et al., 1995); the specific problem was to investigate and make preliminary estimates of the erosion, deposition, transport, and fate of PCBs from a contaminated area in the river (Figure 6.22). Some interesting results of this investigation are presented here.

In the calculations, it was assumed that there was equilibrium sorption of the PCBs to the sediments with a K_p of 2×10^4 L/kg; the nonerosional/nondepositional sediment-water flux of PCBs was not included in order to isolate the flux due to sediment erosion/deposition. The first calculations were to investigate the effects of the magnitude of flow events on the erosion and deposition of sediments and the subsequent transport of PCBs; calculations were therefore made for flow events of 500, 1000, 1500, and 1900 m^3/s. For reference, from 1940 to 1990, the median flow rate was 57 m^3/s and the maximum was 1930 m^3/s. In these first calculations, the following was assumed:

1. Surficial sediments initially in the intensive study area were contaminated with PCBs at a level of 4 µg/g of sediment. This is a reasonable first approximation to the average of the PCB concentrations actually measured at this site.
2. The thickness of this surficial layer was 10 g/cm^2 (on the order of 10 cm), but only where the water depth was less than 3 m (Figure 6.22). This excluded the river channel, where contaminants were generally not found. For the contaminated area, the total amount of PCBs initially in the sediment bed per unit of surface area was then 40 µg/cm^2, whereas the total amount of PCBs initially in the sediment bed was 90 kg.
3. There were sediments coming in from upstream, but they contained no PCBs.

During a flow event, sediments and sorbed PCBs are eroded from the sediment bed. Some of the eroded PCBs are transported and deposited further downstream in the river, and some are transported to Saginaw Bay. From the calculations, it was shown that erosion of PCBs originally in the bed occurs (1) in the shallow nearshore area to a sediment depth generally less than 1 g/cm^2 and (2) at the edge of the channel to sediment depths as great as 30 g/cm^2. The amount of PCBs eroded (in μg/cm^2) for a 1500-m^3/s event is shown in Figure 8.1. The large erosion of PCBs at the edge of the channel is clearly evident. During this event, a total of 28 kg PCBs were eroded and transported downstream; almost all of the eroded PCBs were transported to the bay. A small amount (0.02 kg) was deposited in the wide shallow part of the river, primarily near shore, with the amount of PCBs per unit area increasing toward shore.

A comparison of the amounts of PCBs transported by the different flow events is given in Table 8.1. The amounts transported to the bay increase nonlinearly with the flow rate, from a small amount (0.28 kg) for the 500-m^3/s event to 36.5 kg for the 1900-m^3/s event. However, even for the largest flow event, only about a third of the total PCBs in the bed are eroded. The reason for this is as follows. Because of the currents, the highest shear stresses occur in the deepest water, the channel, whereas the lowest shear stresses occur in the shallow, nearshore areas. Although the shear stresses and erosion rates are high in the channel, no PCBs are present there and therefore no erosion and transport of PCBs occur there. Conversely, in the nearshore region, the shear stresses are low, only small erosion occurs, and little transport of PCBs occurs. The region where the highest erosion and transport of PCBs occurs is on the edge of the channel where PCBs are present and where moderately high shear stresses occur. Here, the depth of

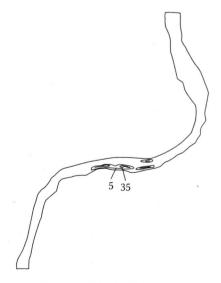

FIGURE 8.1 Amount of PCBs eroded (μg/cm^2) in the Saginaw River for a 1500-m^3/s flow event. (*Source*: From Cardenas and Lick, 1996. With permission.)

TABLE 8.1

PCB Transport in the Saginaw River for Different Flow Events

Flow Event (m³/s)	Transported to Bay (kg)	Deposited Downstream (kg)	Remaining in Bed (kg)
500	0.28	0.02	89.8
1000	11.6	0.06	78.5
1500	28.1	0.09	62.0
1900	36.5	0.11	53.6

erosion and the amount of PCBs eroded and transported depend on the shear stress and hence on the flow rate, at least until the layer of contaminated sediment is eroded. Thereafter, additional PCB erosion and transport is caused only by erosion of surficial layers near shore.

Contaminants initially at the surface of the sediment bed will be buried by sediments depositing during low flows. This process will reduce the subsequent erosion and transport of the contaminated sediments by later and larger flows. To make a preliminary investigation of this process, several calculations were made. In these calculations, it was assumed that (1) as in the first example, a layer of contaminated sediments 10 g/cm² thick was initially present at the surface and had a PCB concentration of 4 μg/g of sediment in the intensive study area; (2) layers of clean sediments were deposited on top of these contaminated sediments for different periods of time at approximately 2 g/cm² per year; and (3) after this deposition, a 1500-m³/s flow event occurred.

The erosion and deposition of sediment for the 1500-m³/s event were the same as in the above example. As far as PCB transport and fate are concerned, the differences in the results described here are that clean overlying sediments must be eroded first before the contaminated sediments can be eroded and transported. Of course, the more clean sediments that are deposited over the contaminated sediments, the less the amount of contaminated sediments that are eroded and transported downstream to the bay.

Calculations were made for no deposition and for deposition time periods of 1, 5, and 20 years. For these scenarios, the masses of PCBs transported to the bay were 28.1, 26.1, 23.1, and 15.2 kg, respectively. There is little difference in PCB transport to the bay without and with 1 year of deposition. Although the newly deposited sediments cover almost all the shallow, nearshore area with a layer of sediment sufficient to eliminate the erosion of contaminated sediments from this area, little erosion occurs there, even without any deposited sediment. Almost all the erosion of contaminated sediments occurs in a narrow region at the edge of the channel, between the deep channel and the shallow, nearshore area. In this region, the currents during a big event are sufficient to erode the newly deposited sediments as well as the older contaminated sediments. Only after 5 years of deposition (approximately 10 g/cm^2) is there a significant decrease in the amount of PCBs resuspended and transported to the bay.

As shown in these calculations, most of the erosion of contaminants occurs at the edge of the channel; that is, the amounts of erosion/deposition vary greatly across a river. This is significant in that, when considering the transport and fate of contaminated sediments and potential remedial actions, it is essential to determine the contaminant concentrations and sediment erosion/deposition rates as a function of distance across the river and, most importantly, at the edge of a channel where the depth and flow velocities may be changing rapidly.

In this investigation, effects of variable sediment properties were not considered. However, sediment properties and hence erosion rates should vary significantly across the channel (because of the changing bathymetry, flow velocities, and deposition of different size particles) as well as with depth (because of flood events). Because of this, variable sediment properties should be considered in a more realistic calculation.

8.1.2 GREEN BAY, EFFECTS OF FINITE SORPTION RATES

To investigate the effects of finite sorption rates on the transport and fate of HOCs in surface waters, calculations were made of the transport and fate of PCBs during storms of different magnitudes on Green Bay (Chroneer and Lick, 1997). Calculations of the hydrodynamics and sediment transport were summarized in Section 6.5 (bathymetry is shown in Figure 6.30) and were the basis for the calculations of PCB transport presented here.

For these calculations, it was assumed that the bottom sediments of the bay were uniformly contaminated with PCBs at a concentration of 1 µg/kg. The overlying water was initially free of PCBs. A moderate wind of 10 m/s from right to left (as in Figure 6.30) for a period of 2 days then caused a resuspension of sediments and contaminants; this was followed by a low wind of 2 m/s for 12 days, during which time the sediments deposited. As the sediments and associated contaminants were resuspended, the contaminants desorbed from the sediments and dissolved in the water. This desorption was quantified by means of the model described in Section 7.2 with the mass transfer coefficient, k, given by Equation 7.34. An average partition coefficient of 10^4 L/kg was assumed. As in Section 7.2, the diffusion coefficient for the HOC within the particle, D, was taken to be 2×10^{-14} cm^2/s. Three sizes of particles were assumed, with diameters of 4, 14, and 29 µm and size fractions of 10, 60, and 30%, respectively. From this, the mass transfer coefficient for each size class was calculated. Calculations were done for finite rates of sorption and also for chemical equilibrium, the more usual assumption in contaminant transport and fate calculations.

Results of these calculations are shown in Figures 8.2 and 8.3. Figure 8.2 shows the changes in C_s and C_w as a function of time at the location in the outer bay denoted by a * in Figure 6.29. Subscripts s and w denote suspended sediment and water, respectively; e and ne denote equilibrium and nonequilibrium, respectively; f, m, and c denote fine, medium, and coarse sediments; and avg

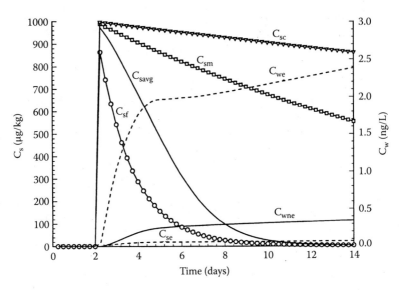

FIGURE 8.2 Green Bay. Contaminant concentrations as a function of time at the point denoted by a * in Figure 6.29. Subscripts s and w denote suspended sediment and water, respectively; e and ne denote equilibrium and nonequilibrium; f, m, and c denote fine, medium, and coarse sediments; and avg denotes the average of all size classes. (*Source*: From Chroneer and Lick, 1997.)

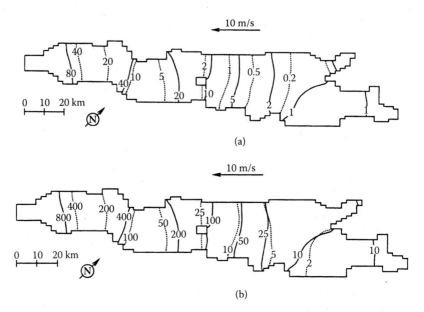

FIGURE 8.3 Concentrations of PCBs in Green Bay at the end of the 14-day event. Solid lines are for the equilibrium calculation, whereas the dashed lines are for the nonequilibrium calculation: (a) C_w in ng/L, and (b) C_s in µg/kg. (*Source*: From Chroneer and Lick, 1997.)

denotes the average of all size classes. For the equilibrium case, $C_{se} = K_p C_{we}$ and C_{we} is therefore proportional to C_{se}. As the sediments are resuspended and also transported to this location from sites in shallower waters near shore, C_{se} and C_{we} increase rapidly at first (due primarily to resuspension) and then more slowly due to transport of sediments and contaminants from the nearshore.

This transport of PCBs is greatly modified by finite sorption rates. In this case, C_{wne} is initially much lower and C_{savg} is much higher than their equilibrium values due to the slow desorption of PCBs from the suspended sediments to the water. In more detail, the fine sediments (C_{sf}) desorb rapidly, whereas the medium (C_{sm}) and coarse (C_{sc}) sediments desorb much more slowly; the medium and coarse sediments lose only a small fraction of their sorbed PCBs during the 14-day event. Because the fine fraction tends to stay in suspension much longer than the medium and coarse fractions, C_{savg} is dominated by the fine fraction and approaches C_{sf} as time increases. For finite rates of sorption and for the first few days, the sediments retain significantly more of their PCBs as they are transported than in the equilibrium case (i.e., $C_{savg} \gg C_{se}$); after deposition, the bottom sediments that are deposited during this time would also have a much higher concentration of PCBs. As time increases, C_{sf} decreases below C_{se} because of the desorption to a low C_{wne}; C_{savg} decreases below C_{se} for the same reason.

The distributions of C_w and C_s in the water of the bay at the end of the 14-day event are shown in Figures 8.3(a) and (b), respectively. Both $C_{wne} < C_{we}$ and $C_{savg} < C_{se}$ throughout the bay, by approximately a factor of two in the inner bay and by a factor of five or more in the outer bay.

At the end of the 14 days, the percentage of PCBs originally resuspended and still remaining in the water (dissolved in the water plus the small amount sorbed to the remaining suspended particles that have not yet deposited) is 27% for the equilibrium case and only 11% for the nonequilibrium case. This percentage depends on the amount of sediment resuspension. At low wind speeds, sediment resuspension is low; a higher percentage of the PCBs desorbs from the resuspended sediments to the water and remains there as the particles settle. At high wind speeds, the resuspension is high; a lower percentage of the PCBs desorbs, and most of the PCBs are therefore still sorbed to the particles and are transported with the particles as they settle to the bottom. Results for winds of 5, 10, and 20 m/s from the northeast for 2 days are summarized in Table 8.2. Of course, although the percentage of resuspended PCBs that remains in the water decreases as the wind speed increases, the total amount of PCBs remaining in the water increases because of the very nonlinear increase of sediment and contaminant resuspension as the wind speed increases.

In these calculations, K_p was assumed to be 10^4 L/kg, a relatively low average value for PCBs. Because desorption rates are inversely proportional to K_p, the transport of PCBs with higher values of K_p, say 10^5 to 10^6 L/kg, would differ considerably from that shown here; because of the much lower desorption rates, there would be much lower values of C_w in the overlying water and higher values of C_s

TABLE 8.2
Percentage of Resuspended Contaminants in Green Bay
That Are Still in the Water at End of Event

	Wind Speed(m/s)		
	5	10	20
Equilibrium	76	27	13
Nonequilibrium	30	11	8

in the overlying water and in the deposited sediments for sediments with high K_p as compared with those with low K_p or with equilibrium partitioning.

PCBs are generally mixtures of PCB congeners, each with a different K_p. These K_p values can differ from one another by more than an order of magnitude, and hence congener desorption rates also can differ by more than an order of magnitude. This dependence on K_p of the congener desorption rate and hence transport in the overlying water also pertains to the nonerosion/nondeposition sediment-water flux (Sections 8.3 and 8.4). Because of this as well as differing solubilities, volatilization rates, and dechlorination rates (all of which depend on K_p), the relative concentrations of PCB congeners will change during transport. This dependence on K_p probably is a significant contributor to the "weathering" of PCBs (e.g., as reported for PCBs in the Hudson River (National Research Council, 2001)).

8.2 THE DIFFUSION APPROXIMATION FOR THE SEDIMENT-WATER FLUX

As a more accurate approximation than the mass transfer approximation of Equation 8.1 or 8.2, the vertical transport of a chemical within the sediment to the overlying water has often been described as simple, or Fickian, diffusion. This approximation is usually only valid for molecular diffusion of an inert, nonreacting substance. Alternately, when chemical reactions are present and are fast (e.g., when adsorption and desorption times are small compared to diffusion transport times in the sediments), then a quasi-equilibrium diffusion approximation can be used. These two limiting approximations are described below and also are compared with results from the mass transfer approximation as described by Equation 8.1 or 8.2.

8.2.1 SIMPLE, OR FICKIAN, DIFFUSION

The basic equations for the diffusion of an inert chemical are essentially the same as those for heat conduction (e.g., Carslaw and Jaeger, 1959) and can be derived in a similar manner. For the one-dimensional, time-dependent diffusive transport of an inert chemical with concentration C(x,t), the flux is given by

$$q = -D \frac{\partial C}{\partial x} \qquad (8.3)$$

where D is a diffusion coefficient (cm^2/s). The conservation equation for C (i.e., the time rate of change of C in an infinitesimal layer due to this flux) is given by

$$\frac{\partial C}{\partial t} = -\frac{\partial q}{\partial x} \qquad (8.4)$$

These two equations can be combined to give

$$\frac{\partial C}{\partial t} = \frac{\partial}{\partial x}\left(D \frac{\partial C}{\partial x}\right)$$

$$= D \frac{\partial^2 C}{\partial x^2} \qquad (8.5)$$

where the second form is valid when D is not a function of x.

To illustrate the application of this equation to a specific problem, consider the one-dimensional, time-dependent transport by diffusion of a dissolved, inert chemical at constant concentration, C_o, in the overlying water into clean sediment. From continuity, the chemical concentration in the sediment at the surface is C_o. The governing equation, initial condition, and boundary conditions can then be written as

$$\frac{\partial C}{\partial t} = D \frac{\partial^2 C}{\partial x^2} \qquad (8.6)$$

$$C(x,0) = 0 \qquad (8.7)$$

$$C(0,t) = C_o \qquad (8.8)$$

$$\lim_{x \to \infty} C(x,t) = 0 \qquad (8.9)$$

where D is assumed constant. The solution to this problem is

$$C = C_o \mathrm{erfc}\left(\frac{x}{2\sqrt{Dt}}\right) \qquad (8.10)$$

where erfc z is the complementary error function (Carslaw and Jaeger, 1959) and $z = x / 2\sqrt{Dt}$. For $D = 1 \times 10^{-5}$ cm^2/s (an approximate value for the molecular diffusion coefficient for many chemicals in water), this solution is shown in Figure 8.4(a) for t = 1, 10, and 100 days. Because C is only a function of z, C can be more economically plotted as a function of z only; this is shown in Figure 8.4(b).

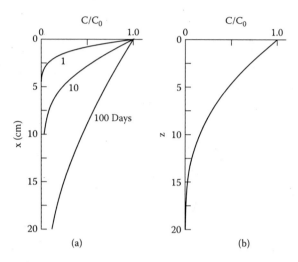

FIGURE 8.4 C/C_0 as a function of (a) depth at different times and (b) $z = x/2(Dt)^{1/2}$.

It is useful to define a decay length as the depth at which $z = 1$. This length is a convenient definition for the approximate distance to which the chemical penetrates with time. For $z = 1$, (1) erfc z = erfc $1 = 0.157$ and C/C_0 has therefore decreased from 1.0 at the surface to 0.157 at $z = 1$, and (2) $x = 2\sqrt{Dt}$; that is, the distance to which the chemical penetrates increases as the square root of time.

At the surface, the chemical flux into the sediment is given by

$$q(0,t) = -D\frac{\partial C(0,t)}{\partial x}$$

$$= \frac{DC_o}{\sqrt{\pi Dt}} \tag{8.11}$$

This flux is singular as $t \to 0$, decreases as $t^{-1/2}$, and goes to zero as $t \to \infty$. The total mass of chemical diffused into the sediment per unit area as a function of time, $Q(t)$, is

$$Q = \int_0^t q(0,t)dt \tag{8.12}$$

For the present problem, it follows from integration of Equation 8.11 that

$$Q = C_o\left(\frac{4Dt}{\pi}\right)^{1/2} \tag{8.13}$$

and grows with time as $t^{1/2}$; it is not singular as $t \to 0$.

8.2.2 SORPTION EQUILIBRIUM

For hydrophobic chemicals, the effects of the adsorption and desorption of chemical between the pore waters and solid particles must be considered. As described in Chapter 7, this sorption process is time dependent and its characteristic times are often comparable to or longer than those for other processes of interest. When this is true, the rate of sorption must be explicitly considered, and the problem becomes mathematically more complex than before. In some cases, sorption rates may be sufficiently fast that the chemical sorbed to the solids is in approximate equilibrium with the chemical dissolved in the pore water. When this occurs, a quasi-equilibrium diffusion approximation is valid and can be used. This approximation is derived as follows.

In the following, HOC transport by both molecular diffusion and bioturbation will be considered for generality. Molecular diffusion will be approximated by a diffusion of the HOC in the pore water with a diffusion coefficient D_m. For simplicity and as a first approximation (see Section 8.4 for a more accurate approximation), bioturbation will be approximated as a diffusion of solids and pore water with an effective diffusion coefficient of D_b. The diffusion coefficient for contaminants sorbed to solids, D_s, is then the same as D_b, whereas the diffusion coefficient for the contaminant dissolved in the pore water, D_w, is the sum of D_m and D_b. In general, D_b is dependent on depth in the sediments; its value is a maximum at the surface and it decreases with depth, with a characteristic length scale on the order of 5 cm for fresh-water organisms and on the order of 10 cm or more for sea-water organisms.

With these approximations and including time-dependent, nonequilibrium sorption as described by Equation 7.31, the one-dimensional, time-dependent mass conservation equation for the contaminant dissolved in water (per unit volume of sediment) is

$$\phi \frac{\partial C_w}{\partial t} - \phi \frac{\partial}{\partial x} \left(D_w \frac{\partial C_w}{\partial x} \right) = (1 - \phi)\rho_s k(C_s - K_p C_w) \qquad (8.14)$$

whereas the conservation equation for the contaminant sorbed to the solids (again, per unit volume of sediment) is

$$(1 - \phi)\rho_s \frac{\partial C_s}{\partial t} - (1 - \phi)\rho_s \frac{\partial}{\partial x} \left(D_s \frac{\partial C_s}{\partial x} \right) = -(1 - \phi)\rho_s k(C_s - K_p C_w) \qquad (8.15)$$

where ϕ is the porosity of the sediments and has been assumed to be constant, and ρ_s is the mass density of the solid particles. The flux of contaminant between the sediments and the overlying water due to diffusion of the dissolved contaminant is given by

$$q(t) = -\phi D_w \frac{\partial C_w}{\partial x}(0, t) \qquad (8.16)$$

It is assumed that there is no flux of contaminant from the solid particles directly into the overlying water.

Equations 8.14 and 8.15 are coupled equations and hence must generally be solved simultaneously. To avoid this complexity, a common approximation is to assume quasi-equilibrium (fast sorption rates) so that $C_s \cong K_p C_w$. By adding Equations 8.14 and 8.15 and then using this assumption, one obtains a diffusion equation with a modified diffusion coefficient:

$$\frac{\partial C_w}{\partial t} = \frac{\partial}{\partial x}\left(D^* \frac{\partial C_w}{\partial x}\right) \qquad (8.17)$$

where

$$D^* = \frac{D_m}{1 + \left(\dfrac{1-\phi}{\phi}\right)\rho_s K_p} + D_b \qquad (8.18)$$

For hydrophobic chemicals with large K_p values and in the absence of bioturbation, $D^* \ll D_m$. If D^* is independent of depth, then Equation 8.17 reduces to

$$\frac{\partial C_w}{\partial t} = D^* \frac{\partial^2 C_w}{\partial x^2} \qquad (8.19)$$

and has the same form as Equation 8.6. In general, the sediment-water flux is still given by Equation 8.16, with no substitution of D^* for D_w.

Although the above three equations are reasonable approximations for many cases, they are only valid when the assumption of quasi-equilibrium is valid, that is, for fast reaction rates when $C_s \cong K_p C_w$. As will be demonstrated in the following sections, these equations are not valid to describe the diffusion of HOCs into and from sediments in many, if not most, realistic conditions.

8.2.3 A MASS TRANSFER APPROXIMATION

Using the mass transfer approximation (Equation 8.2), the transfer of a dissolved chemical at constant concentration, C_{wo}, in the overlying water into a well-mixed sediment layer of thickness h that is initially clean can be described by

$$(1-\phi)\rho_s h \frac{dC_s}{dt} = -\frac{D}{h}\left(\frac{C_s}{K_p} - C_{wo}\right) \qquad (8.20)$$

where $C_s(t)$ is the sorbed chemical concentration in the layer, ϕ is the porosity of the sediment, and ρ_s is the mass density of a solid particle. Because of the assumed large partition coefficient and the implicit assumption of equilibrium partitioning, the amount of chemical dissolved in the pore water can be neglected compared with the amount of chemical sorbed to the solids. The term on the left-hand side is the time rate of change of the mass of chemical per unit area in the layer, whereas the term on the right-hand side is the flux into that layer. It is assumed that there is no flux from the well-mixed layer into the sediments below.

The above equation can be integrated to give

$$C_s = C_{so} (1 - e^{-kt})$$ (8.21)

$$k = \frac{D}{h^2(1-\phi)\rho_s K_p}$$ (8.22)

where $C_{so} = K_p C_{wo}$. From Equations 8.20 and 8.21, the flux is then given as a function of time by

$$q = \frac{D}{h} C_{wo} e^{-kt}$$ (8.23)

A decay time for q can be defined as $kt^* = 1$ (t^* is the time for q to decay to $e^{-1} = 0.368$ of its initial value), or

$$t^* = \frac{1}{k} = \frac{h^2(1-\phi)\rho_s K_p}{D}$$

$$= \frac{h(1-\phi)\rho_s K_p}{H}$$ (8.24)

In the first expression, $t^* \sim h^2$; in the second, $t^* \sim h$. Either way, depending on whether D or H is assumed constant in the modeling, it is clear that h is a crucial parameter for the determination of t^* and hence the time for natural recovery. However, as will be seen in the following sections, the parameter h is difficult to define and even more difficult to accurately quantify.

The related problem of the flux of a chemical from a well-mixed layer of thickness h with an initial chemical concentration C_{so} into clean overlying water can be solved in a similar manner. In this case, the solution for C_s is given by $C_{so}e^{-kt}$; the flux from the sediment is still given by Equation 8.23, with $C_{wo} = C_{so}/K_p$.

As shown in Equation 8.11, the flux due to simple diffusion decays with time as $t^{-1/2}$. By contrast, Equation 8.23 indicates that the flux in the mass transfer approximation decays as e^{-kt}; in addition, this latter decay time, t^*, depends on h, a quantity that cannot be determined from the above equations. The two approximations are inherently different and will give widely different solutions for the flux for large time. This will be discussed more thoroughly in Section 8.5.

8.3 THE SEDIMENT-WATER FLUX DUE TO MOLECULAR DIFFUSION

For HOCs, the sediment-water flux due to molecular diffusion is often significantly modified by finite-rate sorption, with the amount and rate of sorption dependent on the partition coefficient. This has been demonstrated and quantified by experiments and theoretical modeling; some of these efforts will be described here. The most detailed set of experiments were one-dimensional, time-dependent experiments for hexachlorobenzene (HCB) diffusing into and sorbing to a Detroit River sediment (Deane et al., 1999). For these sediments, the measured partition coefficient for HCB was 1.2×10^4 L/kg. The lengths of these experiments were variable, but some continued for up to 512 days. Deane et al. also did tritiated water (THO) experiments in order to (1) illustrate the differences between a purely diffusing (non-sorbing) chemical, THO, and a chemical that diffuses but also strongly sorbs to the sediment, HCB, and (2) obtain parameters for pure diffusion that were then used to more accurately interpret the HCB experiments. These results and analyses by means of numerical models are described first.

For further understanding of the sediment-water flux of HOCs due to molecular diffusion, additional experiments were later done with two different sediments and with HOCs that had a range of K_p values from approximately 10 L/kg to 5×10^4 L/kg (Lick et al., 2006b); these results and analyses are presented next. For remediation purposes, sediment-water fluxes need to be predicted over long periods of time, up to 100 years. On the basis of the experimental work and analyses, numerical calculations were made and are used here to illustrate the characteristic behavior of HOC diffusional fluxes over these long periods of time. For a more general understanding of the molecular diffusion of HOCs, related problems with different boundary and/or initial conditions than those above are also discussed. In particular, results of desorption experiments are presented that, when compared to the results of adsorption experiments, demonstrate the reversibility of the process of molecular diffusion with finite sorption rates.

8.3.1 HEXACHLOROBENZENE (HCB)

8.3.1.1 Experiments

In the experiments described here, HCB diffused from the overlying water into an initially clean sediment. The procedure was such that replicate sediment columns (more accurately called patties because of their small length-to-diameter ratio) were first formed, the upper surfaces of these patties were then exposed to water with a high concentration of dissolved HCB, the HCB diffused into each patty, and the amount of HCB in each patty was measured as a function of depth and time. Radio-labeled HCB was used, and concentrations were measured using liquid scintillation counting. The experimental procedure is only briefly summarized here; Deane et al. (1999) should be consulted for the details.

The sediments used were fine-grained sediments (median particle size of 15 μm) from the Detroit River; the organic carbon content was 3.2%. To form the sediment patties, cylindrical dishes were constructed with an inner diameter of 1.5 cm and a depth of 1 cm. The bottom of each patty was supported by a moveable piston. The piston could be moved at 1-mm increments; 1-mm slices of the patty could then be taken to determine the HCB concentration as a function of depth. Replicate patties were sacrificed at different time intervals; this allowed the determination of HCB concentration as a function of depth at different time intervals. Patties were placed in a 500-mL glass jar with a Teflon-lined lid along with a magnetic stir bar. The jars contained 300 mL HCB-saturated water and a stainless steel source jar containing solid HCB. The HCB source jar and stir bar kept the water well mixed and held the dissolved HCB concentration near saturation over the duration of the experiments. In this way, the upper surfaces of the patties were exposed for long periods of time to water containing HCB at concentrations near saturation.

THO experiments were performed with the same sediments and in a similar manner except that tritiated water was used as the diffusing chemical. Because THO does not sorb to sediments, the quantity measured in the experiments was the THO concentration in the pore water, C_w. HCB has a large K_p and hence sorbs strongly to the solids in the sediments; the amount of HCB dissolved in the water is negligible by comparison with that sorbed to the solids. Because of this, the quantity measured was the concentration of the HCB sorbed to the solids, C_s.

8.3.1.2 Theoretical Models

Although Fickian diffusion is a valid approximation for chemicals diffusing into sediments in some limiting cases (as described in the previous section), the actual transport of either tritiated water or HCB is more complicated than this. Consider first the simpler case of the diffusion of tritiated water. Preliminary comparisons of experimental results with theoretical models demonstrated that, in order to describe this transport accurately, it was necessary to assume that the water-filled pores in the sediments could be divided into two compartments (Coates and Smith, 1964; Van Genuchten and Wierenga, 1976; Nkedi-Kizza et al., 1984; Harmon et al., 1989): (1) the main channels where vertical diffusion occurs and (2) less accessible side pores into which a chemical can diffuse from the main channels but cannot diffuse vertically in these side pores. Mass conservation equations (per unit volume of sediments) to describe this process can be written as

$$\phi v_1 \frac{\partial C_{w1}}{\partial t} = \phi v_1 D \frac{\partial^2 C_{w1}}{\partial x^2} + \phi v_1 k_w (C_{w2} - C_{w1}) \qquad (8.25)$$

$$\phi v_2 \frac{\partial C_{w2}}{\partial t} = -\phi v_1 k_w (C_{w2} - C_{w1}) \qquad (8.26)$$

where C_{w1} = chemical concentration in compartment 1 (the main channels); C_{w2} = chemical concentration in compartment 2 (the side pores); k_w = mass transfer coefficient (cm/s) for transport of the chemical from compartment 1 to compartment 2; and v_1 and v_2 = fractional volumes of sediments (solids and water) in compartments 1 and 2, respectively.

For the case of HCB, or more generally HOC transport, the time-dependent sorption of the HOC to the organic matter in or on the solid particles must be considered. For pure molecular diffusion and subsequent sorption in a single compartment, the conservation equations reduce from Equations 8.14 and 8.15 to

$$\phi \frac{\partial C_w}{\partial t} - \phi D_w \frac{\partial^2 C_w}{\partial x^2} = (1-\phi)\rho_s k(C_s - K_p C_w) \tag{8.27}$$

$$(1-\phi)\rho_s \frac{\partial C_s}{\partial t} = -(1-\phi)\rho_s k(C_s - K_p C_w) \tag{8.28}$$

where it is assumed that D_w is constant.

In the most general case of molecular diffusion with sorption, it is assumed that (1) HCB in solution diffuses through the main channels and then into the side pores; (2) in each compartment, HCB in solution sorbs to the solids in that compartment; and (3) three different size classes of sediments exist, each with a different mass transfer coefficient, k_i, and size fraction, f_i. The resulting mass conservation equations are then

$$\phi v_1 \frac{\partial C_{w1}}{\partial t} = \phi v_1 D_w \frac{\partial^2 C_{w1}}{\partial x^2} + \phi v_1 k_w (C_{w2} - C_{w1})$$

$$+ (1-\phi)v_1 \rho_s \sum_{i=1}^{3} f_i k_i (C_{si1} - K_p C_{w1}) \tag{8.29}$$

$$\phi v_2 \frac{\partial C_{w2}}{\partial t} = -\phi v_1 k_w (C_{w2} - C_{w1}) + (1-\phi)v_2 \rho_s \sum_{i=1}^{3} f_i k_i (C_{si2} - K_p C_{w2}) \tag{8.30}$$

$$(1-\phi)\rho_s f_i v_1 \frac{\partial C_{si1}}{\partial t} = -(1-\phi)\rho_s f_i v_1 k_i (C_{si1} - K_p C_{w1}) \tag{8.31}$$

$$(1-\phi)\rho_s f_i v_2 \frac{\partial C_{si2}}{\partial t} = -(1-\phi)\rho_s f_i v_2 k_i (C_{si2} - K_p C_{w2}) \tag{8.32}$$

where C_{si1} is the chemical concentration in the i-th component of the solids in compartment 1, and C_{si2} is the chemical concentration in the i-th component of the solids in compartment 2.

In the present case of the time-dependent flux of a dissolved HOC in the overlying water into clean bottom sediments, the boundary and initial conditions are that $C_w(0,t) = C_{wo}$ = constant, $C_w(x,0) = 0$, and $C_s(x,0) = 0$; for a sediment of depth h, $\partial C_w(h,t)/\partial x = 0$; and for a sediment of infinite depth, $C_w(x,t) \to 0$ as $x \to \infty$.

8.3.1.3 Diffusion of Tritiated Water

For the diffusion of THO into Detroit River sediments, experimental results for the normalized concentration, C_w/C_{wo}, are shown as a function of depth and at

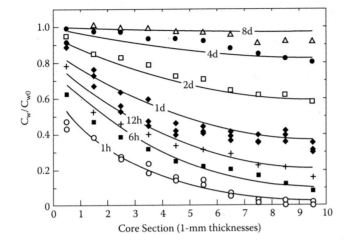

FIGURE 8.5   Experimental and theoretical results for the diffusion of THO into consolidated Detroit River sediment. Concentration of THO in the pore water is normalized by the overlying water concentration. (*Source*: From Deane et al., 1999. With permission.)

different times in Figure 8.5. The results are qualitatively similar to those expected for pure Fickian diffusion. Because the sediment thickness is only 1 cm, the tritiated water rapidly saturates the sediment and reaches an approximate steady state in about 8 days. A numerical calculation was first made with the assumption that the diffusion was Fickian, that is, by means of Equation 8.6. The results were qualitatively correct but did not agree well with the experimental results for all times. Much better agreement was found if it was assumed that the diffusion was governed by Equations 8.25 and 8.26, that is, diffusion into the less accessible pores was significant. However, although the agreement was improved for longer times, the calculated results for the first few hours were still not in good agreement with the experimental results. This was probably an experimental artifact due to the initial placement of the sediment column into the receiving water; this placement inevitably caused small convection currents in the overlying receiving waters that may have caused a slight convective penetration of THO into the sediment column. To correct for this in the calculations, it was assumed that the initial conditions for the calculation were those given by the experimental data at 1 hour. The calculations were then continued by means of Equations 8.25 and 8.26. Results of this latter calculation are shown along with the experimental results in Figure 8.5. There is excellent agreement between the calculations and experimental data for all time. For this calculation, it was assumed that $D = 6 \times 10^{-6}$ cm^2/s, $k_w = 5 \times 10^{-6}$/s, and $v_1 = v_2 = 0.5$. The value of D is approximately the same as that typically found for dissolved chemicals in water.

8.3.1.4   HCB Diffusion and Sorption

For the experiments with HCB diffusing from the overlying water into clean sediments, the measured HCB concentrations on the solids, C_s, as a function of depth

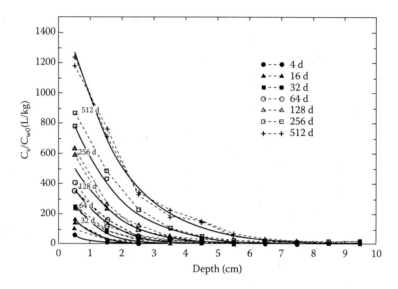

FIGURE 8.6 Experimental and theoretical results for the molecular diffusion of HCB into consolidated Detroit River sediments. C_s/C_{wo} is shown as a function of depth at different times from 4 to 512 days. (*Source*: From Deane et al., 1999. With permission.)

with time as a parameter are shown in Figure 8.6. Due to sorption, the diffusion of HCB into the interior is much slower than that of THO. Despite the length of the experiments (512 days), (1) significant changes in C_s are limited to a few millimeters near the sediment-water interface, and (2) measured values of C_s/C_{wo} near the surface are generally less than 0.1 of their equilibrium value at the surface (where C_s/C_{wo} in equilibrium should equal K_p, i.e., 1.2×10^4 L/kg).

Results of a numerical simulation (which necessarily included nonequilibrium sorption) are shown as the solid lines in Figure 8.6. Parameters assumed were $D = 6 \times 10^{-6}$ cm^2/s; $k_i = 1.6 \times 10^{-8}$, 8×10^{-9}, and 8×10^{-10}/s; and $f_i = 1/3$ for the three size classes with $i = 1, 2$, and 3. The agreement with the experimental results is quite good. In this calculation, the assumptions of three size classes of sediment aggregates (equivalent to diameters of 9, 120, and 400 μm as shown below) as well as diffusion into primary and secondary pores were made. For a calculation with only one size class (120 μm), the agreement was good for intermediate to large times but was not quite as good for small times. A calculation that ignored diffusion into secondary pores modified the results, but not significantly.

Because of experimental limitations, the chemical concentration of HCB in the pore water, C_w, could not be measured. However, this quantity was calculated; values of C_s and K_pC_w (both normalized with respect to C_{wo}) are shown in Figure 8.7 as a function of depth at different times from 4 to 512 days. Both C_s and K_pC_w monotonically increase with time and decrease with depth. It can be seen that (1) C_s and K_pC_w are quite different and are therefore not in local chemical equilibrium with each other, or even close to equilibrium, even after 512 days; and (2) C_w is almost independent of time and is only a function of distance.

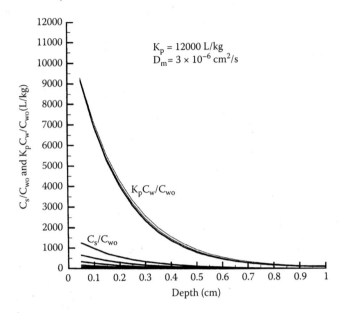

FIGURE 8.7 Calculated results for the molecular diffusion of HCB into consolidated Detroit River sediments. C_s/C_{wo} and $K_p C_w/C_{wo}$ are shown as functions of depth at times of 4, 16, 32, 64, 128, 256, and 512 days. (*Source*: From Deane et al., 1999. With permission.)

When the entire 1-cm column is saturated, C_w should equal C_{wo} and C_s/C_{wo} should equal K_p throughout the sediment column. From the experimental results, it is quite clear that C_s/C_{wo} is far from saturation everywhere. In fact, by means of the model, it can be demonstrated that attainment of 90% of saturation throughout the 1-cm sediment column will take approximately 300 years, or about 10^5 days. This is to be compared with about 8 days for THO saturation (Figure 8.5).

In this calculation, the value of D was taken to be 4×10^{-6} cm²/s, that is, approximately the same as that for THO in the previous calculation and similar to that for many dissolved chemicals in water. It has been suggested that colloids contribute significantly to HOC transport (Thoma et al., 1991). From the present results, the effect of colloids on transport seems to be minimal because the diffusion coefficient necessary for calibration is not significantly greater than those for dissolved chemicals in water.

Although the average disaggregated size of the Detroit River sediments was 15 μm, the effective sizes for sorption can be quite different because, during consolidation, the sediment particles form into larger aggregates of particles. This can be shown as follows. For each size class, an effective mass transfer coefficient is given by Equation 7.34, and therefore

$$k_i = \frac{D}{0.0165d_i^2} \qquad (8.33)$$

From previous experiments and analyses with Detroit River sediments, the value of the diffusion coefficient within the particle, D, is approximately 2×10^{-14} cm^2/s (Lick and Rapaka, 1996). The above equation then gives a relation between the effective diameters for sorption, d_i, and the mass transfer coefficients assumed in the numerical calculations. It follows from the above equation and the assumed values of k_i that the effective particle diameters for the three size classes are 9, 120, and 400 μm. The sediments in the smallest size class are then comparable to the disaggregated sediment particles, whereas the sediments in the larger size classes show the effects of aggregation during consolidation and have much larger sorption equilibration times.

8.3.2 ADDITIONAL HOCs

Experiments and theoretical analyses were done to broaden the HCB investigation so as to include additional HOCs with a wide range of K_p values and with diffusion into two different sediments (Lick et al., 2006b). Experiments were done with three HOCs (a tetrachlorobiphenyl, TPCB; a monochlorobiphenyl, MCB; and pentachlorophenol, PCP) in Detroit River sediments (organic content of 3.2%) and five HOCs (HCB, pyrene, phenanthrene, naphthalene, and benzene) in Lake Michigan sediments (organic content of 1.8%). The procedures to do this were similar to those described above.

8.3.2.1 Experimental Results

The HOCs, K_p values, and the duration of each experiment, T, are given in Table 8.3. The K_p values range from 46,000 to 11.5 L/kg. Results for three representative HOCs

TABLE 8.3
Parameters for Molecular Diffusion Experiments

Chemical	K_p (L/kg)	Length of Experiments, T (days)	Molecular Weight	D_{w0} (10^{-6} cm^2/s)
Detroit River Sediments (3.2% o.c.)				
TPCB	46000	256	292	4.0
HCB	12000	512	285	4.0
MCB	1200	64	113	6.35
PCP	1000	64	266	4.14
Lake Michigan Sediments (1.8% o.c.)				
HCB	9400	96	285	1.12
Pyrene	3700	64	202	1.36
Phenanthrene	2300	64	178	1.44
Naphthalene	80	32	128	1.68
Benzene	11.5	96	78	2.16

are shown in Figure 8.8 for TPCB, K_p = 46,000 L/kg, Detroit River sediments (Figure 8.8(a)); MCB, K_p = 1200 L/kg, Detroit River sediments (Figure 8.8(b)); and naphthalene, K_p = 80 L/kg, Lake Michigan sediments (Figure 8.8(c)). The results for these HOCs as well as for the other HOCs not shown here are qualitatively similar to those for HCB (Figure 8.6); that is, (1) significant changes in C_s are limited to a few millimeters near the sediment-water interface and (2) values of C_s/C_{wo} are generally much less than their equilibrium value of K_p at the surface. In addition,

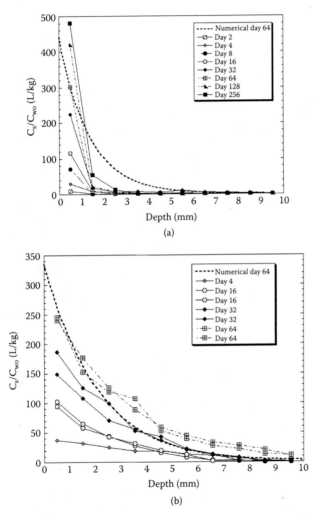

FIGURE 8.8 Experimental and theoretical results for the molecular diffusion of HOCs into consolidated Detroit River sediments. C_s/C_{wo} is shown as a function of depth at different times. (a) TPCB, K_p = 46,000 L/kg. Experimental results are shown at 2, 4, 8, 16, 32, 64, 128, and 256 days; calculated results are shown at 64 days. (b) MCB, K_p = 1,200 L/kg. Experimental results are shown at 4, 16, 32, and 64 days; calculated results are shown at 64 days. (*Source*: From Lick et al., 2006b. With permission.)

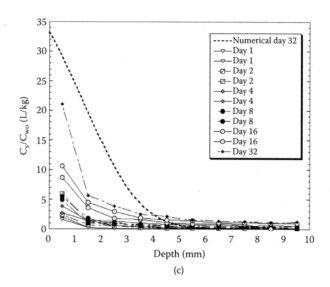

(c)

FIGURE 8.8 (CONTINUED) Experimental and theoretical results for the molecular diffusion of HOCs into consolidated Lake Michigan sediments. C_s/C_{wo} is shown as a function of depth at different times. (c) Naphthalene, K_p = 80 L/kg. Experimental results are shown at 1, 2, 4, 8, 16, and 32 days; calculated results are shown at 32 days. (*Source*: From Lick et al., 2006b. With permission.)

the results show that as K_p decreases, (3) the chemical diffuses more rapidly into the interior, and (4) values of C_s/C_{wo} tend to approach their equilibrium value of K_p at the sediment-water interface more rapidly.

Comparison of the results for HOCs with comparable K_p values in the two different sediments shows that the concentrations are somewhat higher and the HOC has diffused further into the sediments for the Detroit River sediments than for the Lake Michigan sediments. The differences are not unexpected because the two sediments are somewhat different with different amounts and possibly quality and size of organic matter and with somewhat different particle size distributions. However, quantitative reasons for these differences are not known.

8.3.2.2 Theoretical Model

In the modeling of HCB diffusion as described above, one-dimensional, time-dependent diffusion with a finite rate of sorption of HCB between the solid particles and pore water in the sediment was assumed. The general features of the model used in the calculations shown here are essentially the same except that (1) the diffusion coefficient and porosity are assumed to vary with depth (this can be significant near the sediment-water interface) and (2) molecular diffusion occurs in the water in the main channels only; that is, the presence of secondary pores is ignored (this latter process was shown to have a minor effect in the above calculations for HCB).

With these approximations, the one-dimensional, time-dependent conservation equation for the contaminant dissolved in the water (per unit volume of sediment) is

$$\phi\frac{\partial C_w}{\partial t} - \frac{\partial}{\partial x}\left(\phi D_w \frac{\partial C_w}{\partial x}\right) = \sum_i (1-\phi)\rho_s f_i k_i (C_{si} - K_p C_w) \qquad (8.34)$$

whereas the conservation equation for the contaminant sorbed to each size fraction of the solids (again per unit volume of sediment) is

$$(1-\phi)\rho_s f_i \frac{\partial C_{si}}{\partial t} = -(1-\phi)\rho_s f_i k_i (C_{si} - K_p C_w) \qquad (8.35)$$

The flux of contaminant between the sediment and the overlying water is again given by

$$q = -\phi D_w \frac{\partial C_w(0,t)}{\partial x} \qquad (8.36)$$

The spatial variations in the diffusion coefficient and porosity are represented by

$$D_w = D_{w0} - D_{w1}(1 - e^{-\gamma_1 x}) \qquad (8.37)$$

$$\phi = \phi_0 - \phi_1(1 - e^{-\gamma_2 x}) \qquad (8.38)$$

where D_{w0}, D_{w1}, ϕ_0, ϕ_1, γ_1, and γ_2 are constants and are to be determined from the experiments. The variations of D_w and ϕ with depth for different sediments are described below.

8.3.2.3 Numerical Calculations

Based on the above theoretical model, numerical calculations were made for each HOC. In the calculations, several parameters are present that need to be determined. In particular, the effective molecular diffusion coefficient, D_w, and the mass transfer coefficients, k_i, are needed for each chemical and sediment. These can be determined by a comparison of the numerical calculations and experimental results and an adjustment of the parameters in the model until there is agreement between the two.

However, for a uniformly valid analysis, the variations of these parameters from one chemical to another are not arbitrary but are constrained by theory. The basic procedure to determine these parameters was therefore to first adjust the coefficients for HCB for a particular sediment so as to obtain good agreement between the calculated and experimental results. For the other HOCs and the same sediment, the variations in the diffusion and mass transfer coefficients were then determined from basic theory as follows. The coefficient D_w was varied so as to be inversely proportional to the square root of the molecular weight of the chemical (Hirschfelder et al., 1954), whereas the k_i values were varied so as to be proportional to D_w and inversely proportional to K_p, (Equation 7.37). For all calculations, it was assumed that $f_i = 1/3$. When this was done, good agreement between the calculated and experimental results was obtained for all HOCs tested. The agreement was further improved, primarily near the sediment-water interface, by allowing the diffusion coefficient and porosity to vary as in Equations 8.37 and 8.38 by the same amount for each sediment.

With these parameters, the calculated results for TPCB and MCB at day 64 are shown in Figures 8.8(a) and (b) and can be compared there with the experimental results. For all HOCs in Detroit River sediments, reasonably good agreement was obtained. Because only one set of parameters was used for all HOCs, this serves to substantiate the theoretical model and our understanding of the process.

For all HOCs in Lake Michigan sediments, the same procedure was used to determine the parameters. The calculated results at the end of the naphthalene experiment are shown in Figure 8.8(c) and can be compared there with the experimental results. As with HOCs in Detroit River sediments, reasonably good agreement for all HOCs in Lake Michigan sediments was obtained with only one set of parameters.

8.3.3 Long-Term Sediment-Water Fluxes

To illustrate the dependence of the sediment-water flux of an HOC on K_p and time, numerical calculations were made for long times, up to 100 years, and for $K_p = 10^6$, 10^5, 10^4, and 10^3 L/kg. The results for the mass transfer coefficient, q/C_{wo}, are shown in Figure 8.9. For these calculations, the parameters assumed were the same as those for HCB in Detroit River sediments, except that K_p and k_i (~$1/K_p$) were varied. At $t = 0$, all mass transfer coefficients have the same value of 1.44×10^{-5} cm/s or 1.24 cm/day. (For Lake Michigan sediments, this number is essentially the same.) For $t > 0$, each flux decreases with time, with the rate of decrease being greater as K_p decreases. Conversely, as K_p increases, the flux decreases less rapidly with time so that, for example, for $K_p = 1 \times 10^6$ L/kg (a typical value for many PCBs), the flux decreases by less than 30% in 100 years.

8.3.4 Related Problems

The above examples were all concerned with the flux from an HOC dissolved in the overlying water into clean sediments. Related problems and analyses of (1) flux from contaminated bottom sediments to clean overlying water and (2) flux due to a contaminant spill are as follows.

8.3.4.1 Flux from Contaminated Bottom Sediments to Clean Overlying Water

In contrast to the above adsorption experiments, Deane et al. (1999) also performed desorption experiments, that is, HOCs diffusing from contaminated

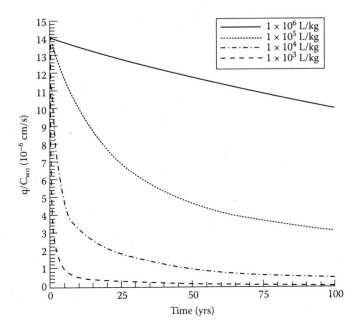

FIGURE 8.9 HOC flux in Detroit River sediments due to molecular diffusion as a function of time, with K_p as a parameter.

bottom sediments to clean overlying water. For both adsorption and desorption experiments with Detroit River sediments, the total amount of HCB adsorbed or desorbed (shown as percent of equilibrium value) during the experiment is shown as a function of time in Figure 8.10. The results are essentially the same for both experiments and demonstrate the reversibility of the process; that is, if the chemical reactions were irreversible, these amounts and hence fluxes would not be the same.

The reversibility of adsorption and desorption also can be demonstrated theoretically. A numerical calculation for the desorption problem with parameters the same as those for the adsorption problem is shown in Figure 8.11; it is the same as the experimental results (Deane, 1998) and is the mirror image of the numerical solution for adsorption (Figure 8.7). This same solution can be found from the adsorption problem (experimental results, calculated solutions, or basic equations) with the substitutions

$$C_w^* = C_{wo} - C_w \qquad (8.39)$$

$$C_s^* = K_p C_{wo} - C_s \qquad (8.40)$$

where C_w^* and C_s^* are the solutions for the desorption problem. Consistent with these substitutions, the initial conditions are that $C_w^*(x,0) = C_{wo}$ and $C_s^*(x,0) = K_p C_{wo}$, whereas the boundary conditions are that $C_w^*(0,t) = 0$ and

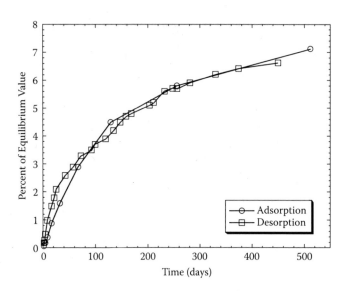

FIGURE 8.10 Comparison of experimental results for HCB adsorption and desorption between Detroit River sediments and overlying water. Percent of equilibrium value is shown as a function of time. (*Source*: From Deane et al., 1999. With permission.)

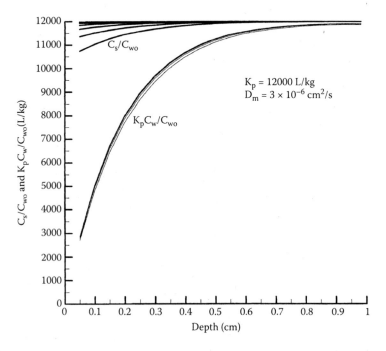

FIGURE 8.11 Desorption of HCB from contaminated sediments to clean overlying water. C_s/C_{wo} and C_w/C_{wo} are shown as functions of time up to 512 days.

$C_w^*(x,t) \to C_{wo}$ as $x \to \infty$. Results for other HOCs for the desorption problem follow in the same way from the previous cases. In particular, the flux can be calculated from Equation 8.36 with the substitution shown in Equation 8.39. Except for a reversal in sign (direction), the flux for different HOCs for the desorption case is therefore exactly the same as that for the adsorption cases shown in Figure 8.9. The fact that the above substitution is valid is another indication that molecular diffusion with finite sorption rates is a reversible process.

8.3.4.2 Flux Due to a Contaminant Spill

The problem of the deposition of a thin layer of contaminated sediments on the surface of clean bottom sediments with clean overlying water is a problem of practical interest and was investigated by means of numerical calculations (Lick et al., 2004b). For this case, the initial conditions are that $C_w(x,0) = C_{wo}$ = constant and $C_s(x,0) = K_p C_{wo}$ for x < h and are zero for x > h, where h is the thickness of the deposited layer. Boundary conditions are $C_w(0,t) = 0$ and $C_w(x,t) \to 0$ as $x \to \infty$. For $K_p = 1.2 \times 10^4$ L/kg and for a deposited layer of 0.1 cm, the numerical results for C_s and $K_p C_w$ are shown in Figure 8.12 as a function of depth and for times up to 512 days. A slow reduction of C_s in the surface layer and a slight increase in C_s at depth as time increases are apparent. C_w is again a function of

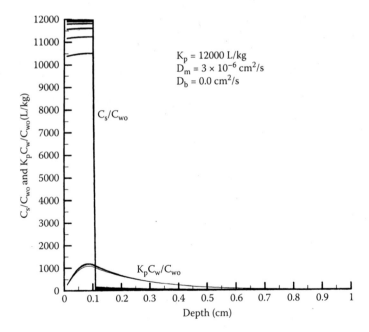

FIGURE 8.12 Desorption of HCB from a thin surficial layer of contaminated sediment. Concentrations are shown as functions of time. (*Source*: From Lick et al., 2004b.)

distance but is almost independent of time. Because of this, the contaminant flux must also be a slowly varying function of time.

Numerical and analytic results indicate that when the thickness of the contaminated layer, h, is greater than a few millimeters, the HOC flux is initially and for many years approximately the same as that for a contaminated layer of semi-infinite thickness (Figure 8.11), with the same maximum value of C_s. This demonstrates that, despite the fact that the deposited layer may be quite thin, on the order of a few millimeters, the flux from this layer to the overlying water is comparable to that for a sediment of semi-infinite depth and will continue at approximately this same level for many years.

8.4 THE SEDIMENT-WATER FLUX DUE TO BIOTURBATION

The effects of bioturbation on the sediment-water flux of an HOC are difficult to understand and quantify due to the large variety and differing amounts of organisms that may be present in surficial sediments, their high variability and activity in space and time, and the different ways that they affect the sediments. This is further complicated by the effects of finite sorption rates. Numerous investigations have been made to generally describe and to some extent quantify the effects of benthic organisms on the physical mixing of sediments (Guinasso and Schink, 1975; Robbins et al., 1977; Fisher et al., 1980; Matisoff, 1982; Wheatcroft et al., 1994; Mohanty et al., 1998) and on contaminant flux (Spaulding, 1987; Reible et al., 1996; U.S. EPA, 2000; Thibodeaux et al., 2001; Sherwood et al., 2002; Luo et al., 2006). A recent survey of work in this area is given by Clarke et al. (2001). Except for Luo et al. (2006), no one has explicitly considered the effects of non-equilibrium sorption.

The effects of benthic organisms on the mixing of sediments are primarily due to feeding and burrowing activities and depend on the organism. Some organisms (such as *Lumbriculus variegatus*, an oligochaete) are primarily vertical burrowers; they ingest sediment at depth and expel the resulting fecal pellets at the sediment-water interface. Other organisms are more horizontal burrowers (e.g., *Chironomus tentans*) or may just mix the sediments in a shallow layer near the sediment-water interface (e.g., *Hyalella azteca*). These three organisms are representative of a wide range of benthic organisms in fresh water; their effects on the flux of HOCs are discussed below. Oligochaetes, because of their vertical burrowing and abundance, are probably the most significant group of organisms in affecting the sediment-water flux of chemicals.

In fresh water, benthic organisms tend to disturb and/or mix the sediments down to depths of 2 to 10 cm. In sea water, vertical burrowers are often much larger than those in fresh water (Oliver et al., 1980; Zwarts and Wanink, 1989) and disturb sediments to deeper depths. As examples, polychaete worms, *Nereis virens*, have been observed at depths of 50 cm (Brannon et al., 1985); echiurea worms at depths of 80 cm (Hughes et al., 1999); and mud shrimps below 1 m (Myers, 1979; Swift, 1978). For marine organisms and for HOCs with large K_p values, measurements of fluxes have not been made.

As an example of how benthic organisms are distributed, consider the survey of organisms in the Great Lakes made by the U.S. Environmental Protection Agency (1999). Densities of organisms were shown to be quite patchy and were generally higher in shallow, nearshore waters (especially near harbors and river mouths) than in deeper parts of the lakes. The amphipod *Diporeia* was generally the most abundant group, except in Lake Erie and parts of Lake Ontario. In these latter two regions, oligochaetes were most abundant and were also the second most abundant group throughout the Great Lakes. Their densities were as follows. In the Central Basin of Lake Erie, local densities of oligochaetes were sometimes quite high, as high as $3.8 \times 10^4/m^2$, but were generally on the order of $10^3/m^2$. In the other Great Lakes, the densities were generally on the order of $10^3/m^2$ or less. In Lake Superior, densities of oligochaetes as well as all other organisms were less than $10^3/m^2$.

A detailed experimental and theoretical study of sediment mixing due to organisms is that by Fisher et al. (1980), who investigated the vertical convection and mixing of sediments by tubificid oligochaetes. This study is described first. Laboratory experiments and modeling of the effects of the three organisms mentioned above on the sediment-water flux of HCB also have been done (Luo et al., 2006) and are described next. On the basis of these results, various effects of bioturbation are then illustrated by means of numerical models.

8.4.1 PHYSICAL MIXING OF SEDIMENTS BY ORGANISMS

In the experiments by Fisher et al. (1980), oligochaetes were placed in a cylinder filled with sediments from Lake Erie. A thin layer of illite containing radioactive ^{137}Cs was then added at the surface. The effects of the organisms as they burrowed in the sediments were followed by measuring the radioactivity of the ^{137}Cs as a function of depth and time.

In modeling sediment mixing due to bioturbation, the process is often described as a time-dependent diffusion process with an effective diffusion coefficient determined by calibration of the model with observations. However, this approximation is not valid to describe the experimental results by Fisher et al. (1980) or the activities of vertical burrowers and feeders in general. Because of this, Fisher et al. used a one-dimensional, time-dependent, convection-diffusion equation to describe the concentration, C, of the radioactive chemical; the resulting equation was

$$\frac{\partial C}{\partial t} + \frac{\partial}{\partial x}(wC) = \delta \frac{\partial^2 C}{\partial x^2} + S \qquad (8.41)$$

where $w(x)$ is the vertical velocity of a sediment layer at depth x, δ is a diffusion coefficient, and $S(x,t)$ is the rate of change of C from a sediment layer due to feeding by the organism. This rate depends on the feeding rate at depth and must also be proportional to C, that is,

$$S(x,t) = -Af(x)C(x,t) \qquad (8.42)$$

where A is a constant of proportionality and f(x) is normalized such that

$$\int_0^\infty f(x)dx = 1 \qquad (8.43)$$

The qualitative dependence of f(x) and the resulting dependence of w(x) must be approximately as shown in Figure 8.13. Specific relations for these quantities are discussed below.

A typical set of experimental results by Fisher et al. (1980) is shown in Figure 8.14. The downward convection of the cesium layer due to feeding by organisms at depth and the deposition of fecal pellets on the surface are clearly evident, as is the broadening of the peak due to diffusive (mechanical mixing) effects. Results of numerical calculations based on the above model are also shown and are in good agreement with the experimental results.

8.4.2 THE FLUX OF AN HOC DUE TO ORGANISMS

The above experiments and modeling were primarily concerned with the physical transport and mixing of sediments due to organisms. Experiments and modeling have also been done that emphasized the vertical transport of an HOC

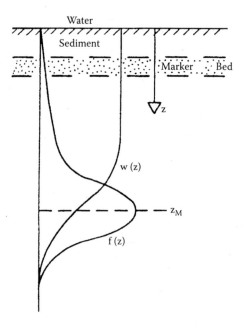

FIGURE 8.13 Diagram of sediment reworking due to benthic organisms. The depth of maximum feeding activity is x_m; f(x) is the vertical distribution of the feeding activity; and w(x) is the vertical velocity of sediment at depth x. (*Source*: From Fisher et al., 1980. With permission.)

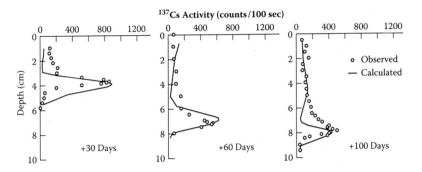

FIGURE 8.14   Observed ^{137}Cs activity in sediment compared with predictions by the convection-diffusion theory. (*Source*: From Fisher et al., 1980. With permission.)

in the sediments and the sediment-water flux of this HOC due to organisms, including the effects of finite-rate sorption dynamics (Luo et al., 2006).

8.4.2.1 Experimental Procedures

The experiments were one-dimensional and time dependent and were meant to determine the transport due to bioturbation (including molecular diffusion) of HCB from the overlying water with a constant concentration of dissolved HCB, C_{wo}, into a clean sediment. The benthic organisms were *Lumbriculus variegatus, Chironomus tentans,* and *Hyalella azteca.* Many of the procedures used in these experiments are similar to those described in the previous section on molecular diffusion. Only a brief summary of these procedures is given here. To better accommodate the organisms, the size of the cylindrical stainless steel dishes was increased compared to that used in the molecular diffusion experiments. Each dish had an interior depth of 15 cm and a surface area of 9.62 cm² (diameter of 3.5 cm). As a substrate for the organisms, fine-grained sediments from Lake Michigan were used. The sediments had a median particle size of 13 μm and an organic carbon content of 1.8%. Radio-labeled HCB was used; HCB concentrations as a function of time and depth in the sediments were measured using liquid scintillation counting.

Approximately 10 cm of wet sediment was added to each dish; this sediment then consolidated in pure water for 10 days before use. After 10 days, ten organisms (a population density of $10^4/m^2$) were added to each dish, and the sediment patty was put into a 1000-mL glass jar. To each jar, 600 mL of the appropriate filtered HCB stock solution was added; source jars of HCB were added to keep the HCB concentration from decreasing significantly during the experiment. At different times, patties were sacrificed and sliced at 0.1-cm intervals to determine the contaminant concentration, C_s, as a function of depth and time. The biological burial rate was determined by putting a layer of glass beads on the sediment surface at the beginning of the experiments; the rate was then determined as the differences in the burial depths of the beads at different times divided by the time interval.

8.4.2.2 Theoretical Model

In describing the sediment-water flux of an HOC due to the activity of *Lumbriculus variegatus*, the processes that are most significant and necessary are (1) sediment convection due to the organisms feeding at depth followed by the transport and deposition of fecal pellets at the sediment-water interface, (2) diffusion of both solid particles and water due to mechanical mixing by the organisms, (3) molecular diffusion of the HOC in the pore water, and (4) finite-rate sorption of the HOC between the solid particles and pore water. The processes necessary to describe the effects of the other two organisms are the same except that vertical sediment convection is less important and can be neglected or approximated as a diffusion process.

With all of the above processes included, the one-dimensional, time-dependent conservation equation for the contaminant dissolved in the pore water (per unit volume of sediment) is

$$\phi \frac{\partial C_w}{\partial t} + \phi \frac{\partial}{\partial x}(w C_w) - \phi \frac{\partial}{\partial x}\left(D_w \frac{\partial C_w}{\partial x}\right) = (1-\phi)\rho_s k(C_s - K_p C_w) + S_w \qquad (8.44)$$

whereas the conservation equation for the contaminant sorbed to the solids (again per unit volume of sediment) is

$$(1-\phi)\rho_s \frac{\partial C_s}{\partial t} + (1-\phi)\rho_s \frac{\partial}{\partial x}(w C_s) - (1-\phi)\rho_s \frac{\partial}{\partial x}\left(D_s \frac{\partial C_s}{\partial x}\right)$$

$$= -(1-\phi)\rho_s k(C_s - K_p C_w) + S_s \qquad (8.45)$$

where S_w and S_s are the rates of loss of C_w and C_s, respectively, from a sediment layer due to feeding by the organism. For simplicity, ϕ is assumed constant, and only one size class of sediments is considered.

The diffusive effects of the organisms are approximated as a diffusion of solids and water with an effective diffusion coefficient of D_b. The diffusion coefficient for contaminants sorbed to solids, D_s, is then equal to D_b, whereas the diffusion coefficient for contaminants dissolved in pore water, D_w, is the sum of D_b and D_m, where D_m is the molecular diffusion coefficient for the HOC in water. Because the activity of the organisms decreases with depth, it is assumed that

$$D_b = D_0 e^{-\gamma x} \qquad (8.46)$$

where D_0 and γ are constants.

The flux of contaminant between the sediment and overlying water is given by

$$q = -\phi D_w \frac{\partial C_w}{\partial x} \qquad (8.47)$$

where all quantities are evaluated at the surface. It is assumed that there is no flux of contaminant from the sediment particles directly into the overlying water.

The above equations combine and extend Equations 8.27 and 8.28 for molecular diffusion with sorption and Equation 8.41 for mechanical mixing by organisms.

The convective velocity, $w(x)$, can be determined as follows (Fisher et al., 1980). In analogy to Equation 8.42, it is assumed that the rate of loss of sediment particles due to feeding by organisms is given by

$$S = -A(1 - \phi)\rho_s f(x) \tag{8.48}$$

where A is a constant and where $f(x)$ satisfies Equation 8.43. For the steady-state movement of the sedimentary particles, the conservation equation is

$$\frac{\partial}{\partial x}\left[(1-\phi)\rho_s w\right] = S \tag{8.49}$$

By substituting Equation 8.48 into Equation 8.49 and integrating the resulting equation over depth, one obtains

$$w(x) = A \int_x^\infty f(x)dx \tag{8.50}$$

From Equation 8.43, it follows that $A = w(0)$ and therefore

$$w(x) = w(0) \int_x^\infty f(x)dx \tag{8.51}$$

where $w(0)$ is the burial rate due to the deposition of fecal pellets and is an experimentally determined quantity.

The specific form of $f(x)$ is assumed to be the same as that suggested by Fisher et al. (1980) and is

$$\frac{f(x)}{N} = a_1 + (1-a_1)\exp[-b(x-x_m)^2] \quad \text{for } x < x_m$$

$$= \exp[-b(x-x_m)^2] \qquad\qquad \text{for } x > x_m \tag{8.52}$$

where a_1 and b are constants, x_m is the depth of maximum feeding, and N is a normalizing factor chosen to satisfy Equation 8.43. The form of $f(x)$ was estimated from observations of organism feeding and is similar to that shown in Figure 8.13.

The fluxes S_w and S_s appearing in Equations 8.44 and 8.45 also must be determined. As in Equation 8.48, it follows that these are given by

$$S_w = -A\phi f(x)C_w \tag{8.53}$$

$$S_s = -A(1-\phi)\rho_s f(x)C_s \tag{8.54}$$

The depositional flux of fecal pellets at the sediment-water interface is the integral of the flux over depth due to feeding, that is,

$$q_s(0,t) = -\int_0^\infty S_s dx$$

$$= A(1-\phi)\rho_s \int_0^\infty f(x)C_s(x,t)dx \tag{8.55}$$

8.4.2.3 Experimental and Modeling Results

For *Lumbriculus variegatus*, experimental results for the concentration of HCB, C_s/C_{wo}, as a function of depth at different time intervals (4, 8, 16, 32, 64, and 96 days) are shown in Figure 8.15(a). Results are qualitatively the same as those for

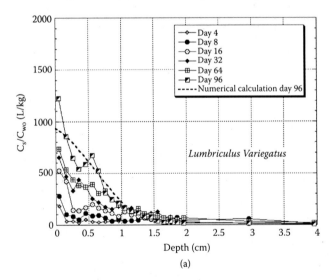

(a)

FIGURE 8.15 Transport of HCB in Lake Michigan sediments due to bioturbation. Sorbed chemical concentration, C_s/C_{wo}, is shown as a function of depth at different times. (a) *Lumbriculus variegatus*; density is $10^4/m^2$. Experimental results are for times of 4, 8, 16, 32, 64, and 96 days, whereas the calculated results are for 96 days. (*Source*: From Luo et al., 2006. With permission.)

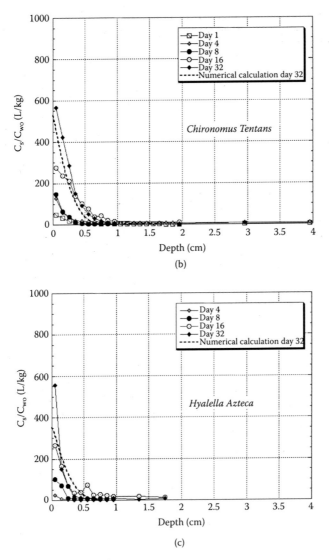

FIGURE 8.15 (CONTINUED) Transport of HCB in Lake Michigan sediments due to bioturbation. Sorbed chemical concentration, C_s/C_{wo}, is shown as a function of depth at different times. (b) *Chironomus tentans*; density is $10^4/m^2$. Experimental results are for times of 1, 4, 8, 16, and 32 days, whereas the calculated results are for 32 days. (c) *Hyalella azteca*; density is $10^4/m^2$. Experimental results are for times of 4, 8, 16, and 32 days, whereas the calculated results are for 32 days. (*Source*: From Luo et al., 2006. With permission.)

HCB as caused by molecular diffusion alone; that is, (1) significant changes in C_s are limited to a shallow layer near the sediment-water interface, and (2) measured values of C_s/C_{wo} are generally much less than their equilibrium value of K_p at the surface ($K_p = 9400$ L/kg). However, by comparison of Figures 8.6 and 8.15(a), it can be seen that, for the present density of organisms ($10^4/m^2$), the effects of bioturbation are much greater than those of molecular diffusion alone; that is, the chemical concentrations are higher and the chemical is transported further into the interior. The chemical concentration decreases smoothly into the interior in an almost exponential manner. Even after 96 days, there is little, if any, evidence of a well-mixed layer due to bioturbation as is sometimes assumed.

For *Chironomus tentans* and *Hyalella azteca*, experimental results for C_s as a function of depth at different times are shown in Figures 8.15(b) and (c), respectively. Results are qualitatively the same as those for *Lumbriculus variegatus* except that the HCB concentrations are lower and the HCB does not penetrate as far.

To demonstrate the behavior of the HCB concentration and flux over long periods of time, numerical calculations were made based on these experiments and the theoretical model described above. In the calculations, various parameters need to be prescribed. For consistency, the quantities D_m (1.2×10^{-6} cm^2/s), φ (0.5), and k (3.1×10^{-8}/s) were kept the same as those same quantities from molecular diffusion experiments and analyses.

For *Lumbriculus variegatus*, the quantities w(0) (8×10^{-8}cm/s), D_b (0.024×10^{-6} cm^2/s), x_m (1.5 cm), and b (1.5/cm^2) were determined from direct observations of the experiments and also by a comparison of the experimental and calculated results. In doing the calculations for this organism, the presence of a thin, very reactive (high sorption rate) layer at the surface was required to obtain good agreement between the calculated and experimental results. The thickness of this layer was on the order of a few millimeters and was due to the fecal pellets deposited on the surface by the oligochaetes. For this layer, a value for k of 282×10^{-8}/s (as compared with a k of 3.1×10^{-8}/s for the interior) was assumed. Because k is inversely proportional to the square of the particle diameter and increases as the porosity increases (Equation 7.37), an explanation for this high value of k is the observed much smaller size and greater porosity of the fecal pellets as compared with those of a sediment aggregate.

With these values for the parameters, calculations of C_s as a function of depth and time were made for *Lumbriculus variegatus*. The calculated results for 96 days are shown in Figure 8.15(a) and compare well with the experimental results at that time. For *Chironomus tentans* and *Hyalella azteca*, the parameters were the same as those for *Lumbriculus variegatus* except that vertical velocities were set to zero and the surficial values for k were assumed to be 40×10^{-8} cm/s and 15×10^{-8} cm/s, respectively. With these values for the parameters, calculated results at 32 days are shown in Figures 8.15(b) and (c) and compare well with the experimental results.

Calculations of the HCB concentration were then made for all three organisms for longer times; results are plotted in Figures 8.16(a), (b), and (c) as a function of depth at the end of each experiment and at times of 1, 3, 5, and 10 years. Figure 8.16(a) shows the sorbed concentrations of HCB, C_s/C_{wo}, for *Lumbriculus variegatus*. As time increases, C_s/C_{wo} near the surface increases but also becomes more uniform with depth; this is due primarily to the feeding of the oligochaetes at depth, the deposition of the fecal pellets at the surface, and

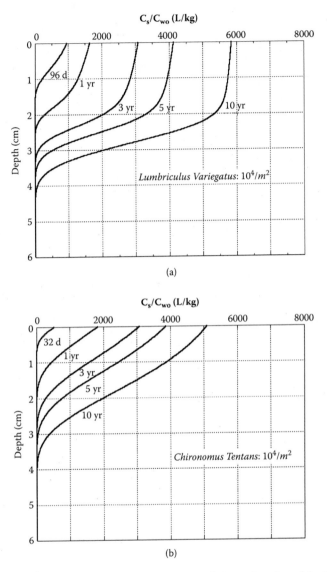

FIGURE 8.16 Calculated results for HCB concentration as a function of depth at times of 96 days and 1, 3, 5, and 10 years. Organism density is $10^4/m^2$. (a) *Lumbriculus variegatus*, and (b) *Chironomus tentans*. (*Source*: From Luo et al., 2006. With permission.)

the resulting downward convection of a sediment layer with the velocity w(x). The maximum feeding rate, $f(x_m)$, occurs at $x = x_m = 1.5$ cm. Below this depth, the feeding rate and hence the vertical velocity decrease rapidly; in addition to the convection, some physical mixing of the sediments by the organisms does occur, but this also decreases with depth. The result is that C_s/C_{wo} decreases rapidly at depths of 2 to 4 cm. A well-mixed layer slowly appears and becomes

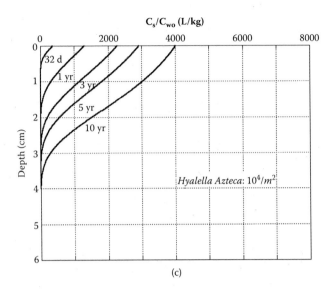

(c)

FIGURE 8.16 (CONTINUED) Calculated results for HCB concentration as a function of depth at times of 96 days and 1, 3, 5, and 10 years. Organism density is $10^4/m^2$. (c) *Hyalella azteca*. (*Source*: From Luo et al., 2006. With permission.)

more distinct as time increases; the thickness of this layer increases with time. The development of this well-mixed layer takes several years.

Near the surface, C_s/C_{wo} is always much less than its chemical equilibrium value of 9400 L/kg and changes relatively slowly as a function of depth. Conversely, in a surficial layer on the order of a few millimeters thick, numerical calculations show that C_w changes rapidly from its value of C_{wo} in the overlying water to its equilibrium value with the solid particles near the surface; that is, C_w is not "well-mixed." As in the pure molecular diffusion of HCB, this rapid change in C_w near the surface is due to the finite-rate sorption of the HCB by the solid particles. Because of this large gradient in C_w near the surface, the transport of HCB near the surface is dominated by diffusion of the dissolved chemical.

For *Chironomus tentans* and *Hyalella azteca*, the results for C_s/C_{wo} are shown in Figures 8.16(b) and (c). Some indication of a well-mixed layer is evident for *Chironomus tentans* after 5 to 10 years, but the effect is much less than that for *Lumbriculus variegatus*. A well-mixed layer is even less noticeable for *Hyalella azteca*.

For *Lumbriculus* at a reduced density of $10^3/m^2$, calculations show that the well-mixed layer does appear with time, but its development is delayed compared with that for a density of $10^4/m^2$. In addition, the chemical concentrations are generally lower and the thickness of the well-mixed layer is less.

In Figure 8.17, the sediment-water flux of HCB as a function of time is shown for all three organisms at densities of $10^4/m^2$, for *Lumbriculus variegatus* at a density of $10^3/m^2$, and for molecular diffusion alone. For *Lumbriculus variegatus* at $10^4/m^2$, q/C_{wo} at t = 0 has a value of about 11.4×10^{-5} cm/s; this decreases with time. For the same organism at a density of $10^3/m^2$, q/C_{wo} at t = 0 has a value of

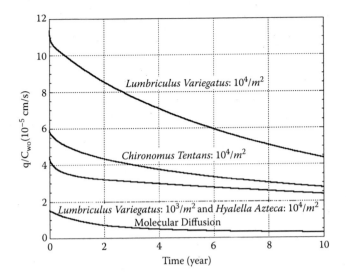

FIGURE 8.17   The sediment-water flux of HCB due to bioturbation as a function of time. Results are shown for all three organisms at densities of $10^4/m^2$, for *Lumbriculus variegatus* at $10^3/m^2$, and for molecular diffusion alone. The results for *Lumbriculus variegatus* at $10^3/m^2$ and *Hyalella azteca* at $10^4/m^2$ are almost the same. (*Source*: From Luo et al., 2006. With permission.)

about 4.5×10^{-5} cm/s; this also decreases with time but at a slower rate than that for $10^4/m^2$. For molecular diffusion alone, the value of C_s/C_{wo} at t = 0 is approximately 1.5×10^{-5} cm/s. By comparison of these three curves, it can be seen that q/C_{wo} is not directly proportional to the concentration of *Lumbriculus* but depends more strongly on the organism concentration at low concentrations than at higher. For the same organism density, the fluxes for the other organisms are less than that for *Lumbriculus*.

As with pure molecular diffusion, the reverse of the above problems — that is, the flux due to bioturbation of an HOC from contaminated sediments to clean overlying water — can be determined from the above solutions by the substitutions $C_w^* = C_{wo} - C_w$ and $C_s^* = K_p C_{wo} - C_s$, where C_w^* and C_s^* are the desired solutions. This is also true for the approximation to bioturbation described below.

8.4.3 MODELING BIOTURBATION AS A DIFFUSION WITH FINITE-RATE SORPTION PROCESS

As indicated above, a diffusion approximation to describe the sediment-water flux due to *Lumbriculus* or other vertical feeders is generally not valid, and a convection-diffusion equation is needed to describe the flux properly. However, for other organisms or for a relatively simple approximation for groups of organisms (including vertical feeders), a diffusion approximation (for both the dissolved chemical and the chemical sorbed to the solids) may be reasonable, as shown above for *Chironomus* and *Hyalella*. When sorption is absent or when sorption equilibrium can be assumed, the simple diffusion approximation (without finite-rate sorption) has been extensively used to describe the flux due to bioturbation. In these cases, effective diffusion coefficients were determined by calibration of the model results with observations. The range of diffusion coefficients typically assumed in the models is 1 to 100 cm^2/yr (Boudreau, 1997), or about 3×10^{-8} to 3×10^{-6} cm^2/s. As an example of bioturbation as approximated by diffusion but with the inclusion of finite sorption rates, the problem of the flux due to a contaminant spill is presented here. The model used is essentially the same as that described above except that the convection term is set to zero.

This is the same problem as that treated in Section 8.3 (molecular diffusion), except that now diffusion of solids and pore water by organisms is included. With $D_b = 1 \times 10^{-5}$ cm^2/s (a large value), numerical solutions for C_s and $K_p C_w$ are shown in Figure 8.18 for times from 4 to 512 days and should be compared with those for molecular diffusion alone in Figure 8.12. The increased diffusion of C_s from its value at the surface (decreasing with time) to lower values in the interior is apparent. For distances greater than about 1 cm, C_s and $K_p C_w$ are approximately equal and are therefore in local chemical equilibrium with each other. However, near

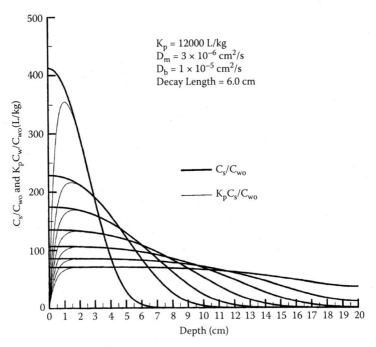

$K_p = 12000$ L/kg
$D_m = 3 \times 10^{-6}$ cm^2/s
$D_b = 1 \times 10^{-5}$ cm^2/s
Decay Length = 6.0 cm

——— C_s/C_{wo}

——— $K_p C_s/C_{wo}$

FIGURE 8.18 Diffusion approximation for transport from a thin surficial layer of contaminated sediment; finite sorption rates. C_s/C_{wo} and $K_p C_w/C_{wo}$ are shown as a function of depth at times of 4, 16, 32, 64, 128, 256, and 512 days. Both C_s/C_{wo} and $K_p C_w/C_{wo}$ decrease with time. $D_b = 1 \times 10^{-5}$ cm^2/s.

the surface, a nonequilibrium layer is apparent where $K_p C_w$ changes rapidly with distance. Both C_s and $K_p C_w$ are changing relatively slowly with time.

8.5 THE SEDIMENT-WATER FLUX DUE TO "DIFFUSION"

In modeling the sediment-water flux of HOCs, the flux due to sediment erosion/ deposition is often described explicitly by means of a sediment transport and HOC sorption model (as in Section 8.1), whereas the remaining components of the flux (molecular diffusion, bioturbation, and groundwater flow) are lumped together and modeled as a "diffusive" flux. This diffusive flux is usually described by means of a mass transfer approximation, Equation 8.1 or 8.2, with the implicit assumptions of a well-mixed contaminant layer of thickness h at the surface of the sediments as well as equilibrium sorption. Conversely, in the previous two sections, more accurate descriptions of the flux due to molecular diffusion and bioturbation were given. On the basis of these latter results, the justification for the use of the mass transfer approximation to predict the sediment-water flux is discussed here (Lick, 2006). Consistent with Sections 8.3 and 8.4 and for HOCs with large partition coefficients, it is argued that a well-mixed layer due to non-erosion/non-deposition processes is slow to form and usually does not exist; that h is difficult to define and even more difficult to quantify; and that, as far as long-term predictions (up to 100 years) of this diffusive flux are concerned, the exact value for an effective h does not matter.

8.5.1 THE FLUX AND THE FORMATION OF SEDIMENT LAYERS DUE TO EROSION/DEPOSITION

In the modeling of sediment transport, because of the nonlinear increase of erosion rates with shear stress, it is clear that big events such as storms and floods, despite their infrequent occurrence, can often cause amounts of sediment transport that are greater than the total of the amounts from smaller erosion/deposition events. After these big events, these large transports are evident as distinct layers in the bottom sediments. This is suggested by the modeling but can be easily seen and verified from the sedimentary record (Sections 1.3 and 6.5). As an example of this record, consider a core from the Kalamazoo River (Figure 3.9(a)). Typical of other cores in the Kalamazoo, there is a distinct 10-cm layer of sand overlaying a layer of silt and clay. In addition, no organisms were found in this core (or in most cores in the Kalamazoo) and therefore could not be the cause of this layering; even if present, organisms would not cause sharp changes in particle size and bulk density with depth as are shown in Figure 3.9(a). For these reasons, this surface layer, as well as most other surface layers in the Kalamazoo, is most certainly due to a depositional event (e.g., the aftermath of a 25-year flood that occurred in 1987) and not due to bioturbation or any other nonerosion/nondeposition process.

As another example, the sediment resuspension and transport in Lake Erie were investigated with the emphasis on the effects of major storms (Section 6.5). As discussed there, the observed steps in the ^{210}Pb data (typically 2 to 10 cm in height) were attributed to distinct depositional layers caused by a series of episodic storms. Distinct sedimentary layers due to erosion by storms and subsequent deposition after the storms have also been observed in the Baltic Sea (Kersten et al., 2005). This layering was deduced from ^{234}Th profiles before and after a storm. From these examples as well as others, it is clear that major storms and large floods are evident as distinct layers in the sedimentary record and also can cause disturbances of sediments to depths much greater than those possible by benthic organisms or molecular diffusion.

As indicated, sedimentary layers due to deposition after a storm or flood are quite common; they are layers of constant properties but are not "well-mixed" layers. The term "well-mixed" as usually used implies that a flux due to mixing is occurring, for example, as in the mixing caused by benthic organisms. In contrast, once deposited, sediments in depositional layers do not move and do not cause a flux of HOCs unless disturbed by erosion or some other flux process.

8.5.2 COMPARISON OF "DIFFUSIVE" FLUXES AND DECAY TIMES

As stated previously, the non-erosion/non-deposition components of the sediment-water flux are often modeled as a "diffusive" flux and quantified by Equation 8.1 or 8.2. In almost all applications of this mass transfer approximation to the modeling of HOC transport, the values for the mass transfer coefficient, H, are estimated from model calibration, not directly from laboratory experiments or field measurements. For HOCs, these values are typically estimated to be on the order of 1 to 10 cm/day, but a range of values from 0.1 to 40 cm/day have been reported [e.g., Fox River (Limno-Tech, 1999; Tracy and Keane, 2000); Grasse River (ALCOA, 1999); Hudson River (QEA, 1999; TAMS, 2000; Erickson et al., 2005); Housatonic River (Weston Solutions, 2004)]. The characteristic decay times, t^*, for these fluxes are given by Equation 8.24 and depend on the layer thickness, h, and K_p. For PCBs in the systems mentioned above, the average value of K_p was usually determined to be about 10^5 L/kg. For this value of K_p, the values for t^* are given in Table 8.4 for h = 2, 10, and 15 cm and for H = 1 and 10 cm/day. The times range from 71 to 5480 years.

TABLE 8.4
Decay Time, t^* (Years), as Determined by the Mass Transfer Approximation

H	H (cm)		
(cm/day)	2	10	15
1	712	3560	5480
10	71.2	356	548

For molecular diffusion alone, the discussion in Section 8.3 indicates that the mass transfer coefficient, H (i.e., the normalized value of the flux), is approximately 1.24 cm/day (independent of K_p but dependent on the sediment) and decays with time at a rate dependent on K_p, as shown in Figure 8.9. For $K_p = 10^5$ L/kg, t* is approximately 45 years. The flux for molecular diffusion alone is essentially the lower bound for the diffusive flux because molecular diffusion is always present, whereas the other components of the flux may not be.

As discussed in Section 8.4, the flux due to bioturbation depends on the type of organism, its number density, the HOC, and the sediment. From Figure 8.17 for HCB (K_p approximately equal to 10^4 L/kg) and for *Lumbriculus* at a number density of 10^4/m^2, H is about 10×10^{-5} cm/s, or 10 cm/day, but decreases to about a third of that after 10 years (i.e., t* is approximately 10 years); for a number density of 10^3/m^2, H is about 4×10^{-5} cm/s or 4 cm/day and t* is about 20 years. For an HOC with $K_p = 10^5$ L/kg, the fluxes at t = 0 would be about the same as the above, but the decay times would be approximately ten times greater, or about 100 to 200 years, depending on the density of *Lumbriculus*. *Lumbriculus* is a relatively small oligochaete. For *Tubifex tubifex,* a larger and more common oligochaete, the flux would be as much as two times greater than that for *Lumbriculus*, whereas the decay time would be halved (Fisher et al., 1980).

In the Great Lakes, oligochaete densities have been measured, were quite patchy, but were generally between 10^3/m^2 and 10^4/m^2; in a few isolated areas, they were as high as 4×10^4/m^2; in Lake Superior, they were on the order of 10^3/m^2 or less (U.S. EPA, 1999). This would indicate that fluxes due to bioturbation in the Great Lakes should generally be about 4 to 10 cm/day, with a few areas as high as 20 to 30 cm/day and some areas (especially in Lake Superior) below 4 cm/day.

Considerable work has been done on the (primarily horizontal) transport of HOCs by groundwater flows. In contrast, the sediment-water flux of an HOC due to groundwater flow as modified by finite rates of sorption has not been investigated or quantified sufficiently (e.g., see review by Medina, 2002). However, as a simple example for discussion, assume equilibrium sorption and that the decay time is given by Equation 8.24. Also assume a layer of contaminated sediments about 1 m thick and a flux similar to that observed, with H approximately 1×10^{-5} to 10×10^{-5} cm/s or about 1 to 10 cm/day (Chadwick and Hawkins, 2004; Chadwick, 2005). For a contaminated sediment layer thickness of 1 m, t* would then be about 3560 years for a flux of 10 cm/day and 35,600 years for a flux of 1 cm/day. For finite-rate sorption, the flux would be less and the decay time would be longer.

8.5.3 Observations of Well-Mixed Layers

A basic assumption in the use of the mass transfer approximation is that there is a well-mixed layer at the surface. But is there such a quantity? Due to big events, erosion/deposition causes distinct sediment layers with constant properties throughout each layer, but these are not "well-mixed" layers, as argued previously. As shown in Figures 8.6 through 8.8, molecular diffusion does not cause a distinct well-mixed layer; the HOC concentration decreases smoothly and almost exponentially with depth, not as a step function as it would in a well-mixed layer. For bioturbation, laboratory data and modeling indicate that HOC concentrations initially decay with depth in an almost exponential manner. It is only after several years and primarily for vertical feeders that a more well-mixed profile develops. In the field, densities and types of organisms are not constant over periods of years but fluctuate locally in space and time, depending on the season (Wheatcroft et al., 1994), big events such as floods and storms, and ecological succession after big events (Rhoads, 1974; Rhoads and Germano, 1986). Because of this, the formation and maintenance of extensive and distinct well-mixed layers due to organisms probably do not occur very often.

Nevertheless, the presence of a well-mixed layer has been inferred from numerous observations of the presence of benthic organisms and their biogenic structures, such as tubes and burrows (e.g., see reviews by Rhoads and Carey, 1997; Boudreau, 1996; Clarke et al., 2001). However, a direct connection between the presence of organisms and the existence of a well-mixed layer has not been demonstrated. The reason is that the mere presence of an organism at a particular depth does not translate into complete or even extensive mixing of the sediments and especially the contaminants at these depths. The relation is tenuous and depends on the number density, size, and activity of the organism as well as the HOC, as demonstrated in Section 8.4.

The depth of the Redox Potential Discontinuity (RPD) is often cited as another indicator of the depth of a well-mixed layer due to bioturbation. The RPD is the transition between oxic and anoxic sediments and can be readily visualized (e.g., see Rhoads and Germano, 1986) as a discontinuity in color between the tan oxic layer and the dark brown to black anoxic layer. This discontinuity occurs at a very low oxygen concentration. Due to molecular diffusion alone, the RPD is on the order of 2 mm. In the presence of benthic organisms, the RPD can descend, usually to several centimeters but occasionally to as much as 10 cm. The presence of the RPD or the uniform color of the sediments above the RPD does not imply that, above the RPD, the sediments are well-mixed or that the oxygen concentration is constant; the depth of the RPD is simply a measure of the distance to which a low concentration of oxygen penetrates into the sediments and is determined by a balance between the downward transport of oxygen by molecular diffusion and bioturbation and the rate of consumption of that oxygen by the viable and nonviable organic matter as well as the inorganic matter in the sediments. Because this rate of consumption of oxygen is much less than the sorption rates of HOCs and because HOCs are conserved as they are transferred between pore water and solid particles (not consumed as for oxygen), the transport rates and the vertical concentration profiles of oxygen and HOCs are generally quite different.

As another argument for a well-mixed layer due to bioturbation, numerous field observations have been made of almost uniform chemical concentrations (such as the radioisotopes ^{210}Pb and ^{137}Cs, or even DDT) in layers near the sediment surface (e.g., see Robbins and Edgington, 1975; Robbins et al., 1978; Fuller et al., 1999; Lewis et al., 2002) and often attributed to mixing by organisms. However, these observations do not prove that the layers are due to organisms. Many, if not most of these layers (such as those discussed above for the Kalamazoo River, Lake Erie, and the Baltic Sea), are due to rapid sediment deposition during and after a storm.

8.5.4 The Determination of an Effective h

From the above and the previous discussions of flux processes, it follows that the presence and properties of any mixed layer depend on the quantity that is being mixed; that is, the mixing of an inert substance is inherently different from that of a rapidly sorbing HOC. In the latter case, as argued above, a well-mixed layer due to non-erosion/non-deposition processes generally does not exist; if it does, its presence would be localized and transient in time, and its thickness would change with time and depend on K_p. For these reasons, it seems quite difficult to define an effective thickness that would be compatible with, and be useful in, the usual model of water quality as implied by Equation 8.1 or 8.2 — that is, an h that is constant with time and has the same value for all diffusive processes (molecular diffusion, bioturbation, and groundwater flow), HOCs, numbers and types of organisms, and magnitudes of groundwater flows.

Nevertheless, this is often done in water quality modeling and may be valid in some cases as a first approximation. If a well-mixed layer is assumed to exist, the next problem is to determine an effective thickness. This parameter is often inferred qualitatively from indirect observations, as indicated above, but is usually quantified by parameterization, that is, by comparison of water quality model results with field measurements and then modifying h (as well as other parameters) until the model and field results agree. Because the value of h has a direct influence on the change with time of the diffusive flux, q, it is this change in q with time that has often been used to determine the value of h by means of parameterization. However, as noted above for an HOC with $K_p = 10^5$ L/kg, the flux due to bioturbation would decay by less than 10% in 10 years, perhaps 15 to 20% in 20 years, and by only half in 100 years. For groundwater flows, t* is even larger, on the order of several thousand years. The largest calibration period may be as much as 20 years but is usually much less. However, during this time, the diffusive flux does not change appreciably, especially when this change is compared with natural variability and errors in field measurements. As a result, h is not a parameter that can be accurately determined by parameterization in this manner. Conversely, h is the most important parameter in determining the change in q with time, a quantity that the water quality model is meant to predict.

Usual assumed values for h are 10 to 15 cm, and considerable argument ensues over which value is better. But, for HOCs, does it matter? As can be seen in Table 8.4, the decay times for h = 10 or 15 cm are hundreds of years and therefore the diffusive flux as described by the mass transfer approximation does not change appreciably over times shorter than this. Within this context, the assumed value for h does not matter.

In summary, for HOCs with large partition coefficients, a well-mixed layer due to non-erosion/non-deposition processes is slow to form and generally does not exist; an effective thickness of this layer is time dependent, depends on K_p, is difficult to define, and is even more difficult to quantify; and, as far as long-term predictions of the HOC diffusive flux are concerned, the exact value for an effective h does not matter. Because of this, the use of Equation 8.1 or 8.2 to predict the sediment-water flux is not recommended. Rather, accurate descriptions of the individual processes (such as those in Sections 8.3 and 8.4) that are significant in causing the sediment-water flux are necessary and should be used in the modeling.

8.6 ENVIRONMENTAL DREDGING: A STUDY OF CONTAMINANT RELEASE AND TRANSPORT

As a conceptually simple problem of contaminant transport and fate, consider the release of a sorbed contaminant from dredged material during and after dredging (Figure 8.19). The purpose of this exercise is to characterize and estimate the magnitudes of various processes that affect this release. For simplicity, the values of many parameters are approximated by typical values or, in some cases, by ranges of values; no attempt is made to determine exact values. Unless specified otherwise, values of most parameters are estimated to be accurate to within factors of two to four. Calculations are approximate; for example, 1 day = 10^5 s.

In the calculations, the following data and values for parameters were assumed (Four-Rs Workshop, 2006). On the average, a dredge removes about 1000 m³ of sediment over an area, A_1, of approximately 2000 m² in 1 day. This is about 10^9 cm³/day and approximately 10^9 g/day of solids. The dredging occurs in a river that has a width, W, of 67 m; a water depth, h, of 10 m; and a flow velocity, v, of 30 cm/s. The dredging lasts for 1000 days. During dredging, 5% of the total mass removed by dredging is relatively coarse and is lost by relatively rapid settling to the bottom in a residual layer within the dredging

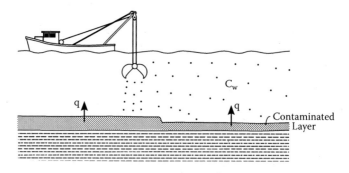

FIGURE 8.19   Release of a sorbed contaminant from dredged material during and after dredging.

area; 1% of the total mass is fine-grained and is lost by drifting downstream with the current, eventually settling and forming a residual layer downstream of the dredging area. The partition coefficient for the contaminant is 10^5 L/kg (a typical average value for PCBs). For convenience and as a reference concentration, the contaminant concentration in the pore waters of the bottom sediment, C_{w0}, is assumed to be 1 μg/L. Because K_p is 10^5 L/kg, the concentration of the contaminant sorbed to the particles, C_{s0}, is 10^5 μg/kg.

As the particles settle through the water column at the dredging site and downstream, contaminants sorbed to the particles may desorb and hence increase the dissolved chemical concentration in the river. In addition, chemicals in the residual layers may desorb and flux to the overlying waters and, in this way, also increase the dissolved chemical concentration in the river. Estimates of the contributions from these two different processes will be made. Sediment erosion is not included so as to isolate these processes.

8.6.1 Transport of Dredged Particles

As noted above, 5% of the total mass dredged is deposited in the dredged area. In 1 day, the depth of sediment dredged is 1000 m^3 divided by the area (2000 m^2), or 50 cm. The dredge makes four passes over the same area for an overall dredging depth of 2 m and a resultant average residual layer thickness, h_1, of 10 cm (5% of 2 m).

From the 1% of the dredged material that drifts downstream, a residual layer is formed over an area, A_2, defined by the width of the river and the travel distance of the particles/flocs, L_s. The travel distance depends on the depth of the river, the flow velocity of the river, and the settling speeds of the sedimentary particles, w_s, and is given by vh/w_s. The time for a particle to settle out of the water, t_s, is given by H/w_s. For a flow velocity of 30 cm/s and a settling speed of 0.01 cm/s (a typical floc), L_s is 30 km, whereas the settling time is 1 day. The area of the downstream residual layer is given by $A_2 = WL_s = 2 \times 10^6$ m^2, whereas the thickness of this layer, h_2, is given by $0.01 \times$ the volume of sediment dredged/A_2, or 0.5 cm.

In a quasi-steady state, the concentration of the suspended sediments, C, depends on a balance between the mass of solids put into suspension during dredging and that lost by deposition. For the coarse-grained fraction of the dredged material at the dredge site, the mass of solids, M_1, put into suspension during dredging is 0.05×10^9 g/day, whereas the loss by deposition is $A_1 w_s C$. If w_s is assumed to be 10^{-2} cm/s, then

$$C = \frac{M_1}{A_1 w_s} = 2.5 \times 10^3 \text{ mg/L} \qquad (8.56)$$

This is a high estimate because the coarse fraction settles as chunks and large aggregates with effective settling speeds much greater than 10^{-2} cm/s.

For the fine-grained fraction that drifts downstream and settles out over an area $A_2 = WL_s$, mass balance gives

$$M_2 = A_2 w_s C = WL_s w_s C = WvhC \qquad (8.57)$$

or

$$C = \frac{M_2}{Wvh} = 0.5 \text{ mg/L} \qquad (8.58)$$

8.6.2 Transport and Desorption of Chemical Initially Sorbed to Dredged Particles

The transport of the chemical that is initially sorbed to the dredged material depends on the partition coefficient and its desorption rate. If desorption rates are fast, then equilibrium partitioning between the chemical sorbed to the solid and that dissolved in the water can be assumed. If desorption rates are low, then the chemical sorbed to the dredged material stays sorbed to the dredged particles as they drift downstream and settle to the bottom.

For the coarse-grained and fine-grained fractions together, the amount of chemical sorbed to the dredged sediment and hence introduced into the water column at the dredge site is 0.06×10^9 g/day $\times C_{s0}$, where C_{s0} is 10^5 μg/kg. When immediate chemical equilibrium is assumed, the sorbed chemical partially desorbs into the water column such that $C_s = K_p C_w$. The chemical is then lost from the water column (1) by settling of the dredged material, $(M_1 + M_2)C_s$, and (2) by downstream convection of the dissolved chemical at a rate of $vhWC_w$. In the quasi-steady state, mass balance then gives that $C_w = 0.23$ μg/L and $C_s = 0.23 \times 10^5$ μg/kg; that is, most of the chemical (77%) desorbs from the dredged material and dissolves into the water.

When desorption rates are slow, almost all the chemical initially sorbed to the dredged material remains sorbed and C_w is practically zero. As the dredged material deposits on the bottom, C_s remains the same as the concentration of the sediments dredged initially, that is, 10^5 μg/kg.

Typical experimental results for the desorption of hexachlorobenzene ($K_p = 10^4$ L/kg), a monochlorobiphenyl ($K_p = 2 \times 10^3$ L/kg), and a hexachlorobiphenyl ($K_p = 6.6 \times 10^4$ L/kg) from Detroit River sediments are shown in Figure 7.5. The rates of desorption depend on K_p and decrease as K_p increases. Because of this, desorption times for a PCB with $K_p = 10^5$ L/kg would be even greater than those for HPCB. The rates of desorption also depend on the particle/floc size, density, and concentration (Chapter 7). The characteristic times for desorption vary significantly, depending on these parameters, but usually are many days or longer and therefore are the same order of magnitude as or greater than the transport time of a particle/floc (the time, t_s, during which most of the particles/flocs settle out of the water column). That is, equilibrium partitioning cannot be assumed. For the dredged material settling in the dredged area, the dredged material is coarse (large chunks and aggregates), desorption is therefore quite slow, and zero desorption (frozen reactions) is probably the best first approximation.

8.6.3 DIFFUSIVE RELEASE OF CONTAMINANT FROM THE RESIDUAL LAYERS

The thickness of the residual layer at the dredging site, h_1, was assumed to be 10 cm. For this layer, the chemical flux from the sediment to the overlying water (assuming negligible concentration of PCB in the overlying water) is approximated by

$$q = HC_w = H\frac{C_s}{K_p} = H^*C_s \tag{8.59}$$

where H and $H^* = H/K_p$ are mass transfer coefficients, and C_w and C_s are chemical concentrations in the bottom sediments. The quantity H is approximately 1 cm/day (molecular diffusion) to 10 cm/day (large numbers of benthic organisms), or possibly somewhat greater for very large numbers of vertical burrowing organisms (Section 8.4). From Equation 8.59, the chemical flux is then

$$q = (1 \text{ to } 10 \text{ cm/day}) \times 1.0 \text{ μg/L}$$

$$= 1 \times 10^{-3} \text{ to } 1 \times 10^{-2} \text{ μg/cm}^2 \text{ day} \tag{8.60}$$

where it has been assumed that $C_w = 1.0$ μg/L. If it is assumed that the flux continues approximately at this same rate, then the time to release all of the chemical from this layer is

$$T = \frac{h_1 CC_s}{q} = 10^5 \text{ to } 10^6 \text{ days} = 274 \text{ to } 2740 \text{ years} \tag{8.61}$$

where C is the concentration of the particles in the bottom sediments and is about 1 g/cm^3.

The amount of chemical released by diffusion over the entire dredged area is given by

$$qA = H \times area \times C_w = \left(1 \text{ to } 10 \frac{cm}{day}\right) \times 0.5 \times 10^6 m^2 \left(\frac{1 \text{ μg}}{L}\right)$$

$$= 0.5 \times 10^7 \text{ to } 0.5 \times 10^8 \text{μg / day} \tag{8.62}$$

The resulting concentration in the overlying water (volume of 2×10^7 m^3/day) is then

$$C_w = 2.5 \times 10^{-4} \text{ to } 2.5 \times 10^{-3} \text{ μg/L} \tag{8.63}$$

For the 10-cm residual layer in the dredged area, the time, T, to release all contaminants by diffusion was estimated at 274 to 2740 years. It follows that, for a 0.5-cm layer, the time T should be on the order of 14 to 140 years. Because the area of the downstream residual layer is four times greater than the area of the residual layer at the dredged site, the release of chemical by diffusion from the downstream residual area is four times that given by Equation 8.62, or 2×10^7 to 2×10^8 µg/day, and also results in four times the concentration given by Equation 8.63, or $C_w = 10^{-3}$ to 10^{-2} µg/L.

If chemical equilibrium is assumed, desorption of the contaminant from the dredged material contributes 0.23 µg/L to the dissolved chemical concentration downstream, whereas the flux from the residual layers contributes 0.00125 to 0.0125 µg/L. These calculations indicate that the direct contribution of the dredged material is significantly greater than that from the residual layers. However, the relative magnitude of the two sources depends on the desorption rate as well as the quantity and activities of the benthic organisms. If desorption rates are slow enough and settling rates are fast enough that the dredged material does not desorb appreciably during its descent (which is probably a very good approximation), then the fluxes from the residual layers are dominant compared with the flux from the settling dredged material. Also, if large numbers of active benthic organisms are present, the sediment-water flux will be greater than that given by Equation 8.62, and the residual layers may again be the major contributors to C_w. Further information on sorption rates and sediment-water fluxes is needed before measured values of C_w can be accurately modeled and extrapolated into the future.

To emphasize this latter point, consider the common situation where C_w has been determined from field measurements and is to be used to calibrate a water quality model. In most cases, (1) the sediment-water flux is not measured directly and (2) either equilibrium sorption is assumed or, if finite-rate sorption is hypothesized, sorption rates for the suspended sediments are not well quantified. Both of these processes significantly affect C_w. Increasing the sediment-water flux or increasing the contaminant desorption rate from the suspended sediments increases C_w. Without additional information, model calibration from measurements of C_w alone cannot determine both of these quantities uniquely. The problem is the same as that described earlier for sediment transport; that is, the problem is inadequately constrained, and non-unique solutions are possible and probable. Because of the uncertainty in the origin of the dissolved chemical, the prediction of C_w as a function of time is therefore uncertain.

Once dredging is completed, the direct contribution of the dredging material to C_w (by desorption of the HOC from settling particles) ends. However, the contributions from the residual layers continue for long times, as indicated above. In fact, the contributions of the residual layers to C_w after dredging are comparable to those from the contaminated sediments before dredging; that is, the effectiveness of environmental dredging alone (without capping, for instance) in reducing exposure risk is negligible.

8.6.4 VOLATILIZATION

In some water quality problems, volatilization of the dissolved chemical from the water to the overlying air can be an important loss mechanism. For the present problem, an estimate of this effect can be made as follows. The water-air mass transfer coefficient, H_a, for a typical PCB ($K_p = 10^5$ L/kg) is on the order of 50 cm/day (e.g., Chapra, 1997). If the dissolved chemical concentration in a volume of water changes due to volatilization alone, then the mass balance equation is

$$h\frac{dC_w}{dt} = H_a\, C_w \qquad (8.64)$$

where h is the depth of the river. For an initial concentration of C_0, the solution to this equation is

$$C_w = C_{w0}\, e^{-\frac{H_a}{h}t} \qquad (8.65)$$

The characteristic decay time is then $t^* = h/H_a$. For the present case with h = 10 m and $H_a = 50$ cm/day, this says that t^* is about 20 days; that is, volatilization will cause a decrease of C_w to e^{-1} of its initial value in about 20 days. At a flow velocity of 30 cm/s (as in the present example), a fluid element will travel approximately 600 km in 20 days. If the river is much shorter than 600 km, volatilization will be negligible. If the length is comparable to or larger than 600 km, volatilization may be significant and must be considered.

Another approach is to compare the air-water flux of the chemical with the sediment-water flux. When the chemical concentration in the air is negligible, the water-air flux is given by

$$q_a = H_a C_w \qquad (8.66)$$

whereas the sediment-water flux due to "diffusion" is

$$q = H(C_{w,sed} - C_w) \qquad (8.67)$$

where $C_{w,sed}$ is a representative dissolved chemical concentration in the sediments. As long as $H_a C_w \ll H C_{w,sed}$, as is often the case for sediments contaminated by chemicals with high K_p, then $q_a \ll q$ and the volatilization flux is small compared with the sediment-water flux due to "diffusion" and can be ignored.

Degradation and other chemical reactions of PCBs are usually considered negligible. This is sometimes justified by stating that this assumption gives an upper limit to PCB concentrations and therefore an error on the safe side. Except for volatilization, no other processes that modify PCB concentrations are usually significant compared to erosion/deposition, non-erosion/non-deposition sediment-water fluxes, chemical sorption, and transport. For other hydrophobic chemicals, chemical transformations may be significant.

8.7 WATER QUALITY MODELING, PARAMETERIZATION, AND NON-UNIQUE SOLUTIONS

In Section 1.2, a preliminary discussion of the problem of parameterization and non-unique solutions in water quality modeling was given. The emphasis was on sediment transport. It was demonstrated there that non-unique solutions could arise when an attempt was made to determine both erosion rates and deposition rates from measurements of suspended sediment concentrations alone without additional information, that is, by determining both E and the settling speed, w_s, from the one equation,

$$C = \frac{E}{p w_s}$$
(8.68)

Non-unique values for E and w_s result from this procedure.

As a more complex illustration of this same problem, the calculation of sediment and contaminant transport in the Fox River (Tracy and Keane, 2000) by two independent modeling groups was also described in Section 1.2. Both groups calibrated their model by means of the suspended solids concentration. For purposes of remediation, one of the most significant quantities to be predicted by each model was the depth of erosion. However, results by the two groups for the amount of sediment eroded at one location for a large, but not maximum, shear stress differed by more than two orders of magnitude! Solutions for other variables were also widely different. As in the above example, the problem is inadequately constrained and an infinite number of solutions are therefore possible.

In the previous section, an idealized problem of environmental dredging was considered with the emphasis on contaminant release and transport. Processes that were significant in increasing the dissolved chemical concentration in the overlying water downstream from the dredging area were (1) the desorption of the HOC at a finite rate from the dredged material that was spilled and settling during the dredging process, and (2) the flux of the HOC from the residual layers formed by the deposition of the dredged material lost during the dredging process. Increasing the contaminant desorption rate from the suspended dredged material or increasing the sediment-water flux increases C_w. Without additional information, calibration of the model from measurements of C_w alone is insufficient to distinguish the effects of these two processes. The problem is the same as that described above and in Section 1.2; that is, the problem is inadequately constrained, and non-unique solutions are possible and quite probable.

The discussion of, and results from, the above example also apply to the problem of the resuspension of HOC-contaminated bottom sediments and the subsequent desorption at a finite rate of the HOC from the suspended sediments to the surrounding water. Calibration of a model is insufficient to distinguish the HOC flux due to erosion/deposition from the HOC sediment-water flux due to non-erosion/non-deposition processes.

As a more complex example (but quite general and typical of water quality models), consider the preliminary modeling of sediment and PCB transport in a contaminated section of the Housatonic River (U.S. EPA, 2006). The EPA report includes references to several other reports that give extensive details as well as peer reviews of the model. The modeling is state of the art and is similar to or better than modeling of PCBs in many other rivers. In the present discussion, the emphasis is on the processes of sediment erosion, sediment deposition, bed armoring, the sediment-water flux due to "diffusion," the finite-rate sorption and partitioning of PCBs and the dependence of these quantities on K_p, and the numerical grid size. All of these have been discussed in detail in previous sections and chapters. Approximations to these processes as used in the model are described below. Although this problem is much more complex than the problems summarized above, it is demonstrated that this problem has essentially the same difficulty as the previous ones; that is, the problem is inadequately constrained and multiple solutions are possible and probable.

8.7.1 PROCESS MODELS

8.7.1.1 Sediment Erosion

Measurements of erosion rates for sediments from the Housatonic River were made by the U.S. Army Corps of Engineers (Gailani et al., 2006) using a flume similar to Sedflume. In their analysis of the results, the Corps and the EPA determined that n in Equation 3.21 for erosion rates was 1.59 for one set of cores and was 0.95 for another set of cores. In all the previous work that was done at UCSB with Sedflume on the determination of erosion rates (the number of cores is on the order of 100), n was typically determined to be about 2 or somewhat greater. The equations described in Section 3.5 were derived and are only applicable to sediments that have the same bulk properties. To determine the parameters in these equations properly, sediments with similar bulk properties (i.e., at the same location and depth in the sediment) must be grouped together (Section 6.4). This procedure was not used by the Corps and EPA, and hence their values for n are probably incorrect and too low. Because of the low values for n, the erosion rates used in the modeling are too low. As a result, during big events, erosion rates may be as large as, or greater than, ten times that predicted in the model. The major effects of this will be on the maximum depth of erosion during storms and the sediment-water flux of PCBs due to erosion/deposition.

8.7.1.2 Sediment Deposition

Deposition rates are probably also too low. In the model, these rates are a calibrated parameter and are determined such that the calculated and measured suspended solids concentrations agree (Section 1.2). If erosion rates are increased, as suggested above, these deposition rates also must be increased to maintain good agreement between calculated and measured suspended solids concentrations.

8.7.1.3 Bed Armoring

Bed armoring is an important process and causes large changes in surficial sediment bed properties and hence large changes in erosion. This occurs, for example, when a layer of coarse sediments (as little as a few particle diameters thick) is deposited on a layer of finer, noncohesive sediments. As the model was configured, any deposited sediments are immediately mixed with a 7-cm surficial layer. Because of this, effective coarsening takes place very slowly (a small amount of added sediment has little effect on the average properties of a 7-cm layer). In reality, this mixing only occurs in a layer a few particle diameters thick, and this thin layer must be present in the model for realistic coarsening to occur.

8.7.1.4 The Sediment-Water Flux of HOCs Due to "Diffusion"

In the model, all of the non-erosion/non-deposition processes were described by means of a "diffusion" model based on the concept of bioturbation and the assumption of a well-mixed layer — that is, the conventional, but not necessarily accurate, approach (Section 8.5). Assumptions were (1) a constant (independent of space and time) sediment-water mass transfer coefficient, H, with a value of 1.5 cm/day, and (2) a surficial, well-mixed layer whose thickness is constant in time but varies spatially from 4 cm upstream to 7 cm downstream. However, there is no evidence that a well-mixed layer exists anywhere in the Housatonic (or anywhere else, as is argued in Section 8.5).

The mass transfer coefficient, H, for the transport of HOCs by molecular diffusion alone is approximately 1.2 cm/day and decreases slowly with time at a rate that decreases as K_p increases (Section 8.3). If there are only a small number of organisms, EPA's assumed value of 1.5 cm/day compares well with this number. However, with benthic organisms present, Luo et al. (2006) give a mass transfer coefficient for HOCs that varies up to 10 cm/day (for benthic organism densities of $10^4/m^2$) and somewhat higher for very dense concentrations of organisms (Section 8.4); these values for H are much higher than those that are assumed in the model.

No consideration is given to groundwater flow, which can be significant, is a convection and not a diffusion process, and does not involve a well-mixed layer of any sort.

8.7.1.5 Equilibrium Partitioning

After sediment particles are resuspended, they will be transported downstream by the current and eventually settle out of the water column. During this time, the contaminant sorbed to the particle will desorb at some finite rate. The time for a particle to settle out of the water column depends on the settling speed and water depth, whereas the distance traveled by the particle before depositing depends on the settling time and current speed. For a reasonable range of settling speeds for fine-grained particles/flocs ($w_s = 2 \times 10^{-3}$ to 1×10^{-1} cm/s) and water depths typical of the Housatonic (h = 1 to 3 m), the settling times ($t = h/w_s$) are in the range of 10^3 s (15 min) to 1.5×10^5 s (1.5 days). For medium- and coarse-grained particles, the settling times are less.

As in the transport problem described in the previous section, PCB desorption times are much greater than settling times. It follows that contaminants on resuspended particles will not desorb completely, or even close to completely, in the water column before the particles settle out of the water column; that is, the chemical sorbed to the suspended particles will not reach chemical equilibrium with the chemical dissolved in the water column. The assumption of equilibrium partitioning is therefore not valid, or even a good approximation, for the sediments in suspension or in the surficial layers of the bottom sediments. This is further complicated by the fact that desorption rates are proportional to K_p, a quantity that depends on the PCB congener.

As in the previous section, the dissolved HOC concentration in the overlying water has two major sources: (1) desorption from resuspended, contaminated sediments and (2) the non-erosion/non-deposition flux of the HOC from the bottom sediments to the overlying water. The assumption of equilibrium partitioning for the suspended sediments negates an accurate determination of dissolved HOC from the first source; this leads, via calibration, to an inaccurate determination of dissolved HOC from the second source.

8.7.1.6 Numerical Grid

With the Housatonic grid, the width of the river is generally approximated as one cell. As in most rivers (see Sections 6.4 and 8.1), there are large differences in erosion between the deeper and the shallower parts of the Housatonic. Predicting the dissimilar amounts of erosion/deposition across a cross-section of the river is crucial in predicting the long-term exposure of PCBs by erosion and/or natural recovery by deposition. As discussed for the Lower Fox River (Section 6.4) and the Saginaw River (Section 8.1), averaging across the cross-section does not describe the erosion/deposition process accurately. A minimum of three cells across the river (two shallow nearshore cells and one deeper center cell) is necessary for quantitative modeling. Inclusion of additional cells where the bathymetry is changing rapidly (e.g., at the steep sides of a channel) is also necessary.

8.7.2 PARAMETERIZATION AND NON-UNIQUE SOLUTIONS

As discussed above, the processes that govern the sediment-water flux of HOCs (sediment erosion, sediment deposition, bed armoring, the sediment-water flux of HOCs due to "diffusion," and the finite-rate sorption and partitioning of HOCs) are described incorrectly and inaccurately in the model. Each of these processes can modify the flux by factors of two to ten. Coarse numerical gridding increases the possibility of error. Nevertheless, EPA documents (U.S. EPA, 2006) indicate that, although there are some discrepancies that cannot be explained by the model despite much effort (e.g., PCB concentrations in the surficial sediments are not predicted correctly, especially downstream), there is reasonably good agreement between the calculated and measured suspended solids concentrations as well as contaminant concentrations in the overlying water. At the same time, sensitivity and uncertainty analyses also seem to say that the model is doing a good job.

How can this be? Are accurate models of sediment and contaminant transport and fate really unnecessary?

The answers to these questions lie in the nonuniqueness of calibrated solutions and the nature of sensitivity and uncertainty analyses. As in the previous examples, calibration of a model does not guarantee that the processes in the model are described properly. A water quality modeler can usually get reasonably good agreement between calculated and observed quantities for a limited time interval and limited conditions, whether or not the fundamental processes are described properly. Another modeler, with quite different descriptions of processes and/or different parameters in his or her model, can get equally good agreement between the calculated and observed quantities. However, future predictions by the different models and modelers will be quite different, as in the examples in this and previous sections. *That is, calibration is necessary but not sufficient.*

Sensitivity and uncertainty analyses are asking the wrong questions. They do not question whether the basic processes are formulated correctly. As an example, equilibrium partitioning is assumed in the above model and in most water quality models. Sensitivity and uncertainty analyses never question this assumption, never demonstrate that it is an inaccurate assumption, nor do they propose a suitable reaction rate.

8.7.3 IMPLICATIONS FOR WATER QUALITY MODELING

If the only purpose of the model is to duplicate known results, then accurate models of sediment and contaminant transport and fate do not matter. Calibration is usually sufficient. However, if the model is to be used for predictive purposes, then accurate process models do matter. In the predictive mode, future conditions (such as sediment properties, contaminant concentrations in the sediments, concentrations and types of benthic organisms, sediment-water fluxes, and flow rates) will be modified (for example, by natural evolution of the body of water and by dredging, capping, or extreme environmental conditions), will change with time, and will be different from those for which the model was calibrated. The basic processes in the present and future are the same. However, their relative effects and significance depend on the modified conditions and will change with time.

Numerous difficulties with the modeling of processes necessary in modeling the transport and fate of HOCs (i.e., sediment erosion, sediment deposition, bed armoring, the finite-rate sorption and partitioning of HOCs and the dependence of these quantities on K_p, and the non-erosion/non-deposition flux) have been described. These processes are all significant and must be understood and quantified for predictive purposes. Calibration by means of measured suspended sediment concentrations and dissolved HOC concentrations in the overlying water alone is insufficient to determine these processes uniquely and accurately. However, all of these processes are now better understood and can be reasonably quantified by means of process experiments and analyses. For an accurate and predictive model, this needs to be done.

References

Accardi-Dey, A. and P.M. Gschwend. 2002. Assessing the combined roles of natural organic matter and black carbon as sorbents in sediments. *Environ. Sci. Technol.*, 36, 21–29.

Accardi-Dey, A. and P.M. Gschwend. 2003. Reinterpreting literature sorption data considering both absorption into organic carbon and adsorption onto black carbon. *Environ. Sci. Technol.*, 37, 99–106.

Ager, D.N. 1981. *The Nature of the Stratigraphical Record.* New York: John Wiley & Sons.

ALCOA. 1999. A Comprehensive Characterization of the Lower Grasse River. ALCOA Report, New York.

Ali, K.H.M. and O. Karim. 2002. Simulation of flow around piers. *J. Hydr. Res.*, 40(2), 161–174.

Allen, J.R.L. 1970. The systematic packing of prolate spheroids with reference to concentration and dilatancy. *Geol. en Mijnbouw*, 49, 13–22.

Arega, F. and E. Hayter. 2007. Skill Assessment of EFDC — a 3D Hydrodynamic and Transport Model — to Simulate Highly Stratified Estuarine Flows. Report, U.S. Environmental Protection Agency, Athens, GA.

Avnir, D., A. Biham, D. Lider, and O. Malcai. 1998. Is the geometry of nature fractal? *Science*, 279, 39–40.

Bagnold, R.A. 1956. The flow of cohesionless grains in fluids. *Phil. Trans. Royal Soc. London A* 249(964), 234–297.

Baker, R.A. 1991. *Organic Substances and Sediments in Water.* Chelsea, MI: Lewis Publishers.

Bedford, K.W. 1985. Selection of Turbulence and Mixing Parameterizations for Water Quality Models. Misc. paper EL-85-2. Vicksburg, MS: U.S. Army Corps of Engineers Waterways Experiment Station.

Been, K. 1981. Non-destructive soil bulk density measurements using X-ray attenuation. *Geotech. Testing J.*, 4, 169–176.

Been, K. and G.C. Sills. 1981. Self-weight consolidation of soft soils: an experimental and theoretical study. *Geotechnique*, 31(4), 519–535.

Berner, R.A. 1980. *Early Diagenesis. A Theoretical Approach.* Princeton, NJ: Princeton University Press.

Blake, A.C., D.B. Chadwick, P.J. White, and C.A. Jones. 2008. Appendix D, User's Guide for Assessing Sediment Transport at Navy Facilities, Technical Report 1960. San Diego, CA: U.S. Navy, Spawar Systems Center.

Blumberg, A.F. 1975. A two-dimensional model for simulation of partially mixed estuaries. In *Estuarine Processes*, Ed. M. Wiley, Vol. II, pp. 323–331. New York: Academic Press.

Blumberg, A.F. 1977. Numerical model of estuarine circulation. *ASCE J. Hydraulic Engr. Div.* 103(HY3), 295–310.

Blumberg, A.F. and D.M. Goodrich. 1990. Modeling of wind-induced destratification in Chesapeake Bay. *Estuaries*, 13(3), 236–249.

Blumberg, A.F. and G.L. Mellor. 1980. A coastal ocean numerical model. In *Mathematical Modeling of Estuarine Physics. Proceedings of an International Symposium*, Eds. J. Sunderman and K.P. Holz, Vol. II, pp. 323–331. Berlin: Springer-Verlag.

Blumberg, A.F. and G.L. Mellor. 1987. A description of a three-dimensional coastal ocean circulation model. In *Three-Dimensional Coastal Ocean Models*, Ed. N.S. Heaps, 4, 1–16. Washington: American Geophys. Union.

Blumberg, A.F., B. Galperin, and D.J. O'Conner. 1992. Modeling vertical structure of open-channel flows. *J. Hydraul. Eng., ASCE,* 118(8), 1119–1134.

Blumberg, A.F., R.P. Signell, and H.J. Jenter. 1993. Modeling transport processes in the coastal ocean. *J. Marine Env. Eng.,* 1, 3–52.

Blumberg, A.F., L.A. Khan, and J.P. St. John. 1999. Three-dimensional hydrodynamic model of New York Harbor region. *J. Hydraul. Eng.,* 125(8), 799–816.

Borah, D., C. Alonso, and S. Prasad. 1983. Routing graded sediments in streams: formulations. *J. Hydraul. Eng.,* 108(12), 1486–1502.

Borglin, S., R. Jepsen, A. Wilke, and W. Lick. 1996. Parameters affecting the desorption of hydrophobic organic chemicals from suspended sediments. *Environ. Toxicol. Chem.,* 15, 2254–2262.

Borrowman, T.D., E.R. Smith, J.Z. Gailani, and L. Caviness. 2005. Erodibility Study of Passaic River Sediments Using ASACE Sedflume. Vicksburg, MS: U.S. Army Corps of Engineers.

Boudreau, B.P. 1997. *Diagenetic Models and Their Implementation.* Berlin: Springer-Verlag.

Brannon, J.M., R.E. Hoeppel, T.C. Sturgis, L. Smith, and D. Gunnison. 1985. Effectiveness of Capping in Isolating Contaminated Dredged Material from Biota and the Overlying Water. Technical Report D-85-10. Vicksburg, MS: U.S. Army Corps of Engineers Waterways Experiment Station.

Bretschneider, C.L. 1958. Revision in wave forecasting; deep and shallow water. *Proc. 6th Conf. Coastal Engineering, ASCE*, pp. 30–67.

Breusers, H.N.C. and A.J. Raudkivi. 1991. Scouring, LAHR/AIRH Hydraulic Structures Design Manual, 2, Rotterdam/Brookfield: Balkema.

Buchak, E.M. and J.E. Edinger. 1984a. Generalized, Longitudinal-Vertical Hydrodynamics and Transport: Development, Programming, and Applications, Document 84-18-R. Vicksburg, MS: U.S. Army Corps of Engineers Waterways Experiment Station.

Buchak, E.M. and J.E. Edinger. 1984b. Simulation of a density underflow into Wellington Reservoir using longitudinal-vertical numerical hydrodynamics, Document 84-18-R. Vicksburg, MS: U.S. Army Corps of Engineers Waterways Experiment Station.

Burban, P.-Y., W. Lick, and J. Lick. 1989. The flocculation of fine-grained sediments in estuarine waters. *J. Geophys. Res.,* 94(C6), 8323–8330.

Burban, P.-Y., Y. Xu, J. McNeil, and W. Lick. 1990. Settling speeds of flocs in fresh and sea waters. *J. Geophys. Res.,* 95(C10), 18213–18220.

Camp, T.R. and P.C. Stein. 1943. Velocity gradients and internal work in fluid motion. *J. Boston Soc. Civil Eng.,* 30(4), 219–237.

Cardenas, M. and W. Lick. 1996. The transport of sediments and hydrophobic contaminants in the lower Saginaw River. *J. Great Lakes Res.,* 22, 669–682.

Cardenas, M., J. Gailani, C.K. Ziegler, and W. Lick. 1995. Sediment transport in the lower Saginaw River. *Marine Freshwater Res.,* 21, 257–274.

Carslaw, H.S. and J.C. Jaeger. 1959. *Conduction of Heat in Solids.* Oxford: Oxford University Press.

CFD 2000. CFD2000 Theoretical Background, Manual, CFD2000 Computational Fluid Dynamics Package, Pacific-Sierra Research Co.

Chadwick, B. 2005. Coastal Contaminant Migration Monitoring Assessment Report. San Diego, CA: SPAWAR Systems Center.

Chadwick, B. and A. Hawkins. 2004. Monitoring of Water and Contaminant Migration at the Groundwater-Surface Water Interface Report. San Diego, CA: SPAWAR Systems Center.

Chapra, S.C. 1997. *Surface Water Quality Modeling*. Boston: McGraw-Hill.

Cheng, N.S. 1997. Simplified settling velocity formula for sediment particle. *J. Hydraul. Eng., ASCE*, 123(2), 149–152.

Cheng, R.T., C.H. Ling, J.W. Gartner, and P.F. Wang. 1999. Estimates of bottom roughness length and bottom shear stress in South San Francisco Bay, California. *J. Geophys. Res.*, 104(C4), 7715–7728.

Chepil, W.S. 1959. Equilibrium of soil grains at the threshold of movement by wind. *Soil Sci. Soc. Proc.*, 23, 422–428.

Chien, N. and Z. Wan. 1999. *Mechanics of Sediment Transport*. ASCE Press.

Chin, C.O. and Y.M. Chiew, 1992. Effect of bed surface structure on spherical particle stability. *J. Waterway, Port, Coastal Ocean Eng.*, 119(3), 231–242.

Christoffersen, J.B. and I.G. Jonsson. 1985. Bed friction and dissipation in a combined current and wave motion. *Ocean Eng.*, 12(5), 387–423.

Chroneer, Z. and W. Lick. 1997. Parameters affecting the transport and fate of hydrophobic contaminants in surface waters. *Computing Environ. Res. Mgmt.*, Air and Water Mgmt. Assoc., pp. 183–197.

Chroneer, Z., M. Cardenas, J. Lick, and W. Lick. 1996. Sediment and contaminant transport in Green Bay. In *Estuarine & Coastal Modeling*, Ed. M.L. Spaulding, pp. 313–324.

Clark, M.N. 1982. Discussion of forces acting on floc and strength of floc. *ASCE, J. Environ. Eng. Division*, 108(EE3), 592–594.

Clarke, D.G., M.R. Palermo, and T.C. Sturgis. 2001. Subaqueous Cap Design: Selection of Bioturbation Profiles, Depths, and Rates, DOER Technical Notes Collection (ERDC TN-DOER-C21). Vicksburg, MS: U.S. Army Corps of Engineers Research and Development Center.

Clarke, J.F. and M. McChesney. 1964. *The Dynamics of Real Gases*. London: Butterworths.

Coates, J.T. and A.W. Elzerman. 1986. Desorption kinetics for selected PCB congeners from river sediments. *J. Contam. Hydrol.*, 1, 191–210.

Coates, K.H. and B.D. Smith. 1964. Dead-end pore volume and dispersion in porous media. *Soc. Petrol. Eng., J.* 4, 73–84.

Cole, T.M. and E.M. Buchak. 1995. CE-QUAL-2: A Two-Dimensional, Laterally-Averaged, Hydrodynamic and Water Quality Model. Report EL-95-1, Vicksburg, MS: U.S. Army Corps of Engineers Waterways Experiment Station.

Cornelissen, G., O. Gustafsson, T.D. Bucheli, M. Jonker, A.A. Koelmans, and P.C.M. Van Noort. 2005. Extensive sorption of organic compounds to black carbon, coal, and kerogen in sediments and soils: mechanisms and consequences for distribution, bioaccumulation, and biodegradation. *Environ. Sci. Technol.*, 39(18), 6881–6895.

Csanady, G.T. 1973. *Turbulent Diffusion in the Environment*. Boston: Reidel.

Damgaard, J.S., R.J.S. Whitehouse, and R.L. Soulsby. 1997. Bed-load sediment transport on steep longitudinal slopes. *J. Hydraul. Eng.* 123(12), 1130–1138.

Daraghi, B. 1989. The turbulent flow around a circular cylinder. *Exp. Fluids*, 8,1–12.

Deane, G.F. 1998. The Diffusion and Sorption of Hydrophobic Organic Chemicals in Consolidated Sediments and Saturated Soils. Ph.D. thesis, Santa Barbara, CA: University of California.

Deane, G., Z. Chroneer, and W. Lick. 1999. The diffusion and sorption of hexachlorobenzene in consolidated sediments and saturated soils. *J. Environ. Eng.*, 125, 689–696.

Deng, G.B. and J. Piquet. 1992. Navier-Stokes computations of horseshoe vortex flows. *Int. J. Numerical Meth. Fluids,* 15, 99–124.

Diplas, P. and A.N. Papanicolaou, 1997. Batch analysis of slurries in zone settling regime. *J. Environ. Eng.,* 123, 7, 659–667.

DiToro, D.M., C.S. Zarba, D.J. Hansen, W.J. Berry, R.C. Swartz, C.E.Cowan, S.P. Pavlou, H.E. Allen, N.A. Thomas, and P.R. Paquin. 1991. Technical basis for the equilibrium partitioning method for establishing sediment quality criteria. *Environ. Toxicol. Chem.,* 11(12), 1541–1583.

Donev, A., I. Cisse, and P.M. Chaikin. 2004. Improving the density of jammed disordered packings using ellipsoids. *Science,* 303, 990–993.

Drake, D.E. and D.A. Cacchione. 1992. Wave-current interaction in the bottom boundary layer during storm and non-storm conditions: observations and model predictions. *Continental Shelf Res.,* 12, 1331–1352.

Ducker, W.A., T.J. Senden, and R.M. Pashley. 1991. Direct measurement of colloidal forces using an atomic force microscope. *Nature,* 353, 239–241.

Ducker, W.A., T.J. Senden, and R.M. Pashley. 1992. Measurement of forces in liquids using a force microscope. *Langmuir,* 8, 1831–1836.

Dyer, K.R. 1986. *Coastal and Estuarine Sediment Dynamics.* New York: John Wiley & Sons.

Einstein, H.A. 1950. The Bed Load Function for Sediment Transportation in Open Channel Flows. U.S. Dept. of Agriculture, Soil Conservation Service Technical Bull. No. 1026, September.

Einstein, H.A. and N.L. Barbarossa. 1952. River channel roughness. *Trans. Am. Soc. Civ. Eng.,* 117117, 1121–1132.

Eisma, D. 1986. Flocculation and de-flocculation of suspended matter in estuaries. *Neth. J. Sea Res.,* 20,183–199.

Endicott, D. and M. DeGraeve. 2006. A Review of the Available Literature Relating to the Contamination of the Lower Passaic River. Report. Traverse City, MI: Great Lakes Environmental Center.

Engelund, F. 1981. The Motion of Sediment Particles on an Inclined Bed. Progress Rep. No. 53, Technical University of Denmark, Inst. Hydrodynamics and Hydr. Engrg., Lyngby, Denmark.

Engelund, F. and E. Hansen. 1967. *A Monograph on Sediment Transport in Alluvial Streams.* Copenhagen, Denmark: Teknisk Vorlag.

Erickson, M.J., C.L. Turner, and L.J. Thibodeaux. 2005. Field observations and modeling of dissolved fraction sediment-water exchange coefficients for PCBs in the Hudson River. *Environ. Sci. Technol.,* 39(2), 549–556.

Etter, R.J., R.P. Hoyer, E. Partheniades, and J.F. Kennedy. 1968. Depositional behavior of kaolinite in turbulent flow. *J. Hydraul. Div., ASCE,* 94(HY6), 1439–1452.

Fenton, J.R. and J.E. Abbott. 1977. Initial movement of grains on a stream bed: the effect of relative protrusion. *Proc. Roy. Soc. London Ser. A,* 352(1671), 523–537.

Fisher, J.B., W. Lick, P.L. McCall, and J.A. Robbins. 1980. Vertical mixing of lake sediments by tubificid oligochaetes. *J. Geophys. Res.,* 85, 3997–4006.

Fitch, B. 1983. Kynch theory and compression zones. *AIChE J.,* 29(6), 940–947.

Four-Rs Workshop. 2006. Resuspension, Release, Residual, and Risk, Sponsored by the U.S. Environmental Protection Agency and the U.S. Army Corps of Engineers, Vicksburg, MS.

Friedlander, S.K. 1977. *Smoke, Dust, and Haze.* New York: John Wiley & Sons.

Froelich, D.C. 1989. HW031-D Finite Element Surface Water Modeling System: Two-Dimensional Flow in a Horizontal Plane, User's Manual. U.S. Federal Highway Administration Report No. FHWA-RD-88-177.

Fukuda, M. and W. Lick. 1980. The entrainment of cohesive sediments in fresh water. *J. Geophys. Res.*, 85, 2813–2824.

Fuller, C.C., A. van Geen, M. Baskaran, and R. Anima. 1999. Sediment chronology in San Francisco Bay, California defined by ^{210}Pb, ^{234}Th, ^{137}Cs, and ^{239,240}Pu. *Mar. Chem.*, 64, 7–27.

Gailani, J.Z., S.J. Smith, M.G. Channell, G.E. Banks, and D.B. Brister. 2006. Sediment Erosion Study for the Housatonic River. Massachusetts, Report, U.S. Army Corps of Engineers, Vicksburg, MS.

Galperin, B., L.M. Kantha, S. Hassis, and A. Rosati. 1988. A quasi-equilibrium turbulent energy model for geophysical flows. *J. Atmosph. Sci.*, 45, 55–62.

Gessler, J. 1967. The Beginning of Bedload Movement of Mixtures Investigated as Natural Armoring in Channels. W.M. Keck Laboratory of Hydraulics and Water Resources, California Institute of Technology, Translation T-5.

Gibson, R.E., G.L. England, and M.J.L. Hussey. 1967. The theory of one-dimensional consolidation of saturated clays. *Geotechnique*, 17, 261–273.

Ghosh, U., J.R. Zimmerman, and R.G. Luthy. 2003. PCB and PAH speciation among particle types in contaminated harbor sediments and effects on PAH bioavailability. *Environ. Sci. Technol.*, 37, 2209–2217.

Glenn, S.M. and W.D. Grant. 1987. A suspended sediment correction for combined wave and current flows, *J. Geophys. Res.*, 92, 8244–8246.

Gotthard, D. 1997. Three-Dimensional, Non-Destructive Measurements of Sediment Bulk Density Using Gamma Attenuation, M.S. thesis, Department of Mechanical and Environmental Engineering, University of California, Santa Barbara, CA.

Graf, W.H. and B. Yulisiyanto. 1998. Experiments on flow around a cylinder; the velocity and vorticity fields. *J. Hydraul. Res.*, 36(4), 637–653.

Grant, W.D. and O.S. Madsen. 1979. Combined wave and current interaction with a rough bottom. *J. Geophys. Res.*, 84(C4), 1797–1808.

Grass, A.J. 1970. The initial instability of fine bed sand. *J. Hydr. Div., ASCE*, 96(3), 619–632.

Graton, L.C. and H.J. Fraser. 1935. Systematic packing of spheres with particular relation to porosity and permeability. *J. Geol.*, 43, 785–809.

Guinasso, N.L. and D.R. Schink. 1975. Quantitative estimates of biological mixing rates in abyssal sediments. *J. Geophys. Res.*, 80, 3032–3044.

Gularti, R.C., W.E. Kelly, and V.A. Nacei. 1980. Erosion of cohesive sediments as a rate process. *Ocean Eng.*, 7, 539–551.

Gust, G. and V. Muller. 1997. Interfacial hydrodynamic and entrainment functions of currently used erosion devices. In *Cohesive Sediment, 4th Nearshore and Estuarine Cohesive Sediment Transport Conference*. New York: John Wiley & Sons.

Guthrie-Nichols, E., A. Grasham, C. Kazunga, R. Sangaiah, A. Gold, J. Bortiatynski, M. Salloum, and P. Hatcher. 2003. The effect of aging on pyrene transformation in sediments. *Environ. Sci. Technol.*, 22(1), 40–49.

Guy, H.P., D.B. Simons, and E.V. Richardson. 1966. Summary of Alluvial Channel Data from Flume Experiments, 1956–1961, Geological Survey, Professional Paper 462-1, Washington, D.C.

Hakanson, L. and M. Jansson. 2002. *Principles of Lake Sedimentology*. Caldwell, NJ: The Blackburn Press.

Hamblin, P.F. 1971. An investigation of horizontal diffusion in Lake Ontario. *Proc. 14th Conf. Great Lakes Res.*, 570–577. Int. Assoc. Great Lakes Res.

Hamrick, J.M. 1992a. A Three-Dimensional Environmental Fluid Dynamics Computer Code: Theoretical and Computational Aspects. The College of William and Mary, Virginia Institute of Marine Science, Special Report 317, 63 pp.

Hamrick, J.M. 1992b. Estuarine environmental impact assessment using a three-dimensional circulation and transport model. In *Estuarine and Coastal Modeling 2*, Ed. M.L. Spaulding, pp. 292–303, ASCE, New York.

Hamrick, J.M. 1994. Linking hydrodynamic and biogeochemical transport models for estuarine and coastal waters. In *Estuarine and Coastal Modeling 3*, Ed. M.L. Spaulding, pp. 591–608, ASCE, New York.

Hamrick, J.M. and T.S. Wu. 1997. Computational design and optimization of the EFDC/HEM3D surface water hydrodynamic and eutrophication models. In *Next Generation Environmental Models and Computational Methods*, Eds. G. Delich and M.F. Wheeler, pp. 143–156. Society of Industrial and Applied Mathematics, Philadelphia.

Haney, R.L. 1991. On the pressure gradient force over steep topography in sigma coordinate ocean models. *J. Phys. Oceanogr.*, 21, 610–619.

Harmon, T.C., H.P. Ball, and P.V. Roberts. 1989. Nonequilibrium transport of organic contaminants in groundwater. In *Reactions and Movement of Organic Chemicals in Soils,* Eds. B.L.Sawhney and K. Brown. Madison, WI: Soil Science Society of America.

Hayter, E.J., M. Bergs, R. Gu, S. McCutcheon, S.J. Smith, and H.J. Whiteley. 1999. HSCTM-2D, a Finite Element Model for Depth-Averaged Hydrodynamics, Sediment, and Contaminant Transport. Technical Report, U.S. Environmental Protection Agency Environmental Research Laboratory, Athens, GA.

Heinrich, J., W. Lick, and J. Paul. 1981. The temperatures and currents in a stratified lake: a two-dimensional analysis. *J. Great Lakes Res.*, 7, 264–275.

Heinrich, J.C., J.F. Paul, and W. Lick. 1983. Validity of a two-dimensional model for variable-density hydrodynamic circulation. *Mathematical Modeling,* 4, 323–337.

Hellou, J., S. Steller, V. Zitko, J. Leonard, T.G. Milligan, and P. Yeats. 2002. Distribution of PACs in surficial sediments and bioavailability to mussels, *Mytilus edulis* of Halifax Harbor. *Mar. Env. Res.*, 53, 357–379.

Hiemenz, P.C. 1986. *Principles of Colloid and Surface Chemistry.* New York: Marcel Dekker.

Hirschfelder, J., C. Curtis, and R. Bird. 1954. *Molecular Theory of Gases and Fluids.* New York: John Wiley & Sons.

Hughes, D.J., R.J.A. Atkinson, and A.D. Ansell. 1999. The annual cycle of sediment turnover by the echiuran worm *Maxmuelleria lankesteri* (Herdman) in a Scottish sea loch. *J. Exper. Mar. Biol. Ecol.*, 238, 209–223.

Hunt, R.J. 1984. Particle aggregate breakup by fluid shear. In *Estuarine Cohesive Sediment Dynamics*, Ed. A.J. Mehta, pp. 85–109. New York: Springer-Verlag.

Ijima, T. and F.L. Tang. 1966, Numerical calculations of wind waves in shallow water. *Proc. 10th Coastal Eng. Conf.*, pp. 38–45.

Israelachvili, I. 1992. *Intermolecular and Surface Forces.* San Diego: Academic Press.

Ives, K.J. 1978. Rate theories. In *The Scientific Basis of Flocculation*, Ed. K.J. Ives, pp. 37–61. Alphen aan den Rijn, The Netherlands: Sijthoff and Noordhoff International Publishers, B.V.

James, S.C., C. Jones, and J.D. Roberts. 2005. Consequence Management, Recovery and Restoration after a Contamination Event. Report SAND 2005-6797, Sandia National Laboratories, Albuquerque, NM.

Jepsen, R. and W. Lick. 1996. Parameters affecting the adsorption of PCBs to suspended sediments. *J. Great Lakes Res.*, 22, 341–353.

Jepsen, R. and W. Lick. 1999. Nonlinear and interactive effects in the sorption of hydrophobic organic chemicals by sediments. *Environ. Toxicol. Chem.*, 18, 1627–1636.

Jepsen, R., S. Borglin, W. Lick, and D. Swackhamer. 1995. Parameters affecting the adsorption of hexachlorobenzene to natural sediments. *Environ. Toxicol. Chem.*, 14, 1487–1497.

Jepsen, R., J. Roberts, and W. Lick. 1997. Effects of bulk density on sediment erosion rates. *Water, Air, Soil Pollut.*, 99, 21–31.

Jepsen, R., J. Roberts, and W. Lick. 1998a. Long Beach Harbor Study, Report, Department of Mechanical and Environmental Engineering, University of California, Santa Barbara, CA.

Jepsen, R., J. Roberts, W. Lick, D. Gotthard, and C. Trombino. 1998b. New York Sediment Study, Report, Department of Mechanical and Environmental Engineering, University of California, Santa Barbara, CA.

Jepsen, R., J. McNeil, and W. Lick. 2000. Effects of gas generation on the density and erosion of sediments from the Grand River. *J. Great Lakes Res.*, 209–219

Jin, L., J. McNeil, W. Lick, and J. Gailani. 2000. Effects of Bentonite on Sediment Erosion Rates. Report, Dept. of Mech. and Env. Eng., University of California, Santa Barbara, CA.

Jin, L., J. McNeil, W. Lick, and J. Gailani. 2002. Effects of Bentonite on the Erosion Rates of Quartz Particles. Report, Dept. of Mech. and Environ. Eng., University of California, Santa Barbara, CA.

Johnson, B.H., K.W. Kim, R.E. Hurth, B.B. Hsieh, and H.L. Butler. 1993. Validation of Three-Dimensional Numerical Hydrodynamic, Salinity, and Temperature Model of Chesapeake Bay. Tech. Report HL-91-7, Vicksburg, MS: U.S. Army Corps of Engineers Waterway Experiment Station.

Jones, C. and J. Gailani. 2008. Discussion of "Comparison of two techniques to measure sediment erodibility in the Fox River, Wisconsin." *J. Hydraul. Eng.*, 134, 898.

Jones, C. and W. Lick. 2000. Effects of bed coarsening on sediment transport. *Estuarine and Coastal Modeling, VI*, 915–930.

Jones, C. and W. Lick. 2001a. SEDZLJ, A Sediment Transport Model, Report, University of California, Santa Barbara, CA.

Jones, C. and W. Lick. 2001b. Sediment erosion rates: their measurement and use in modeling. ASCE Conference on Environmental Dredging.

Kang, S.W., Y.P. Sheng, and W. Lick. 1982. Wave action and bottom shear stresses in Lake Erie. *J. Great Lakes Res.*, 8(3), 482–494.

Karickhoff, S.N. and K.R. Morris. 1985. Sorption dynamic of hydrophobic pollutants in sediment: suspension. *Environ. Toxicol. Chem.*, 4, 469–479.

Kersten, M., T. Leipe, and F. Tauber. 2005. Storm disturbance of sediment contaminants at a hot-spot in the Baltic Sea assessed by [234]Th radionuclide tracer profiles. *Environ. Sci. Technol.*, 39, 984–990.

Kohnke, H. 1968. *Soil Physics*. New York: McGraw-Hill.

Kraaij, R.H., J. Tolls, D. Sijm, G. Cornelissen, A. Heikens, and A. Belfroid. 2002. Effects of contact time on the sequestration and bioavailability of different classes of hydrophobic organic chemicals to benthic oligochaetes (tubificids). *Environ. Toxicol. Chem.*, 21(4), 752–759.

Krone, R.B. 1962. Flume Studies of the Transport of Sediments in Estuarial Shoaling Processes. Report, Hydraulic Engineering Laboratory and Sanitary Engineering Research Laboratory, University of California, Berkeley, CA.

Lavelle, J. W., H.O. Mofjeld, and E.T. Baker. 1984. An in situ erosion rate for a fine-grained marine sediment. *J. Geophys. Res.*, 89, 6543–6552.

Lee, D.Y., S.W. Kang, and W. Lick. 1981. The entrainment and deposition of fine-grained sediments. *J. Great Lakes Res.*, 7, 224–233.

Lewis, P.J. 1987. Severe Storms Over the Great Lakes: A Catalogue Summary for the Period 1957 to 1985. Canadian Climate Center, Report No. 87–13.

Lewis, R.C., K.H. Coale, B.D. Edwards, M. Marot, J.N. Douglas, and E.J. Burton. 2002. Accumulation rate and mixing of shelf sediments in the Monterey Bay National Marine Sanctuary. *Mar. Geol.*, 181, 157–169.

Lick, W. 1992. The importance of large events. *Reducing Uncertainty in Toxic Mass Balance Models,* Great Lakes Monograph No. 4, State University of New York at Buffalo.

Lick, W. 2006. The sediment-water flux of HOCs due to "diffusion" or is there a well-mixed layer? If there is, does it matter?. *Environ. Sci. Technol.*, 40(18), 5610–5617.

Lick, W. and H. Huang. 1993. Flocculation and the Physical Properties of Flocs. In *Nearshore and Estuarine Cohesive Sediment Transport*, Ed. A. Mehta. New York: Springer-Verlag.

Lick, W. and S.W, Kang. 1987. Entrainment of sediments and dredged materials in shallow lake waters. *J. Great Lakes Res.*, 13(4),619–627.

Lick, W. and J. Lick. 1988. On the aggregation and disaggregation of fine-grained sediments. *J. Great Lakes Res.*, 14(4), 514–523.

Lick, W. and J. McNeil. 2001. Effects of sediment bulk properties on erosion rates. *Science of the Total Environment*, 266, 41–48.

Lick, W. and V. Rapaka. 1996. A quantitative analysis of the dynamics of the sorption of hydrophobic organic chemicals to suspended sediments. *Environ. Toxicol. Chem.*, 15, 1038–1048.

Lick, W., J. Lick, and C.K. Ziegler. 1992. Flocculation and its effects on the vertical transport of fine-grained sediments. *Hydrobiologia,* 235, 1–16.

Lick, W., H. Huang, and R. Jepsen. 1993. The flocculation of fine-grained sediments due to differential settling. *J. Geophys. Res.*, 98, 10279–10288.

Lick, W., J. Lick, and C.K. Ziegler. 1994a. The resuspension and transport of fine-grained sediments in Lake Erie. *J. Great Lakes Res.*, 20, 599–612.

Lick, W., C.K. Ziegler, J. Lick, and A. Joshi. 1994b. Effects of flocculation on particle transport. In *Estuarine Coastal Modeling*, Ed. by M.L. Spaulding, pp. 172–186.

Lick, W., J. McNeil, Y.J. Xu, and C. Taylor. 1995. Measurements of the Resuspension and Erosion of Sediments in Rivers. Report, University of California, Santa Barbara.

Lick, W., Z. Chroneer, and V. Rapaka. 1997. Modeling the dynamics of the sorption of hydrophobic organic chemicals to suspended sediments. *Water, Air, Soil Pollut.,* 99, 225–235.

Lick, W., C. Jones, Z. Chroneer, and R. Jepsen. 1998. A predictive model of sediment transport. *Estuarine Coastal Modeling*, pp. 389–399.

Lick, W., J. Lick, L. Jin, J. McNeil, and J. Gailani. 2002. Effects of Clay Mineralogy on Sediment Erosion Rates. Report, University of California, Santa Barbara, CA.

Lick, W., L. Jin, and J. Gailani. 2004a. Initiation of movement of quartz particles. *J. Hydraul. Eng.,* 130, 755–761.

Lick, W., J. Lick, and C. Jones. 2004b. Effects of Finite Sorption Rates on Sediment-Water Fluxes of Hydrophobic Organic Chemicals. Report, University of California, Santa Barbara, CA.

Lick, W., J. Lick, L. Jin, and J. Gailani. 2006a. Approximate equations for sediment erosion rates. In *Estuarine and Coastal Fine Sediment Dynamics.* Amsterdam: Elsevier.

Lick, W., X. Luo, G. Deane, and C. Jones. 2006b. Sediment-water fluxes of hydrophobic organic chemicals due to molecular diffusion. In *Estuarine and Coastal Modeling IX, ASCE*, pp. 448–467.

Limno-Tech, Inc. 1999. Development of an Alternative Suite of Models for the Lower Fox River. Limno-Tech, Inc., Ann Arbor, MI.

Little, W.C. and P.G. Mayer. 1972. The Role of Sediment Gradation on Channel Armoring. Publication No. ERC-0672, School of Civil Engineering, Georgia Institute of Technology, Atlanta, GA.

Litton, G.M. and T. Olson. 1993. Colloid deposition rates on silica bed media and artifacts related to collector surface preparation methods. *Environ. Sci. Technol.*, 27, 185–193.

Lodge, P. 2006. Effects of Surface Slope on Erosion Rates of Quartz Particles. M.S. thesis, University of California, Santa Barbara, CA.

Lohmann, R., J.K. MacFarlane, and P.M. Gschwend. 2005. Importance of black carbon to sorption of native PAHs, PCBs, and PCDDs in Boston and New York harbor sediments. *Environ. Sci. Technol.*, 39, 141–148.

Lunsman, T.D. and W. Lick. 2005. Sorption of hydrophobic organic chemicals to bacteria. *Environ. Toxic. Chem.*, 24, 2128–2137.

Luo, X., W. Lick, and C. Jones. 2006. Modeling the sediment-water flux of hydrophobic organic chemicals due to bioturbation. In *Estuarine Coastal Modeling IX, ASCE*, pp. 468–485.

Luque, R.F. 1972. Erosion and Transport of Bed-Load Sediment. Ph.D. thesis, Delft University of Technology, Meppel, The Netherlands.

Luque, R.F. and R. van Beek. 1976. Erosion and transport of bedload sediment. *J. Hydraul. Res.*, 14(2), 127–144.

MacIntyre, S., W. Lick, and C.H. Tsai. 1990. Variability of entrainment of cohesive sediments in freshwater. *Biogeochemistry*, 9, 187–209.

Mandelbrot, B.B. 1977. *The Fractal Theory of Nature*. New York: W.H. Freeman and Company.

Martin, J.L. and S.C. McCutcheon. 1999. *Hydrodynamics and Transport for Water Quality Modeling*. New York: Lewis Publishers.

Massion, G. 1982. The Resuspension of Uniform-Sized Fine-Grained Sediments, M.S. thesis, University of California, Santa Barbara, CA.

Matisoff, G. 1982. Mathematical models of bioturbation. In *Animal Sediment Relations*. Eds. P.L. McCall and M.S. Tevasz. New York: Plenum Press.

Matsuo, T. and H. Unno. 1981. Forces acting on floc and strength of floc. *J. Environ. Eng. Div. Am. Soc. Civ. Eng.*, 107(EE3), 527–545.

May, R.M. 2004. Uses and abuses of mathematics in biology. *Science*, 303, 790–793.

McNeil, J. and W. Lick. 2002a. Erosion Rates and Bulk Properties of Sediments in Lake Michigan. Report, Department of Mechanical and Environmental Engineering, University of California, Santa Barbara, CA.

McNeil, J. and W. Lick, 2002b. Erosion Rates and Bulk Properties of Sediments from the Kalamazoo River, Report, University of California, Santa Barbara, CA.

McNeil, J. and W. Lick. 2004. Erosion rates and bulk properties of sediments from the Kalamazoo River. *J. Great Lakes Res.*, 30, 407–418.

McNeil, J., C. Taylor, and W. Lick. 1996. Measurement of the erosion of undisturbed bottom sediments with depth. *J. Hydraul. Eng.*, 122, 316–324

McNeil, J., L. Jin, and W. Lick. 2001a. Erosion Rates and Bulk Properties of Grasse River Sediments and a Proposed Capping Material. Report, University of California, Santa Barbara, CA.

McNeil, J., L. Jin, and W. Lick. 2001b. Effects of Gas on Sediment Consolidation and Erosion of a Capping Material. Report, University of California, Santa Barbara, CA.

Medina, A.A. 2002. Surface water–ground water interactions and modeling applications. In *Environmental Modeling and Management*, Ed. C.C. Chien, Today Media, Inc.

Mehta, A.J., 1973. Depositional Behavior of Cohesive Sediments. Ph.D. thesis, University of Florida, Gainesville, FL.

Mehta, A.J., 1994. Characterization of cohesive soil erosion with special reference to the relationship between erosion shear strength and bed density. Report UFL/COEL/MP-91-4, Coastal and Oceanographic Engineering Department, University of Florida, Gainesville, FL.

Mehta, A.J., T.M. Parchure, J.G. Dixit, and R. Ariathurai. 1982. Resuspension potential of deposited cohesive sediment beds. In *Estuarine Comparisons*, Ed. V.S. Kennedy, pp. 591–609. New York: Academic Press.

Mei, C.C. 1983. *The Applied Dynamics of Ocean Surface Waves*. New York: John Wiley & Sons.

Mellor, G.L. and T. Yamada. 1982. Development of a turbulence closure model for geophysical fluid problems. *Rev. Geophys. Space Phys.*, 20, 851–875.

Mellor, G.L., T. Ezer, and L.Y. Oey. 1994. The pressure gradient conundrum of sigma coordinate models. *J. Atmos. Oceanic Technol.*, 11(4), 1126–1134.

Mellor, G.L., L.Y. Oey, and T. Ezer. 1998. Sigma coordinate pressure gradient errors and the seamount problem. *J. Atmosph. Oceanic Technol.*, 15, 1122–1131.

Melville, B. 1999. Bridge Scour: A Compendium of Technical Papers by Researchers at University of Auckland. Compendium 592, Dept. of Civil and Resource Engr., University of Auckland, New Zealand.

Meyer-Peter, E. and R. Muller. 1948. Formulas for bed-load transport. *Proceedings of the International Association for Hydraulic Structures Research*, pp. 39–64. Report of Second Meeting, Stockholm.

Miller, M., I.N. McCave, and P.D. Komar. 1977. Threshold of sediment motion under unidirectional currents, *Sedimentology*, 24, 507–528.

Mitchell, J.K. 1993. *Fundamentals of Soil Behavior*. New York: John Wiley & Sons.

Miyagi, N., M. Kimura, H. Shojie, A. Saina, C.M. Ho, S. Tang, and Y.C. Tai. 2000. Statistical analysis on wall shear stress of turbulent boundary layer in a channel flow using micro-shear stress imager. *Int. J. Heat Fluid Flow*, 21, 576–581.

Mohanty, S., D.D. Reible, K.T. Valsaraj, and L.J. Thibodeaux. 1998. A physical model for the simulation of bioturbation and its comparison to experiments with oligochaetes. *Estuaries*, 21(2), 255–262.

Munk, W.H. and E.R. Anderson. 1948. Notes on the theory of the thermocline. *J. Mar. Res.*, 1, 276–295.

Murty, T.S. and R.J. Polavarapu. 1975. Reconstruction of some of the early storm surges on the Great Lakes. *J. Great Lakes Res.*, 1, 116–129.

Myers, A.C. 1979. Summer and winter burrows of a mantis shrimp, *Squilla empusa*, in Narragansett Bay, Rhode Island. *Estuar. Coast. Shelf Sci.*, 8, 87–98.

Nairn, B.J. 1998. Incipient Transport of Silt-Sized Sediments. Report No. KH-R-59, W.M. Keck Laboratory of Hydraulics and Water Resources, California Institute of Technology, Pasadena, CA.

National Climatic Data Center. 1990. Climatic Summaries for NDBC Buoys and Stations, Update 1, National Oceanic and Atmospheric Administration.

National Research Council. 2001. *A Risk-Management Strategy for PCB-Contaminated Sediments*. National Academy Press, Washington, D.C.

Nkedi-Kizza, P., J.W. Biggar, H.M. Selim, J.M. Davidson, and D.R. Nielsen. 1984. On the equivalence of two conceptual models for describing ion exchange during transport through an aggregated oxisol. *Water Resour. Res.*, 20, 1123–1130.

Obi, S., K. Inoue, T. Furakawa, and S. Masuda. 1996. Experimental study on the statistics of wall shear stress in turbulent channel flow. *Int. J. Heat Fluid Flow*, 17, 187–192.

O'Connor, D.J. 1988. Models of sorptive toxic substances in freshwater systems. *J. Environ. Eng.*, 114(3), 507–551.

Okubo, A. 1971. Oceanic diffusion diagrams. *Deep Sea Res.*, 18, 789–802.

Oliver, J.S., P.N. Slattery, L.W. Hulberg, and J.W. Nybakken. 1980. Relationships between wave disturbance and zonation of benthic invertebrate communities along a high-energy subtidal beach in Monterey Bay. *California Fishery Bulletin*, 78, 437–454.

Owen, M.W. 1975. Erosion of Avonmouth Mud. Hydraulics Research Station Report INT 150.

Paphitis, D. 2001. Sediment movement under unidirectional flows: An assessment of empirical threshold curves. *Coast Eng.*, 43, 227–245.

Parker, D.S. 1982. Discussion of forces acting on floc and strength of floc. *J. Environ. Eng. Div., ASCE*, 108(EE3), 594–598.

Parker, G., C. Paola, and S. Leclair. 2000. Probabilistic Exner sediment continuity equation for mixtures with no active layer. *J. Hydraul. Eng.*, 126(11), 818–826.

Partheniades, E. 1965. Erosion and deposition of cohesive solid. *J. Hydraulics Div. ASCE*, 91, 105–139.

Partheniades, E., 1986. The present knowledge and needs for future research on cohesive sediment dynamics. *Proceedings of the 3rd International Symposium on River Sedimentation*, pp. 3–25.

Paul, J.F. and W. Lick. 1974. A numerical model of thermal plumes and river discharges. *Proc. 17th Conf. Great Lakes Res.*, Int. Assoc. Great Lakes Research, pp. 445–455.

Paul, J.F., and W. Lick. 1985. A Numerical Model for Three-Dimensional, Variable-Density Hydrodynamic Flows, Report, U.S. Environmental Protection Agency, Duluth, MN.

Paul, J., P. Kasprzyk, and W. Lick. 1982. Turbidity in the western basin of Lake Erie. *J. Geophys. Res.*, 5779–5784.

Pickens, K., Z. Chroneer, P. Patel, and W. Lick. 1994. The formation of a turbidity maximum in an estuary. In *Estuarine & Coastal Modeling*, Ed. M.L. Spaulding, pp. 187–201.

Postma, H. 1967. Sediment Transport and Sedimentation in the Estuarine Environment. In *Estuaries*, Ed. G.H. Lauff, pp. 158–179. American Association for the Advancement of Science.

QEA. 1999. PCBs in the Upper Hudson River, Report to the General Electric Company, QEA LLC, Montvale, NJ.

Saffman, P.G. and J.T. Turner. 1956. On the collision of drops in turbulent clouds. *J. Fluid Mech.*, 1, 16–30.

Raveendran, P. and A. Amirtharajah. 1995. Role of short-range forces in particle detachment during filter backwashing. *J. Environ. Eng.*, 121(12), 860–868.

Ravens, T.M. 2007. Comparison of two techniques to measure sediment erodibility in the Fox River, Wisconsin. *J. Hydraul. Eng., ASCE* 133(1), 111–115.

Reible, D.D., V. Papov, K.T. Valsaraj, L.J. Thibodeaux, K. Lin, M. Dikshit, M.A. Todure, and J.W. Fleeger. 1996. Contaminant fluxes from sediment due to *tubificid oligochaete* bioturbation. *Wat. Res.*, 30, 704–714.

Rhoads, D.C. 1974. Organism-sediment relations on the muddy seafloor. In *Oceanography and Marine Biology Annual Reviews*, Ed. H. Barnes, 12, 263–300.

384 Sediment and Contaminant Transport in Surface Waters

Rhoads, D.C. and D.A. Carey. 1997. Capping dredged materials in the New York Bight: evaluation of the effects of bioturbation. Science Applications International Report No. 374; Report No. 39 of the New York Mud Dump Series, U.S. Army Corps of Engineers, New York, NY.

Rhoads, D.C. and J.D. Germano. 1986. Interpreting long-term changes in benthic community structure: a new protocol. *Hydrobiologia,* 145, 291–304.

Richards, A.F., T.J. Hirst, and J.M. Parks. 1974. Bulk density-water content relationship in marine silts and clays. *J. Sediment Petrol.,* 44, 1004–1009.

Richardson, J.E. and V.G. Panchang. 1998. Three-dimensional simulation of scour inducing flow at bridge piers. *J. Hydraul. Eng.,* 124, 530–540.

Robbins, J.A. and D.N. Edgington. 1975. Determination of recent sedimentation rates in Lake Michigan using Pb-210 and Cs-137. *Geochim. Cosmochim. Acta,* 39, 285–304.

Robbins, J.A., J.R. Krezoski, and S.C. Mozley. 1977. Radioactivity in sediments of the Great Lakes: post-depositional redistribution by deposit feeding organisms. *Earth Planet. Sci. Lett.,* 36, 325–333.

Robbins, J.A., D.N. Edgington, and A.L.W. Kemp. 1978. Comparative ^{210}Pb, ^{137}Cs, and pollen geochronologies of sediments from Lakes Ontario and Erie. *Quat. Res.,* 10, 256–278.

Roberts, J., R. Jepsen, and W. Lick. 1998. Effects of particle size and bulk density on the erosion of quartz particles. *J. Hydraul. Eng.,* 124, 1261–1267.

Roberts, J., C. Jones, and S. James. 2006. Report on Baseline Conditions, Cedar Lake, Illinois. Report to USACE, Chicago District. Sandia National Laboratories.

Saffman, P.G. 1965. The lift on a small sphere in slow shear flow. *J. Fluid Mech.,* 22(2), 385–400.

Saffman, P.G. and J.T. Turner. 1956. On the collision of drops in turbulent clouds. *J. Fluid Mech.,* 1, 16–30.

Sawhney, B.L. and K. Brown. 1989. *Reactions and Movement of Organic Chemicals in Soils.* Madison, WI: Soil Science Society of America.

Schlichting, H. 1955. *Boundary-Layer Theory.* New York: McGraw-Hill.

Schubel, J.R. 1974. Effects of tropical storm Agnes on the suspended solids of the Northern Chesapeake Bay. In *Suspended Solids in Water,* Ed. R.J. Gibbs, 4, 113–132. Plenum Marine Science.

Scott, G.D. 1960. Packing of equal spheres. *Nature,* 188, 908–909.

Sea Engineering. 2004. Hydrodynamic Field Measurement and Modeling Analysis of the Lower Fox River, Report Sea Engineering, Inc., Santa Cruz, CA.

Sherwood, C.R., D.E. Drake, P.L. Wiberg, and R.A. Wheatcroft. 2002. Prediction of the fate of pp'-DDE in sediments on the Palos Verdes shelf, California, USA. *Continental-Shelf Res.,* 22, 1025–1058.

Simons, T.J. 1974. Verification of numerical models of Lake Ontario. Part I. Circulation in spring and early summer. *J. Phys. Oceanogr.,* 4, 507–523.

Smart, G.M. 1984. Sediment transport formula for steep channels. J. *Hydraul. Eng., ASCE* 110(3), 267–276.

Smith, J.M., D.T. Resio, and A.K. Zundel. 1999. STWAVE: Steady-State Spectral Wave Model; Report 1: User's Manual for STWAVE Version 2.0, Instructional Report CHL-99-1, Vicksburg, MS: U.S. Army Corps of Engineers Research and Development Center.

Smoluchowski, M. 1917. Versuch einer Mathematischen Theorie der Koagulations-Kinetik Kolloid Losungen. *Z. fur Physikalische Chem.,* 92, 129–168.

Sorensen, R.M. 1978. *Basic Coastal Engineering.* New York: John Wiley & Sons.

Soulsby, R.L. 1997. *Dynamics of Marine Sands.* London: Thomas Telford.

Soulsby, R.L., L. Hamm, G. Klopman, D. Myrhaug, R.R. Simons, and G.P. Thomas. 1993. Wave-current interactions within and outside the bottom boundary layer. *Coastal Eng.,* 21, 41–69.

Soulsby, R.L. and R.J.S. Whitehouse. 1997. Threshold of sediment motion in coastal environments, *Proc. Combined Australian Coast. Eng. and Ports Conf.,* 1, 149–154, University of Canterbury, New Zealand.

Spaulding, M.L. 1987. Selected studies on PCB transport in New Bedford Harbor, ASA86-18, Applied Science Associates, Narragansett, RI.

Spaulding, M.L., 2006. Estuarine and Coastal Modeling. *Proceedings of the Ninth International Conference,* ASCE.

Stumm, W. and J.J. Morgan. 1996. *Aquatic Chemistry.* New York: Wiley Interscience.

Styles, R. and S.M. Glenn. 2000. Modeling stratified wave and current bottom boundary layers on the continental shelf. *J. Geophys. Res.,* 105(10), 24119–24139.

Sudarsan, R., C. Jones, and W. Lick. 2003. Numerical Modeling of Clear Water Scour around Cylinders. Report, University of California, Santa Barbara, CA.

Swift, D.J. 1978. Survey of sport shellfishing potential in San Francisco Bay, in Southern San Francisco Bay, and Northern San Mateo Counties. Final Report to the San Francisco Wastewater Program, City and County of San Francisco, 77 p.

TAMS. 2000. Hudson River PCBs Reassessment RI/FS Phase 3 Report: Feasibility Study, Report to the U.S. Environmental Protection Agency by TAMS Consultants, Inc.

Tetra Tech. 1999. A Theoretical Description of Toxic Contaminant Transport Formulations Used in the EFDC Model. Tech. Memo. TT-EFDC-99-2, Tetra Tech, Inc., Fairfax, VA.

Tetra Tech. 2000. A Theoretical Description of Sediment Transport Formulations Used in the EFDC Model. Tech. Memo. TT-EFDC-00-1, Tetra Tech, Inc., Fairfax, VA.

Thibodeaux, L.J., K.T. Valsaraj, and D.D. Reible. 2001. Bioturbation-driven transport of hydrophobic organic chemicals from bed sediment. *Environ. Eng. Sci.,* 18, 215–223.

Thoma, G.J., A.C. Koulermos, K.T. Valsaraj, D.D. Reible, and L.J. Thibodeaux. 1991. The effects of pore-water colloids on the transport of hydrophobic organic compounds from bed sediments. In *Organic Substances and Sediments in Water,* Ed. H.A. Baker. Chelsea, MI: Lewis Publishers.

Thomann, R.V. and D.M. DiToro. 1983. Physico-chemical model of toxic substances in the Great Lakes, *J. Great Lakes Res.,* 9(4), 474–496.

Thomann, R.V. and J.A. Mueller. 1987. *Principles of Surface Water Quality Modeling and Control.* New York: Harper Collins.

Thomas, R.L., J.M. Jaquet, A.L.W. Kemp, and C.F.M. Lewis. 1976. The surficial sediments of Lake Erie. *J. Fish. Res. Board. Can.,* 33, 385–403.

Thomas, W.A. and W.H. McAnally. 1985. User's Manual for the Generalized Computer Program System-Open Channel Flow and Sedimentation TABS-2. Vicksburg, MS: U.S. Army Corps of Engineers Waterways Experiment Station.

Tooby, P.F., G.L. Wick, and J.D. Isaacs. 1977. The motion of a small sphere in a rotating velocity field: a possible mechanism for suspending particles in turbulence. *J. Geophys. Res.,* 82, 2096–2100.

Toorman, E.A. 1996. Sedimentation and self-weight consolidation: general unifying theory. *Geotechnique,* 46(1), 103–113.

Toorman, E.A. 1999. Sedimentation and self-weight consolidation: constitutive equations and numerical modeling. *Geotechnique,* 49(6), 709–726.

Tracy, J.P. and C.M. Keane. 2000. Peer Review of Models Predicting the Fate and Export of PCBs in the Lower Fox River Below DePere Dam, American Geological Institute, Alexandria, VA.

Tsai, C.H. and W. Lick. 1986. A portable device for measuring sediment resuspension. *J. Great Lakes Res.*, 12(4), 314–321.

Tsai, C.H. and W. Lick. 1988. Resuspension of sediments from Long Island Sound. *Water Sci. Technol.*, 20(6/7), 155–164.

Tsai, C.H., S. Iacobellis, and W. Lick. 1987. The flocculation of fine-grained lake sediments due to a uniform shear stress. *J. Great Lakes Res.*, 3, 135–146.

Tye, R., R. Jepsen, and W. Lick. 1996. Effects of colloids, flocculation, particle size, and organic matter on the adsorption of hexchlorobenzene to sediments. *Environ. Toxicol. Chem.*, 15, 643–651.

U.S. Army Coastal Engineering Research Center, 1973. Shore Protection Manual, Vol. 1. Fort Belvoir, VA.

U.S. EPA. 1999. Benthic invertebrates. www.epa.gov/glnpo/monitoring/indicators/benthic/paper.htm.

U.S. EPA. 2000. Aquatox Release 1: A Simulation Model for Aquatic Ecosystems, EPA 823-F-00-015.

U.S. EPA. 2006. GE/Housatonic River Site in New England, Rest of River, Reports. www.epa.gov/NE/ge/thesite/restofriver/reports/model_validation_finalcomments/252998.pdf.

U.S. Weather Service. 1976. Climatological data. Michigan Annual Summary, 1914–present. National Weather Service, National Oceanographic and Atmospheric Administration, Dept. of Commerce, Washington, D.C.

Van Genuchten, M.T. and P.J. Wierenga. 1976. Mass transfer studies in sorbing porous media. I. Analytic solutions. *Soil Sci. Soc. Am. J.*, 40, 473–483.

Van Niekerk, A., K. Vogel, R. Slingerland, and J. Bridge. 1992. Routing of heterogeneous sediments over movable bed: model development. *J. Hydraul. Eng.*, 110(10), 246–263.

Van Rijn, L.C. 1993. *Principles of Sediment Transport in Rivers, Estuaries, and Coastal Seas.* Amsterdam, The Netherlands: Aqua Publications.

Verwey, E.J.W. and T.G. Overbeek. 1948. *Theory of the Stability of Lyophobic Colloids.* Amsterdam: Elsevier.

Walling, D.E. and B.W. Webb. 1981. The reliability of suspended sediment load data. *Proc. Symp. Erosion and Sediment Transport Measurement*, IAHS Publication No. 133, pp. 177–194.

Wang, K.P., Z. Chroneer, and W. Lick. 1996. Sediment transport in a thermally stratified bay. *Estuarine Coastal Modeling*, pp. 466–477.

Weston Solutions. 2004. Modeling Study of PCB Contaminants in the Housatonic River. Report, Weston Solutions, Inc., West Chester, PA.

Wheatcroft, R.A., I. Olmez, and F.X. Pink. 1994. Particle bioturbation in Massachusetts Bay. Preliminary results using a new deliberate tracer technique. *J. Mar. Res.*, 52, 1129–1150.

White, S.J., 1970. Plane bed thresholds of fine-grained sediments. *Nature*, 228, 152–153.

Whitehouse, R.J.S. 1991. Slope-inclusive bedload transport: experimental assessment and implications of bedform development, Euromech 262. In *Sand Transport in Rivers, Estuaries, and the Sea*, Eds. R.L. Soulsby and R. Bettess. Balkema, Rotterdam, The Netherlands.

Whitehouse, R.J.S. 1998. *Scour at Marine Structures.* London: Thomas Telford.

Whitehouse, R.J.S., R.L. Soulsby, and J.S. Damgaard. 2000. Discussion, inception of sediment transport on steep slopes. *J. Hydraul. Eng., ASCE*, 126, 553–555.

Wilson, B.W. 1960. Note on surface wind stress over water at low and high speeds. *J. Geophys. Res.*, 65, 3377–3382.

Winterwerp, J.C. 2007. On the sedimentation rate of cohesive sediment. In *Estuarine and Coastal Fine Sediments Dynamics*. Amsterdam: Elsevier.

Witt, O. and B. Westrich. 2003. Quantification of erosion rates for undisturbed contaminated cohesive sediment cores by image analysis. *Hydrobiologia*, 494, 271–276.

Wright, S. and G. Parker. 2004. Flow resistance and suspended load in sand-bed rivers: simplified stratification model. *J. Hydraul. Eng.*, 130(8), 796–805.

Wu, S. and P.M. Gschwend. 1986. Sorption kinetics of hydrophobic organic compounds to natural sediments and solids. *Environ. Sci. Technol.*, 20, 717–725.

Wu, T.S., J.M. Hamrick, S.C. McCutcheon, and R.B. Ambrose. 1997. Benchmarking the EFDC/HEM3D surface water hydrodynamic and eutrophication models. In *Next Generation Environmental Models and Computational Methods*, Eds. G. Delic and M.F. Wheeler. Society of Industrial and Applied Mathematics, Philadelphia, PA.

Wu, W., S.Y. Wang, and Y. Jia. 2000. Nonuniform sediment transport in alluvial rivers. *J. Hyd. Res.*, 38(6), 427–434.

Yang, C.T. 1996. *Sediment Transport, Theory and Practice*. New York: McGraw-Hill.

Yen, C.L. and K.T. Lee. 1995. Bed topography and sediment sorting in channel bend with unsteady flow. *J. Hydraul. Eng.*, 121(8), 591–599.

You, Z.J. 2000. A simple model of sediment initiation under waves. *Coastal Eng.*, 41, 399–412.

Zabawa, C.F. and J.R. Schubel. 1974. Geologic effects of tropical storm Agnes on Upper Chesapeake Bay. *Maritime Seds.*, 10, 79–84.

Ziegler, C.K. and W. Lick. 1986. A Numerical Model of the Resuspension, Deposition, and Transport of Fine-Grained Sediments in Shallow Water. Report, University of California, Santa Barbara, CA.

Zwarts, L. and J.H. Wanink. 1989. Siphon size and burying depth in deposit and suspension feeding benthic bivalves. *Mar. Biol.*, 100, 227–240.

Index

A

Aggregation
  basic theory of, 104–108
  collision frequency, 104–106
  particle interactions, 106–108
Albite density, 39
Amorphous organic matter, 292, 311
Angle of repose, particle size and, 100, 102
Annular flumes, 46–50
Approximate uniformly valid equation,
  125–126
Armistice Day storm, 1940, Lake Erie,
  263–264

B

Bacteria, effect on erosion rates, 80–81
Bays, sediment transport modeling, 261–271
Bed armoring, 149
  effect on sediment transport, 220
  parameterization, water quality modeling,
    non-unique solutions, 368
Bedload
  erosion into, 223–225
  transport, 220–226
Benthic organisms, effect on erosion rate,
  80–81
Bentonite, 68, 72–75, 80, 88–90, 92, 379
  properties of, 68
Benzene, 334, 375
Bioturbation, sediment-water flux, 342–355
  finite-rate sorption process, modeling
    bioturbation as diffusion with,
    353–355
  hydrophobic organic chemical flux from
    organisms, 344–353
  experimental modeling, 348–353
  experimental procedures, 345
  theoretical model, 346–348
  organisms, physical mixing of sediments
    by, 343–344
Bottom shear stress, 182–187
  current modeling, effects of currents,
    185–187
  effects of currents, 182–187
  waves, effects of, 185–187

Boulder sizing, 22
Boundary conditions, current modeling,
  176–179
Bridge piers, cylindrical, flow around partially
  submerged, 206–210
Bulk density
  bottom sediments, 37–44
  measurements, 39–41
  variations in, 41–44
  effect on erosion, 68–69, 87–88
Bulk properties, effects on erosion rates, 67–81
  bacteria, 80–81
  benthic organisms, 80–81
  bulk density, 68–69
  erosion rate comparisons, 79–80
  gas, 77–79
  mineralogy, 72–75
  organic content, 75–76
  particle size, 70–72
  salinity, 76

C

Calcite, 35, 39, 42
  density, 39
  Kalamazoo River sediment, 35
Carbonaceous geosorbents, 84, 114–115, 121,
  125–126, 137–140, 176, 292, 311
CG. See Carbonaceous geosorbents
Chemical properties, effects on adsorption, 311
Chlorite, 33, 35
  Kalamazoo River sediment, 35
Classification, sediment particle size, 21–23
Clay
  density, 39
  Kalamazoo River sediment, 35
  minerals
    effects on erosion, 88–90
    erosion rate equation, 92
  sizing, 22
Coarse clay sizing, 22
Coarse silt sizing, 22
Coarse stand sizing, 22
Coarsening in straight channel, 227–229
Cobble sizing, 22
Cohesive forces, effects on erosion, 85–87